TOWARDS THE ENVIRONMENTAL MINIMUM

Pervasive environmental harm that disproportionately impacts vulnerable members of society is left largely unregulated across the globe despite existing legal commitments to human rights and environmental protection in many states. To address this shortcoming, Stefan Theil proposes a new normative framework for environmental protection through human rights law. In clear and accessible prose, he demonstrates how such a human rights-based approach can strengthen environmental protection without requiring radical departures from established protection regimes and legal principles. The environmental minimum developed in the book translates the general and abstract commitments of states into specific and practical measures that protect the natural environment. The framework develops the doctrine of international, regional, and domestic courts, analysed through an innovative approach that improves contextual awareness. This book is thus a valuable resource for lawyers, social scientists, political theorists, environmental and human rights advocates.

Stefan Theil is a postdoctoral research fellow at the Faculty of Law, University of Oxford. He teaches and conducts research in public and constitutional law, environmental law, European Union law, as well as human rights, with a particular focus on freedom of expression. He writes and blogs frequently on current legal developments, often drawing comparisons between legal systems and reflecting on deepening European integration.

Towards the Environmental Minimum

ENVIRONMENTAL PROTECTION THROUGH HUMAN RIGHTS

STEFAN THEIL
University of Oxford

CAMBRIDGE
UNIVERSITY PRESS

CAMBRIDGE
UNIVERSITY PRESS

Shaftesbury Road, Cambridge CB2 8EA, United Kingdom

One Liberty Plaza, 20th Floor, New York, NY 10006, USA

477 Williamstown Road, Port Melbourne, VIC 3207, Australia

314–321, 3rd Floor, Plot 3, Splendor Forum, Jasola District Centre, New Delhi – 110025, India

103 Penang Road, #05–06/07, Visioncrest Commercial, Singapore 238467

Cambridge University Press is part of Cambridge University Press & Assessment, a department of the University of Cambridge.

We share the University's mission to contribute to society through the pursuit of education, learning and research at the highest international levels of excellence.

www.cambridge.org
Information on this title: www.cambridge.org/9781108799744

DOI: 10.1017/9781108891769

First published 2021
First paperback edition 2026

A catalogue record for this publication is available from the British Library

ISBN 978-1-108-83514-5 Hardback
ISBN 978-1-108-79974-4 Paperback

For my parents and Lauren

Contents

PART IV BEYOND HUMAN RIGHTS LAW

Figures

Acknowledgements

This book represents the culmination of many years of research, beginning with my doctoral work at the University of Cambridge and concluding at the Bonavero Institute, University of Oxford. The last few months have been particularly challenging in light of the restrictions imposed by the novel coronavirus. The uncertainty and pressures that the pandemic has exerted on everyday life have given what was intended to be a celebratory moment a distinctly sombre overlay. Life, it seems, will be different once we emerge from this crisis. Should my book provide some humble contributions towards a kinder, more generous, and environmentally just world, then I shall consider the effort worthwhile.

In such times we realise what is truly important in our lives, and I have been humbled by the generous support from my loved ones and many extraordinary friends and colleagues. Their advice and guidance were the essential ingredients that allowed me to complete this project. My fledgling career would not have been possible without my academic mentors, and I endeavour to follow their example in instructing the next generation of students.

I am grateful to the Faculty of Law, University of Cambridge, and my supervisors, Markus Gehring and Stephanie Palmer, who, above all, gave me the confidence to pursue my research when it was still in its infancy. Thanks are also due to my examiners Eyal Benvenisti and Ole Pedersen for their helpful comments during and after the viva. Many friends went far beyond the call of friendship in their constructive advice and countless discussions over the years. I am especially indebted to John Adentire, Will Bateman, Amelia Papworth, and Barry Solaiman for reading the manuscript in its entirety and providing crucial feedback. I will always be grateful for their support, and the intellectual renaissance I was privileged to experience with them in Cambridge.

Of my academic mentors, I owe a particular debt of gratitude to Jeff King. Jeff has guided me throughout my academic career and offered crucial advice at virtually every major fork in the road. Research assistance and collaboration with Jeff on various projects taught me invaluable lessons on approaching complex legal problems with rigorous theoretical, doctrinal, and empirical analysis. His seminal work to date, *Judging Social Rights* (Cambridge University Press, 2012), was a key inspiration for this book and provided important guidance.

I would like to thank the inaugural director of the Bonavero Institute, Kate O'Regan, and the Faculty of Law, University of Oxford for taking a chance on a junior academic at the close of his doctoral studies. The community in Oxford proved a stimulating and supportive environment for research that I have come to treasure no less than my time at 'the other place.' Special thanks are due to Liora Lazarus who provided crucial guidance on my work and career throughout her time at the institute. Many outstanding colleagues and friends have contributed their insights in conversations and discussions over the years that helped shape the ideas in this book: Jeremias Adams-Prassl, Simon Alison, Kaitlin M Ball, Nick Barber, John Bell, Dominic Burbidge, Oliver Butler, Paul Craig, Colm O'Cinneide, Ya Lan Chang, Michael Dafel, Pavlos Eletheriadis, Franziska Fischer, Kent McFadzien, Liz Fisher, Lilia Giugni, Andrew Higgins, Hayley Hooper, Miles Jackson, Aileen Kavanagh, Knut Karnapp, George Letsas, Daniella Lock, Richard Martin, Kaara Martinez, Ana Julia Mauricio, Michael Molavi, Shelby Newsad, Maha Rafi Atal, Farhan Samanani, Rajiv Shah, Arjun Sharath, Ewan Smith, Lukas Schäffner, Eugenio Velasco Ibarra, Jorge Viñuales, and Alison Young. Their work consistently inspires me and discussions with them broaden my horizons. They are the fabric of a vibrant community of scholars without which no serious intellectual project can succeed. Elinor Buys provided outstanding research assistance on the final manuscript, especially in updating the dataset. Thanks are also due to Finola O'Sullivan, Tom Randall, Gemma Smith, and the team from Cambridge University Press, as well as the anonymous reviewers for their helpful and constructive feedback.

I am grateful to my siblings, Daniel and Christian, who in their own unique ways reminded me that the world out there is bigger than it can appear from an Oxbridge library. My deepest gratitude goes to my parents, Angela and Udo. Their unconditional love and support gave me the confidence to overcome a learning disability, while their education and encouragement broadened my horizons beyond what many are privileged to experience.

The most important contribution to the book was made by my partner, Lauren. Her love, humour, patience, advice, and unwavering kindness kept me sane during the lows of my academic career and made the high points all the more enjoyable. Not a single page would be the same without the contribution of these wonderful people in my life.

Abbreviations

ACHPR	African Charter on Human and Peoples' Rights
ACHR	American Convention on Human Rights
AfComHR	African Commission on Human and Peoples' Rights
AfCtHR	African Court of Human and Peoples' Rights
American Declaration	American Declaration on the Rights and Duties of Man
BGBl	*Bundesgesetzblatt* – German Federal Gazette
BGH	*Bundesgerichtshof* – German Federal Court of Justice
BGHZ	*Entscheidungen des Bundesgerichtshofes in Zivilsachen* – Official Law Report of the German Federal Court of Justice for Private Law Matters
BNatSchG	*Bundesnaturschutzgesetz* – German Federal Nature Conservation Act
BVerfG	*Bundesverfassungsgericht* – German Federal Constitutional Court
BVerfGE	*Entscheidungen des Bundesverfassungsgerichts* – Official Law Report of the German Federal Constitutional Court
BVerfGK	*Bundesverfassungsgericht (Kammerentscheidung)* – German Federal Constitutional Court (Chamber judgment)
BVerwG	*Bundesverwaltungsgericht* – German Federal Administrative Court

BVerwGE	*Entscheidungen des Bundesverwaltungsgerichts* – Official Law Report of the German Federal Administrative Court
CFC	chlorofluorocarbons
CoE	Council of Europe
COP	United Nations Climate Change Conference of Parties
dB	decibel, a logarithmic unit used to express the ratio of two values of a physical quantity
dBA	a decibel measurement of sound, achieved through use of the A weighting filter, which approximates a human ear's response to sound
dBA Leq	average continuous sound level on a logarithmic scale, given in decibel
dBA Lmax	highest RMS sound pressure level within the measuring period, given in decibel
ECHR	European Convention on Human Rights
ECJ	European Court of Justice
EComHR	European Commission of Human Rights
ECtHR	European Court of Human Rights
EFSA	European Food Safety Authority
EU	European Union
GG	*Grundgesetz* – German Basic Law
Grand Chamber	Grand Chamber of the European Court of Human Rights
HRA	UK Human Rights Act 1998
HRC	UN Human Rights Council
IAComHR	Inter-American Commission on Human Rights
IACtHR	Inter-American Court of Human Rights
ICCPR	International Covenant on Civil and Political Rights
ICESCR	International Covenant on Economic, Social and Cultural Rights
ICJ	International Court of Justice
ICJ Rep	International Court of Justice Reports of Judgments, Advisory Opinions and Orders
NGO	non-governmental organisation
NO_2	nitrogen dioxide
OAU	Organisation of African Unity

Plenary	Plenary of the European Court of Human Rights (abolished)
PM10	fine particulate matter (10 µm or less)
PM2.5	fine particulate matter (2.5 µm or less)
RMS	root mean square of a time-varying quantity, obtained by squaring the amplitude at each instant, obtaining the average of the squared values over the interval of interest, and then taking the square root of this average
SDGs	Sustainable Development Goals
UKHL	United Kingdom House of Lords
UKSC	Supreme Court of the United Kingdom
UN	United Nations
UN Rep Int'l Arb Awards	Reports of International Arbitral Awards
UNTS	United Nations Treaty Series
VCLT	Vienna Convention on the Law of Treaties
VwGO	*Verwaltungsgerichtsordnung* – German Code of Administrative Court Procedure
VwVfG	*Verwaltungsverfahrensgesetz* – German Administrative Procedure Act
WHO	World Health Organization
ZACC	South African Constitutional Court

1

Introduction

1.1 KEY ISSUES

Media outlets have been awash in recent years with reports of environmental degradation and pollution, chiefly focusing on major cities in the developing world. Images of air pollution from New Delhi and Beijing are infamous, as are unprecedented levels of contamination of major rivers and waterways, such as the Ganges.[1] These reports are often accompanied by accounts of ineffective public administration, corruption, unrestricted economic development, and urbanisation.[2]

Contrary to the prevailing narrative, serious environmental harm is not solely, nor perhaps even primarily, a phenomenon in developing countries. Despite the comparatively detailed regulatory regimes that operate in Europe, millions of Europeans are routinely exposed to the most serious public health hazard of the twenty-first century, one that has been described as a greater threat than Ebola or AIDS.[3] Largely unbeknownst to the wider public, the air is contaminated with highly dangerous pollutants that cause as many as 3.7

[1] Minsi Zhang, Yu Song, and Xuhui Cai, 'A Health-Based Assessment of Particulate Air Pollution in Urban Areas of Beijing in 2000–2004' (2007) 376 *Science of the Total Environment* 100; Steve Hamner and others, 'The Role of Water Use Patterns and Sewage Pollution in Incidence of Water-Borne/Enteric Diseases Along the Ganges River in Varanasi, India' (2006) 16 *International Journal of Environmental Health Research* 113.

[2] Hbib Sekrafi and Asma Sghaier, 'Examining the Relationship between Corruption, Economic Growth, Environmental Degradation, and Energy Consumption: A Panel Analysis in MENA Region' [2016] *Journal of the Knowledge Economy* 1; Matthew A. Cole, 'Corruption, Income and the Environment: An Empirical Analysis' (2007) 62 *Ecological Economics* 637; R. J. Smith and others, 'Governance and the Loss of Biodiversity' (2003) 426 *Nature* 67.

[3] *Global Environment Outlook – Geo-6 Assessment for the Pan-European Region* (United Nations Environment Programme, 2016) 61; 'WHO Chief: Air Pollution Link to 600,000 Deaths in Children', *BBC News* (6 March 2017) www.bbc.co.uk/news/health-39175410, accessed 1 April 2021.

million deaths annually in the world and 480,000 in Europe, according to estimates by the World Health Organization (WHO).[4] Those disproportionately affected by ambient air pollution are already vulnerable segments of society: children, the elderly, and the poor.[5]

The front line of this crisis lies in major cities such as London, as much as in New Delhi and Beijing, where the pollution has reached dramatic proportions. Overall, it was estimated in successive studies that in any given year, depending on the methodology employed, at least 3,537 and as many as 9,416 deaths in Greater London were attributable to long-term exposure to air pollution.[6] The extremely small particles from fossil fuel power plants, combustion engines, tyres, and brakes contained in the air settle in the lungs and brains of individuals, and have been linked to various respiratory illnesses such as cancers and Alzheimer's disease.[7] Despite these serious health implications, regulatory and enforcement responses from governments have been sluggish and ineffective: more than 95 per cent of the urban population in Europe remains exposed to air pollution above WHO guidelines.[8] Indeed, one does not have to be a radical ecologist to appreciate that a staggering number of everyday human activities result in hazardous environmental harm, driven by an ideology that prioritises short-term wealth accumulation over long-term and sustainable economic development.

The argument that the thriving of humankind is fundamentally linked to preserving the environment and ecosystems has been advanced in the field of environmental ethics for decades.[9] Regardless of whether one ultimately assigns an instrumental or an intrinsic value to the natural environment, it is

4 For detailed definitions of harmful substances, *see* 'Ambient (Outdoor) Air Quality and Health', *World Health Organization* (2014) www.who.int/mediacentre/factsheets/fs313/en/, accessed 1 April 2021; *Burden of Disease from Ambient Air Pollution for 2012* (World Health Organization, 2014) www.who.int/phe/health_topics/outdoorair/databases/AAP_BoD_results_March2014.pdf, accessed 1 April 2021, 1.

5 *Global Environment Outlook*, 9.

6 Heather Walton and others, *Understanding the Health Impacts of Air Pollution in London* (King's College London – Environmental Research Group, 2015) 40; A. M. Gowers, B. G. Miller, and J. R. Stedman, *Estimating Local Mortality Burdens Associated with Particulate Air Pollution* (Public Health England, 2014) 21.

7 Barbara A. Maher and others, 'Magnetite Pollution Nanoparticles in the Human Brain' (2016) 113 *Proceedings of the National Academy of Sciences* 10797.

8 *Global Environment Outlook – Geo-6 Assessment for the Pan-European Region*, 61.

9 Arne Naess, 'The Shallow and the Deep, Long-Range Ecology Movement: A Summary' (1973) 16 *Inquiry* 95; John Passmore, *Man's Responsibility for Nature* (2nd ed., Duckworth 1980); Andrew Light and Eric Katz, 'Introduction: Environmental Pragmatism and Environmental Ethics as Contested Terrain' in Eric Katz and Andrew Light (eds.), *Environmental Pragmatism* (Routledge 1996) 1; Avner de Shalit, *Why Does Posterity Matter* (Routledge 1994).

abundantly clear that the current levels of environmental harm cannot continue without putting the long-term survival of humankind in serious doubt.[10] The most prominent example is anthropocentric climate change, which has attracted mass protests and shapes the political activism of an entire generation. Beyond this headline-grabbing policy debate, countless environmental degradations and pollutants remain chronically under-regulated.

This is perplexing given that many national constitutions and international treaties recognise the value of a sound environment, and at times contain express provisions mandating its protection: most prominently in the form of individual rights to a healthy environment or constitutional mission statements that commit states to environmental protection. Widespread breaches of these legal norms present a fundamental challenge to the seriousness of such commitments, especially as the meaningful enjoyment of a host of human rights is intimately linked to and presupposes basic environmental conditions. A right to protest government policy is an empty guarantee if the air that protesters breathe is contaminated to a point where venturing outdoors entails serious health risks.[11] Property rights safeguarding home ownership are meaningless if states permit airplanes to pass directly overhead, and a state that orchestrates a crackdown on independent journalism does no less violence to human rights than a state that permits industries to contaminate drinking water with impunity.[12] Given these pervasive threats, the apparent failure to translate general and abstract commitments to human rights and environmental protection into specific and practical measures that safeguard the environment is a serious shortcoming. In part, this failure is fuelled by the lack of attention that human rights law scholarship pays to the implications of environmental harm, and the ad hoc doctrinal approach taken by many international and domestic courts.

Orthodox legal thinking has considerable reservations about attaching the status of human rights to environmental interests. Such scepticism was

[10] Dale Jamieson, *Reason in a Dark Time – Why the Struggle Against Climate Change Has Failed – and What It Means for Our Future* (Oxford University Press 2014) 237.

[11] Oliver Wainwright, 'Inside Beijing's Airpocalypse – A City Made "Almost Uninhabitable" by Pollution' *The Guardian* (16 December 2014) www.theguardian.com/cities/2014/dec/16/beijing-airpocalypse-city-almost-uninhabitable-pollution-china, accessed 1 April 2021.

[12] Harry Cockburn, 'Turkey Coup: Erdogan Bans All Academics from Leaving Country as Government Crackdown Intensifies' *Independent* (20 July 2016) www.independent.co.uk/news/world/europe/turkey-coup-erdogan-academics-ban-leaving-country-government-crackdown-latest-a7146591.html, accessed 1 April 2021; *Stockton and others* v. *United Kingdom* App no 36053/97 (Commission, 15 January 1998) [liquid aluminium entered water distribution network]; *Otgon* v. *The Republic of Moldova* App no 22743/07 (ECtHR, 25 October 2016) [ruptured sewage pipeline contaminated drinking water supply].

admittedly more fashionable in earlier scholarship and court decisions, but its echoes remain audible in critiques that question the utility and importance of environmental human rights, our ability to define them with sufficient clarity, and concerns as to how we balance their demands with competing interests.[13] These critiques often connect with wider concerns: for instance, the capacity of human rights to bring about substantial policy improvements and social change, and the suitability of courts as forums for the effective and legitimate enforcement of such rights, especially when their decisions conflict with those of democratic legislatures. Accordingly, critics instead emphasise procedural over substantive guarantees, and a gradual implementation model based on economic, social, and cultural rights as a template for environmental protection through human rights.[14]

While complete opposition has become rare, human rights and environmental law scholarship nonetheless tend to focus only in passing on environmental human rights and their implications.[15] Even environmentally progressive accounts often present environmental harm as a human rights issue only in specific contexts such as climate change,[16] and the associated

[13] Philippe Sands and others, *Principles of International Environmental Law* (4th ed., Cambridge University Press 2018) chapter 17; Alan Boyle, 'Human Rights and the Environment – Where Next?' in Ben Boer (ed.), *Environmental Law Dimensions of Human Rights* (Oxford University Press 2015) 219; Conor Gearty, 'Do Human Rights Help or Hinder Environmental Protection?' (2010) 1 *Journal of Human Rights and the Environment* 7.

[14] Stuart Bell, Donald McGillivray, and Ole W. Pedersen, *Environmental Law* (8th ed., Oxford University Press 2013) 15; Margaret DeMerieux, 'Deriving Environmental Rights from the European Convention for the Protection of Human Rights and Fundamental Freedoms' (2001) 21 *Oxford Journal of Legal Studies* 521; B. Weston and David Bollier, 'Toward a Recalibrated Human Right to a Clean and Healthy Environment: Making the Conceptual Transition' (2013) 4 *Journal of Human Rights and the Environment* 116; James Harrison, 'Reflections on the Role of International Courts and Tribunals in the Settlement of Environmental Disputes and the Development of International Environmental Law' (2013) 25 *Journal of Environmental Law* 501.

[15] William A. Schabas, *The European Convention on Human Rights – A Commentary* (Oxford University Press 2015) 387–8; Philip Alston, Ryan Goodman, and Henry J. Steiner, *International Human Rights: Text and Materials* (Oxford University Press 2013) 1536; Olivier De Schutter, *International Human Rights Law – Cases, Materials, Commentary* (2nd ed., Cambridge University Press 2014) 372–3, 457; Rhona K. M. Smith, *Textbook on International Human Rights* (6th ed., Oxford University Press 2014) 395, remarkably does not mention the environmental cases, nor do Ilias Batekas and Lutz Oette, *International Human Rights Law and Practice* (Cambridge University Press 2013) 516–7; Merris Amos, *Human Rights Law* (2nd ed., Hart Publishing 2014) 425, 508–9, 515.

[16] Catalina Devandas Aguilar and others, *The Effects of Climate Change on the Full Enjoyment of Human Rights* (United Nations Human Rights Council, 2015); Anna Grear, 'Towards "Climate Justice"? A Critical Reflection on Legal Subjectivity and Climate Injustice: Warning Signals, Patterned Hierarchies, Directions for Future Law and Policy' (2014) 5 *Journal of Human Rights and the Environment* 103; Eric A. Posner, 'Climate Change and International

refugees,[17] as well as natural resources such as water.[18] Overall, the scholarship thus conveys the image of a fragmented relationship, a patchwork of largely unrelated issues that do not lend themselves to a comprehensive and coherent normative or indeed doctrinal approach.

In my view, this image is misleading and based on an undertheorised relationship between human rights and environmental harm. Many important questions remain unsettled, including to what extent environmental harm can be accommodated by human rights and more significantly, how human rights protection regimes can provide effective safeguards in practice. The prospect of a practical normative framework for courts would be fanciful if appropriately addressing environmental harm required a radical departure from fundamental characteristics of human rights law, or the practice and procedures of courts. Against prevailing scepticism, this book argues that there is a sufficient basis for a comprehensive framework that theorises the relationship between human rights and environmental harm while remaining practically viable in the context of existing protection regimes: that framework is the environmental minimum.

In simplified terms, the environmental minimum framework facilitates the practical protection of the environment through existing human rights law. On the one hand, it provides individuals with a claim against states to ensure basic regulations of harmful degradation and pollution (protective function), while on the other hand providing a justification for environmental regulations when they are challenged based on human rights (limitation function). The environmental minimum endeavours to give specific guidance to courts based on the abstract commitments to environmental protection that human rights presuppose, and which constitutional rights, mission statements, and international treaties express in general terms. This book defends two connected claims: that the environmental minimum represents a normatively

Human Rights Litigation: A Critical Appraisal' (2007) 155 *University of Pennsylvania Law Review* 1925.

17 Bonnie Docherty and Tyler Giannini, 'Confronting a Rising Tide: A Proposal for a Convention on Climate Change Refugees' (2009) 33 *Harvard Environmental Law Review*; Daniel A. Farber, 'Basic Compensation for Victims of Climate Change' (2007) 155 *University of Pennsylvania Law Review* 1605.

18 Daphina Misiedjan, *Towards a Sustainable Human Right to Water: Supporting Vulnerable People and Protecting Water Resources* (Intersentia 2019); Malcolm Langford and Anna F. S. Russell, *The Human Right to Water: Theory, Practice and Prospects* (Cambridge University Press 2017); Astrida Neimanis, 'Alongside the Right to Water, a Posthumanist Feminist Imaginary' (2014) 5 *Journal of Human Rights and the Environment* 5; Paula Gerber and Bruce Chen, 'Recognition of the Human Right to Water – Has the Tide Turned?' (2011) 36 *Alternative Law Journal* 21; John Scanlon, Angela Cassar, and Noémi Nemes, 'Water as a Human Right?' [2004] IUCN Environmental Policy and Law Paper.

consistent and attractive as well as a practically viable framework for conceptualising the relationship between human rights and the environment.

The normative appeal derives from the compatibility of the environmental minimum with the normative claims of environmental human rights. The framework presents a principled account that incorporates the core characteristics of human rights and is consistent with key principles in domestic and international law, including the rule of law and *pacta sunt servanda*. The key normative claim is this: states must provide principled regulation of environmental harms in good faith, based on the best available scientific evidence, and ensure their consistent practical enforcement. This applies to any state that takes its human rights obligations and the rule of law seriously. This normative argument is developed in Part I of this book.

The environmental minimum framework is practically viable because it accounts for the extensive and rich environmental case law of international (UN, Africa, American, European) and domestic courts, most notably the European Court of Human Rights (ECtHR). The environmental minimum translates the general and abstract commitment to human rights and environmental protection into actionable standards of review that chime with and incrementally develop the modern doctrinal approach. This practicality argument preoccupies Part II of this book which conducts an in-depth analysis of the environmental minimum in the context of the ECHR.

Part III considers the potential contributions of the environmental minimum in other international protection regimes (ACPHR, ACHR, ICCPR, ICESCR) and for environmental constitutionalism through the lens of two case studies, namely South Africa and Germany. Part IV explores what the environmental minimum framework can achieve beyond human rights law, notably by alleviating three pathologies that plague domestic and international environmental regulation, as well as clarifying and enhancing the claims and enforcement mechanisms of international environmental law (transboundary environmental harm and Sustainable Development Goals).

The following section of the introduction outlines definitions of key terms used throughout the book before turning to a concise summary of the distinct elements of the overall argument.

1.2 DEFINITIONS

1.2.1 *Environment*

The term 'environment' describes 'all of the external abiotic and biotic factors, conditions, and influences that affect the life, development, and survival of an organism or a community.'[19] The term thus encompasses all natural resources, such as air, land, and water, as well as the natural and the human-made environment, as both play important roles in human well-being.[20] The book is however concerned only with harm to the natural environment, that which is comprised of the interaction of all living species, climate, weather, and natural resources that affect human survival and economic activity.[21] This definition excludes social environments, for instance a workplace, adversely affected through a loss of trust amongst co-workers. The overarching term of environmental harm encompasses both degradation and pollution.

Degradation is defined as the 'depletion or destruction of a potentially renewable resource such as air, water, soil, forest, or wildlife, by using it at a rate faster than it can be naturally renewed.'[22] Pollution is defined as the 'process of contaminating or polluting', for instance of air and water.[23] In simplified terms, degradation takes something away from the natural environment, for instance by depleting a source of water, whereas pollution adds something undesirable to the environment, for instance arsenic to drinking water. This understanding therefore excludes the degradation of human-made environments, such as those found within buildings and those created through infrastructure, but notably includes their pollution: while a poor state of maintenance of a bridge is not relevant with respect to environmental degradation, as it does not entail the depletion of resources, it may well constitute pollution if the bridge is hazardous to wildlife. Both environmental degradation and pollution will be collectively referred to as 'environmental harm' in this book.

1.2.2 *Human Rights*

This book draws on the account of James Nickel, who succinctly formulates the key characteristics of human rights in contemporary legal thinking. It is undoubtedly worthwhile to engage with other accounts of human rights,[24] but

[19] 'Environment' in Chris Park (ed.), *A Dictionary of Environment and Conservation* (Oxford University Press 2007).

[20] Adopted from the 'Declaration of the United Nations Conference on the Human Environment', 16–23 June 1972, UN Doc.A/CONF 48/14/Rev.1 [1].

[21] D. L. Johnson and others, 'Meanings of Environmental Terms' (1997) 26 *Journal of Environmental Quality* 581.

[22] 'Environmental Degradation' in Park (ed.), *A Dictionary of Environment and Conservation.*

[23] 'Pollution' in Park (ed.), *A Dictionary of Environment and Conservation.*

[24] See, for instance, Charles R. Beitz, *The Idea of Human Rights* (Oxford University Press 2011) and his practical conception; Amartya Sen, 'Elements of a Theory of Human Rights' (2004) 32 *Philosophy & Public Affairs* 315 and his freedom-based conception; Rex Martin, *A System of*

a comprehensive discussion would go beyond the requirements of the argument presented here. Save for a few basic assumptions, the environmental minimum does not rely on any particular notion of human rights.[25]

According to Nickel, human rights are rights which contain numerous and specific claims focusing on freedoms, protections, and benefits (as opposed to few and general statements of principles), held by rightsholders (all living persons) against addressees (those with duties and responsibilities towards rightsholders, typically states). They set minimum standards (as opposed to exhaustively formulating an ideal),[26] which are independent of recognition or enactment,[27] but that make no claim to absolute priority, only to relative priority over other considerations ('high-priority norms').[28] This understanding of human rights has three immediate consequences for the idea of protecting the environment through human rights.

First, because the rightsholders are individual humans, human rights do not protect from environmental harm that is not established with sufficient certainty and linked to detrimental impacts on humans. This excludes from consideration poorly evidenced environmental harm which does not directly impact humans: for instance, electromagnetic waves from cell phone towers, or harm that occurs naturally regardless of human activities, for instance a volcanic eruption.

The 'specific claims' characteristic meanwhile excludes broad expressions of principle, such as those found in constitutional mission statements proclaiming the importance of environmental protection. These cannot be considered a sufficient basis for protection because it is not clear whether individuals can make legal claims and typically only commit the state to take environmental policy considerations into account.[29] Flexibility and

Rights (Oxford University Press 1997) and his moral conception; in the specific context of environmental rights, see Tim Hayward, *Constitutional Environmental Rights* (Oxford University Press 2005) 43, he draws considerably on the views of Jürgen Habermas, 'Remarks on Legitimation through Human Rights' (1998) 24 *Philosophy & Social Criticism* 157, 161 and sees recognition as a necessary, but on its own insufficient, guide to the normative content of human rights.

25 See Chapter 2.

26 Minimalism also features in other contemporary human rights accounts, see James Griffin, *On Human Rights* (Oxford University Press 2008) 33; Beitz, *The Idea of Human Rights*, 141–2.

27 A key difference to the account by Martin, *A System of Rights* and Philip Alston, 'Conjuring up New Human Rights: A Proposal for Quality Control' (1984) 78 *American Journal of International Law* 607.

28 James W. Nickel, *Making Sense of Human Rights* (2nd ed., Blackwell 2007) 9–10; a key difference to the account of Sen, 'Elements of a Theory of Human Rights'.

29 See, for instance, Article 20a Basic Law for the Federal Republic of Germany (adopted 8 May 1949, in force 23 May 1949, as last amended 29 September 2020) (GG), discussed in Chapter 9.

interpretational leeway are only compatible with the concept of human rights to the extent that obligations remain reasonably determinable in the context of specific cases. Human rights require more than expressions of principle and must endeavour to offer tangible protections to rightsholders.

The minimum standard is a key conceptual feature of human rights, as they do not purport to provide optimal protection,[30] nor enshrine any particular comprehensive notion of justice.[31] Likewise, protections from environmental harm are not intended to describe, nor can they provide, comprehensive protection of what may constitute an ideal environment. Such an attempt would not only conflict with the concept of human rights but also fail to recognise that human rights can make significantly diverging and conflicting normative claims. As many sceptics of environmental human rights point out, reasonable people disagree over what constitutes an ideal environment and the means by which it should be achieved.[32] The scope for agreement is reduced the further one ventures away from the lowest common denominators that necessarily underpin any vision of environmental protection.

Finally, the high-priority characteristic of human rights requires a justification for the prioritisation of environmental protection over other desirable interests. We must explain why environmental interests belong to the category of things that human rights protect: what distinguishes them, for instance, from business opportunities in the extraction of natural gas resources through fracking, or majority preferences in the distribution of resources, for instance in allocating public funding to football clubs over comparatively less popular theatres, opera houses, and museums?[33]

Once this priority of environmental interests is established, we further require an idea of how environmental interests can be balanced with competing interests. This is because recognising a human rights-based interest in environmental protection does not equate to absolute priority over other considerations. Not every marginal furtherance of environmental protection can justify limitations of other human rights: for instance, expropriating land in order to increase the size of a nature reserve requires carefully weighing environmental protection against property ownership. An adjudicative body must resolve cases where human rights are invoked on both sides of an issue: to strike an appropriate balance, determine compensation and necessary remedies. This is particularly salient in the context of environmental protection, as it has the potential to

30 Nickel, *Making Sense of Human Rights*, 36.
31 Jeff King, *Judging Social Rights* (Cambridge University Press 2012) 21.
32 Boyle, 'Human Rights and the Environment – Where Next?', 219.
33 King, *Judging Social Rights*, 21.

conflict with a wide range of competing interests based on human rights, notably general economic development: this has been a source of particular tension with respect to international efforts to combat climate change.[34]

1.2.3 *Environmental Rights*

Based on this understanding of the environment, environmental harm, and human rights, there are several types of environmental rights that should be distinguished.

Procedural environmental rights typically focus on status obligations granting individuals participatory rights, such as access to information during administrative or regulatory proceedings, or standing to raise environmental claims in court. The prime international example is found in the Aarhus Convention, which commits states to providing access to information and to justice and public participation in environmental decision-making processes.[35] Domestic examples include the many forms of participation rights granted to non-governmental organisations (NGOs) and procedural rights to pursue public interest litigation, for instance under § 61 of the Federal Nature Protection Act in Germany.[36]

Substantive environmental rights contain positive and negative obligations, focusing on guaranteeing an environment of a certain quality or the prevention of certain environmental harms. They are often codified in domestic constitutions, contained within ordinary legislation, or form a part of international agreements, for instance regional human rights protection regimes. Their scope is potentially wide-ranging and may encompass protections of the natural environment beyond the immediate needs of humans. Typical examples are section 24 of the Constitution of South Africa as well as Article 24 (Right to a general satisfactory environment) of the ACHPR and Article 11 of the Protocol of San Salvador to the ACHR.

As we have seen, human rights are rights based on considerations of political morality, and individuals enjoy them independent of express recognition or codification.[37] Consequently, environmental human rights describe

34 For a discussion of climate change as the pivotal transboundary environmental harm, see Chapter 11.

35 Convention on Access to Information, Public Participation in Decision-Making and Access to Justice in Environmental Matters (adopted 28 June 1998, in force 30 October 2001), 2161 UNTS 447 (Aarhus Convention).

36 Bundesnaturschutzgesetz of 29 July 2009 (BGBl. I S. 2542), last amended through Article 5 of the Law of 25 February 2021 (BGBl. I S. 306).

37 Nickel, *Making Sense of Human Rights*, 9; King, *Judging Social Rights*, 21.

a subset of such rights that focus on preserving and limiting harm to the natural environment. The key difference to procedural and substantive environmental human rights is the independence from recognition, the focus on immediate threats to humans only, and the more limited scope of protection. Environmental human rights stretch only as far as necessary to safeguard the fundamental commitments that underpin human rights and are therefore more circumspect in their protection of the natural environment than many procedural and substantive environmental rights. Nonetheless, elements of environmental human rights are found in both procedural and substantive environmental rights, and are broadly recognised at the UN Human Rights Council (HRC), as well as reflected in the doctrine of courts and tribunals at the domestic and international levels.[38] Indeed, it is difficult to make sense of the environmental case law of many courts and tribunals without inferring a commitment to the notion of environmental human rights.

With the key definitions established, the following section offers an outline of the overall argument in favour of the environmental minimum: it sketches the moral case for environmental human rights and how their normative claims and practical appeal are realised in the environmental minimum. This is intended both as a primer and a convenient reference point for the following chapters.

1.3 THE ARGUMENT IN A NUTSHELL

The environmental minimum provides a framework for a human rights-based review of environmental harm. The core contribution of the environmental minimum is that it translates general and abstract commitments to human rights and environmental protection into specific and practical measures to protect the environment, informed by generally accepted scientific standards.

The normative argument in favour of the environmental minimum framework centres on its consistency with core characteristics of human rights, specifically its compatibility with the normative claims of environmental human rights, and its alignment with fundamental legal principles, namely the rule of law and *pacta sunt servanda*, the principle that every treaty in force

[38] See *25th Session of the Human Rights Council (3–28 March 2014) Resolution – Human Rights and the Environment* (UN A/HRC/25/L31); and the reports of the special rapporteur, most recently, John H. Knox, *Report of the Special Rapporteur on the Issue of Human Rights Obligations Relating to the Enjoyment of a Safe, Clean, Healthy and Sustainable Environment* (UN A/HRC/34/49, 2017).

is binding upon the parties to it and must be performed by them in good faith (Article 26 Vienna Convention on the Law of Treaties).[39] The framework distinguishes between human rights, understood as an expression of idealised values and norms derived from considerations of political morality, and human rights *law*, the adjudicated domestic and international protection regimes that entail practical limitations and trade-offs.

What renders the environmental minimum preferable over other conceivable incarnations of environmental human rights is its incremental and minimalist approach. The environmental minimum does not purport to facilitate the comprehensive protection of the environment in human rights law that might be justified as a matter of political morality. Crucially, the framework only becomes operational if rightsholders succeed in establishing a specific risk to a human right under an existing protection regime. To a significant extent, the environmental minimum is thus compatible with the doctrinal position adopted under the ECHR and other international and domestic protection regimes. However, the framework does not rely on any specific human rights, nor does it seek to introduce novel rights unfamiliar to the respective human rights regime.

Section 1.3.1 covers the groundwork for the environmental minimum developed in Part I of this book: it illustrates the moral case for environmental human rights. This notably includes explaining how commitments to human rights necessarily entail some environmental safeguards and responding to sceptical accounts that centre on vagueness and the potential for conflict between rights.

Section 1.3.2 then sketches the principles of the environmental minimum developed in Chapter 3 and how these derive from (and are ultimately justified through) the normative claims of environmental human rights.

Section 1.3.3 provides an overview of Part II of this book, where the environmental minimum framework is applied to analyse the doctrinal approach of the ECHR to environmental harm. It shows significant areas of compatibility with the framework alongside some limitations under the current doctrinal approach.

Section 1.3.4 sketches the approach taken in other international and regional protection regimes and introduces the two case studies of domestic environmental constitutionalism, namely South Africa and Germany, that comprise Part III of this book.

[39] Vienna Convention on the Law of Treaties (adopted 23 May 1969, in force 27 January 1980) 1155 UNTS 331 (VCLT).

Finally, Section 1.3.5 outlines the potential contributions of the environmental minimum beyond human rights law discussed in Part IV, most notably for the purposes of environmental regulation and international environmental law.

1.3.1 *Environmental Human Rights*

This book argues that human rights must contain a sub-category of rights that provide protections from environmental harm. These environmental human rights follow from the insight that degradation and pollution of a sufficient severity threaten the enjoyment, and hence ultimately the fundamental commitments, that underpin all human rights.[40] At the very least, environmental human rights seek protections to safeguard human dignity and autonomy, and the environmental minimum represents an attractive avenue towards implementing such protections through human rights protection regimes. Nonetheless, the environmental minimum does not suggest that it comprehensively reflects the wide range of possible normative claims made by environmental human rights.

The argument is significantly narrower: the environmental minimum constitutes an internally coherent and plausible interpretation of the protections necessary to safeguard commitments to human dignity and autonomy in the face of environmental harm.

The persistent and rightly identified difficulties of agreeing on a precise definition of an ideal environment should not distract us from the wholly achievable task of agreeing on basic safeguards, a lowest common denominator that nonetheless holds the promise of tangible improvements.[41] While it is true that even at this basic level there is scope for conflict with competing interests,[42]

40 See the dataset gathered for this book.

41 Scepticism is nonetheless commonplace, *see* Patricia W. Birnie, Alan E. Boyle, and Catherine Redgwell, *International Law and the Environment* (3rd ed., Oxford University Press 2009) 281, 302; James W. Nickel, 'The Human Right to a Safe Environment: Philosophical Perspectives on Its Scope and Justification' (1993) 18 *Yale Journal of International Law* 281, 285; Ulrich Beyerlin and Thilo Marauhn, *International Environmental Law* (Hart – C. H. Beck – Nomos 2011) 392; Christian Tomuschat, *Human Rights – Between Idealism and Realism* (2nd ed., Oxford University Press 2008) 57; Francesco Francioni, 'International Human Rights in an Environmental Horizon' (2010) 21 *European Journal of International Law* 41, 54.

42 Such concerns have in my view been exaggerated, *see* Alston, 'Conjuring up New Human Rights: A Proposal for Quality Control', 614; Michael Kloepfer, 'Droht der Autoritäre Ökologische Staat?' (1993) 48 *Universitas* 236, 240; Chris Miller, *Environmental Rights: Critical Perspectives* (Routledge 1998) xi.

human rights law offers a well-established mechanism to balance competing demands and resolve any conflicts.

Some critics reject human rights as inherently incapable of providing the necessary protection of the natural environment and climate,[43] or as unduly dismissive of the rights of the broader ecosystem, for instance the protection of (sentient) animals and the legal personhood of nature.[44] These criticisms are understandable but ultimately misguided. A focus on humans is the inevitable consequence of a human rights law-based approach, and minimalism is inherent to the concept of human rights because rights often make conflicting demands. Environmental human rights scepticism is often nested within a broader debate in environmental ethics between those who argue in favour of the intrinsic value of the environment (intrinsic approach) and those who seek environmental protection primarily due to its utility to humans (anthropocentric approach). However, both the anthropocentric and intrinsic approaches make significantly overlapping normative claims and even beyond their points of agreement, the differences are not necessarily irreconcilable.[45] Both approaches accept that environmental harm should be minimised because of its detrimental impacts on a complex system of interconnection between organisms, of which humans form an integral part. As we explore in Part I of

[43] Catherine Redgwell, 'Life, The Universe and Everything: A Critique of Anthropocentric Rights' in Michael R. Anderson and Alan E. Boyle (eds.), *Human Rights Approaches to Environmental Protection* (Clarendon 1996) 71; Francioni, 'International Human Rights in an Environmental Horizon'; Anna Grear, 'Towards New Legal Futures? In Search of Renewing Foundations' in Anna Grear and Evadne Grant (eds.), *Thought, Law, Rights and Action in the Age of Environmental Crisis* (Edward Elgar 2015); Evadne Grant, 'Re-imagining Adjudication: Human Rights Courts and the Environment' in Grear and Grant (eds.), *Thought, Law, Rights and Action*; Pierre-Marie Dupuy and Jorge E. Viñuales, *International Environmental Law* (Cambridge University Press 2015) 300.

[44] Christopher D. Stone, 'Should Trees Have Standing?' (1972) 45 *Southern California Law Review* 491; Susan L. Goodkin, 'The Evolution of Animal Rights' (1986–1987) 18 *Columbia Human Rights Law Review* 259; Peter Singer, *Animal Liberation: A New Ethics for Our Treatment of Animals* (Harper Collins 1976); Tom Regan, *The Case for Animal Rights* (Routledge 1983); Tom Regan and Peter Singer, *Animal Rights and Human Obligations* (Prentice-Hall 1976); Cynthia Giagnocavo and Howard Goldstein, 'Law Reform or World Reform: The Problem of Environmental Rights' (1989) 35 *McGill Law Journal* 345; Massimiliano Montini and Francesca Volpe, 'Rethinking Regulation for Promoting an Ecologically Based Approach to Sustainability' [2016] C-EENRG Working Papers 1, 24; Elaine C. Hsiao, 'Whanganui River Agreement' (2012) 42 *Environmental Policy and Law* 371.

[45] Hayward, *Constitutional Environmental Rights*, 34–5; de Shalit, *Why Does Posterity Matter*, 7–8; Andrew Light, 'Compatibilism in Political Ecology' in Eric Katz and Andrew Light (eds.), *Environmental Pragmatism* (Routledge 1996) 161 [uses the terms environmental materialists and environmental ontologists to distinguish between anthropocentric and environment centred approaches, respectively]; Bryan G. Norton, *Toward Unity among Environmentalists* (Oxford University Press 1991) chapter 10.

this book, this basic agreement provides the foundation for some level of reconciliation and jointly supported normative claims.

1.3.2 *Principles of the Framework*

The principles of the environmental minimum are modelled on the core characteristics of human rights and the normative claims of environmental human rights.

The threats to human dignity and autonomy that environmental harm represents are reflected in the specific risk standard, which requires the degradation or pollution to affect a specific, previously recognised human right. These specific human rights anchor the broader practical argument in favour of the framework. Primarily, the environmental minimum normatively demands good faith engagement of states with scientific evidence of the threats to humans posed by environmental harm and mandates the establishment, as well as the consistent enforcement, of regulations that address specific environmental risks to human rights.[46]

A specific risk is established where a reasonable hypothesis is brought forward, based on scientific evidence,[47] that links an environmental harm to a threat towards a specific human right. The specific risk standard is best understood as a preliminary filter which excludes from consideration environmental harm that either does not threaten established human rights or has an insufficient basis in science. An example of the former is the annual culling of badgers in the United Kingdom, which is intended to reduce the spread of bovine tuberculosis.[48] The cull as such does not impact on human rights, notwithstanding its environmental impact and broader potential for harm and unintended consequences. An example of an insufficient scientific basis are the unsubstantiated claims of harm to humans through the fluoridation of drinking water.[49] Even though this is a recurring complaint from certain pressure groups, no adverse health effects have

[46] For the purposes of this book, 'regulation' refers to any primary legislation, subordinate, or delegated legislation, and administrative action.

[47] Adapted from Matthew T. Wansley, 'Regulation of Emerging Risks' (2016) 69 *Vanderbilt Law Review* 401, 431.

[48] Christl A. Donnelly and others, 'Positive and Negative Effects of Widespread Badger Culling on Tuberculosis in Cattle' (2006) 439 *Nature* 843.

[49] J. Fawell and others, *Fluoride in Drinking-Water*, World Health Organization (2006) www.who.int/water_sanitation_health/publications/fluoride_drinking_water_full.pdf, accessed 1 April 2021; *Water Fluoridation – Health Monitoring Report for England 2014* (Public Health England, 2014).

been proven to be caused through such policies. Should a specific risk be established, the environmental minimum moves on to examine whether the environmental harm has been appropriately regulated by the state. This is determined through minimum standards. Minimum standards are subdivided into three categories: legal standards, established standards, and emerging standards.

Domestic and international obligations that require a state to regulate environmental harm comprise the legal standards: here the environmental minimum chiefly requires consistent enforcement, in line with the rule of law, and the principle of *pacta sunt servanda* in an International Law setting. However, legal standards must be adopted and updated in good faith in line with available evidence. Bad faith would primarily attach to regulations that fail to consider relevant information, seek to circumvent generally accepted environmental protection standards of the relevant scientific community,[50] or unduly favour competing interests over environmental concerns. Bad faith regulations are not recognised as legal standards, and hence lead to the application of either established or emerging standards, dependent on the underlying level of scientific certainty.

In the absence of legal standards, the environmental minimum substitutes established standards of environmental protection: namely, those standards generally accepted in the relevant scientific community. Experts at international organisations, such as the WHO, compile, weigh, and evaluate scientific evidence to arrive at standards and guidelines for regulatory interventions.[51] Should environmental harm exceed the level permissible under relevant standards and guidelines, it would result in a violation of the human right identified at the specific risk stage. Nonetheless, regulators remain free to balance environmental concerns with competing interests, and prescribe a different standard of protection in good faith by adopting legal standards.

Finally, emerging standards apply where uncertainty prevails to a point where the relevant scientific community does not generally accept any standard on a given environmental harm. In those cases, it is generally left to regulators to determine whether intervention is required at all, and how it should be balanced with competing interests. An exception to this rule applies in cases where a risk of severe environmental degradation or pollution exists, and the harm assessment formula requires intervention. The harm assessment formula mandates intervention if the harm caused by regulatory inaction

50 Adapted from: King, *Judging Social Rights*, 246.

51 I reference further possible sources for such evidence throughout the book, but they cannot be considered exhaustive.

(assessed by assuming that the environmental harm is harmful as outlined in the reasonable hypothesis) outweighs the harm caused by regulatory action (assessed by instead assuming that the environmental harm is not harmful to humans).

The conceptually related precautionary principle has been criticised for making partially simplistic and unobjectionable, and partially incoherent, normative claims when employed to reach decisions on potentially harmful conduct. Critiques such as those from Cass Sunstein expose several pertinent concerns,[52] even though they are at least partially based on an unreasonable expectation of coherence. The strongest rejoinder to Sunstein is that the precautionary principle fulfils different functions within different contexts, and that once it is sufficiently specified it can make valuable contributions to regulatory practice. The environmental minimum can alleviate some of the valid concerns identified by Sunstein, as conflicts between different courses of regulatory action and their associated risks can be resolved in the framework of human rights through a harm-based approach adopted by courts.

Objections to the adjudication of environmental regulations are best addressed against the backdrop of one of the most influential advocates of judicial restraint in the face of policy decisions: Lon Fuller and his seminal article on polycentricity.[53] The core rejoinder is that polycentricity ultimately draws any decision-making mechanism into question. Courts appear no more implicated than other decision-making bodies, and have in fact proven themselves reasonably competent resolvers of polycentric issues over centuries, especially in the common law legal tradition. The concept of polycentricity as stated by Fuller is at odds with its selective invocation in the context of resource allocation, policy questions and human rights, and its conspicuous absence in discussions on taxation,[54] competition, and anti-trust law. Crucially, the remedies offered by Fuller cannot account for instances where legislatures and market-based mechanisms have demonstrably failed to secure acceptable resolutions of polycentric issues: that is apparent as much in the landmark case of *Brown* v. *Board of Education*,[55] as it is in the regulatory failures to address air pollution in Europe.

Owing to its flexibility, compliance with the environmental minimum does not determine whether a state has comprehensively lived up to human rights

52 Cass R. Sunstein, *Laws of Fear: Beyond the Precautionary Principle* (Cambridge University Press 2005).

53 Lon L. Fuller and Kenneth I. Winston, 'The Forms and Limits of Adjudication' (1978) 92 *Harvard Law Review* 353.

54 Jeff King, 'The Pervasiveness of Polycentricity' [2008] Public Law 101.

55 *Brown* v. *Board of Education of Topeka* 347 US 483 (1954).

obligations. Rather, the framework is best seen as an expression of environmental human rights that tackles specific threat scenarios arising from environmental harm in a manner consistent with the requirements and doctrine of human rights protection regimes.

In that sense, the environmental minimum does not purport to replace, but rather to complement, existing remedies under public and private law. Furthermore, the environmental minimum is not a mechanism for resolving contentious scientific debates, or second-guessing regulatory decisions taken in good faith: it does not render value judgments beyond the general need to address environmental harm in order to uphold commitments to human rights. The tree diagram in Figure 1.1 summarises the general structure of the framework developed more fully in Chapter 3.

1.3.3 *Practical Viability*

There are a number of reasons why the ECHR is a good test case for the practical viability of the environmental minimum. Chiefly, it stands out among other regional human rights protection regimes, and indeed surpasses many domestic constitutions, with respect to the breadth and depth of its environmental case law. Cases from other protection regimes that have dealt with environmental harm are comparatively rare, and the doctrinal position is often less sophisticated. Partially, this is a consequence of raw case figures: with well over 800 million potential applicants, and an average of approximately 12.8 environmentally themed applications and 9.9 decisions annually between 1985 to 2014, the ECtHR quite simply has more occasion to engage with environmental harm.[56] Quite apart from the raw figures, the ECHR has several advantages as a case study.

First, the ECHR offers a comparatively hard sell in terms of conceptualising environmental harm in the context of human rights. It contains no express right to environmental protection, and reforms of the Convention that would have introduced such a right have been rejected on several occasions. Any argument for the environmental minimum must hence rely on the set of civil and political rights that have already been recognised. Incidentally, these rights are also more established and pervasive in domestic and international protection regimes. While a focus on civil and political rights makes the argument more challenging, it simultaneously makes it comparatively easier to generalise the conclusions.

[56] See the figures calculated for both category one and category two cases in the dataset published alongside this book.

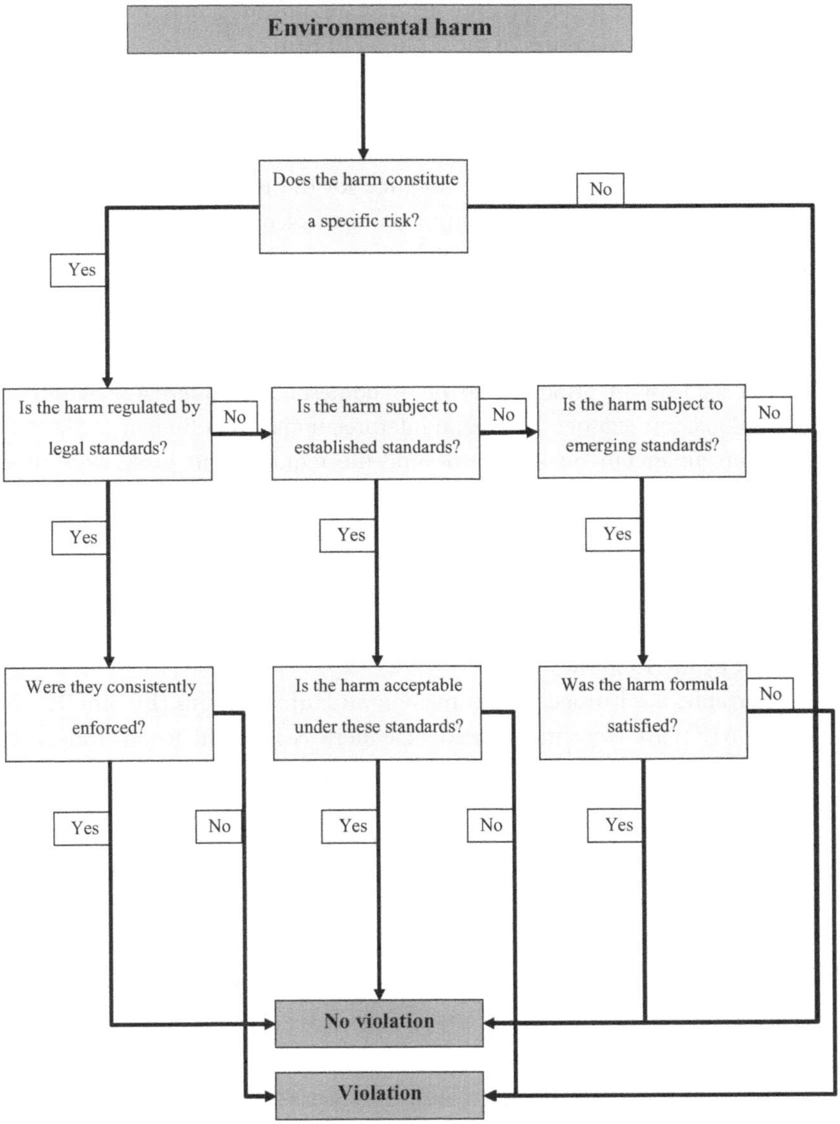

FIGURE 1.1 The operation and structure of the environmental minimum

Second, the ECHR is directly relevant to 47 distinct European jurisdictions. Crucially, these include the current members of the European Union, which are already committed to a comparatively detailed system of environmental regulations under EU law. Hence, the ECHR provides the opportunity to investigate what improvements the environmental minimum can hope to achieve in sophisticated regulatory regimes and provide a reasonable comparison with non-EU members of the Council of Europe.[57]

Third, the ECHR can account well for domestic human rights protections in legislation and constitutions that run parallel to and interact with the Convention. The ECHR is generally acknowledged by the domestic legal systems of contracting states as a resource for the interpretation of domestic rights and for providing a minimum standard of protection. In many ways, demonstrating that the environmental minimum is viable in the context of the ECHR renders it indicative of a genuine European consensus that is presumptively compatible with all domestic rights protection regimes. On this basis, it is reasonable to argue that the environmental minimum can be applied, at least in principle, to other domestic and international human rights protection regimes: be it as an interpretational guideline, or as a starting point for enhancements that go beyond the requirements prescribed under the ECHR.

The investigation into the ECHR case law has unearthed important findings. First, the extent and scope of engagement with environmental harm are consistently underestimated in contemporary scholarship, and second, the environmental minimum is largely compatible with the doctrinal position adopted by the ECtHR.[58]

Ostensibly, the European Commission on Human Rights (EComHR) and the ECtHR have long maintained that there is no right to environmental protection under the Convention. This claim is widely stated as a starting point in contemporary scholarship.[59] Indeed, there is no provision in the Convention (or its additional Protocols) that provides a right to environmental protection, or so much as mentions the natural environment. It is equally correct that this was a key consideration for the EComHR in dismissing early environmentally themed applications. However, this historic attitude is just that: historic. There is a rich development of environmental case law that

[57] A data disaggregation by individual countries can be found in the country analysis section of the dataset.

[58] Notable exceptions include Dinah Shelton, 'Human Rights and the Environment: What Specific Environmental Rights Have Been Recognized' (2006) 35 *Denver Journal of International Law and Policy* 129 and on procedural safeguards specifically Chris Hilson, 'Risk and the European Convention on Human Rights: Towards a New Approach' [2009] Working Paper.

[59] Juliane Pätzold, 'Artikel 8 EMRK' in Ulrich Karpenstein and Franz C. Meyer (eds.), *Europäische Menschenrechtskonvention – Kommentar* (2nd ed., C. H. Beck 2015) 263 [38]; D. J. Harris and others, *Harris, O'Boyle & Warbrick: Law of the European Convention on Human Rights* (3rd ed., Oxford University Press 2014) 341.

exceeds the passing mention it receives in most contemporary scholarship.[60] The treatment of pertinent cases is often superficial and lacking in an appreciation of the broader context.[61] The environmental cases are in fact more nuanced than is commonly acknowledged, and the doctrine developed far surpasses the prevailing image of a patchwork focused on procedural protections.[62]

When instead conducting a comprehensive evaluation of the case law, one is struck by the rich developmental history and broader context of earlier and subsequent cases that is crucial to a sound doctrinal appraisement. Acknowledging the predecessors and evaluating the effects especially of ostensible landmark cases on subsequent decisions is not just a matter of good record-keeping but promises a clearer and more accurate picture of the law. This is where a methodology that is based on both doctrinal and data analysis of decisions can achieve significant improvements.

This book engages in an exhaustive analysis of the case law, comprising a comprehensive search of the ECHR decision database, the manual examination of over 3,000 cases, and the categorisation of environmentally themed applications according to the type of environmental harm alleged. This work is reflected in 23 data points per judgment that form the basis of further analysis. Even those authors claiming to provide a comprehensive overview of the case law fall well short of recognising the over 350 cases identified in the course of this research.[63] This analysis leads to the conclusion that the doctrinal requirements set out by the ECtHR support the environmental minimum to a significant extent, especially with respect to:

- being content with a reasonable hypothesis to find an infringement, provided a specific risk is established;
- accepting minimum standards, particularly legal standards as a basis for the review of state action; and
- acknowledging that environmental motives cannot justify disproportionate infringements of Convention rights.

[60] See, for instance, Alston and others, *International Human Rights*: Text and Materials, 1536; De Schutter, *International Human Rights Law – Cases, Materials, Commentary*, 372–3, 457; Smith, *Textbook on International Human Rights*, 395, Batekas and Oette, *International Human Rights Law and Practice*, 516–17; Amos, *Human Rights Law*, 425, 508–9, 515.

[61] See, for instance, Gearty, 'Do Human Rights Help or Hinder Environmental Protection?', 18; R. R. Churchill, 'Environmental Rights in Existing Human Rights Treaties' in Michael R. Anderson and Alan E. Boyle (eds.), *Human Rights Approaches to Environmental Protection* (Clarendon 1996) 89.

[62] Philippe Sands and others, *Principles of International Environmental Law* (3rd ed., Cambridge University Press 2012) 788; Bell and others, *Environmental Law*, 15.

[63] See the dataset gathered for this book.

Areas of tension include a reluctance by the ECtHR to rely on established and emerging standards in the absence of legal standards, and certain aspects of the rules of evidence. Finally, the ECtHR does not unequivocally rule that inconsistent enforcement of legal standards invariably amounts to a violation of the Convention. There was a tendency in early cases to make questionable use of the margin of appreciation doctrine and accept unsubstantiated economic benefits as a justification for environmental harm.

1.3.4 *Beyond the European Convention*

Beyond the European Convention, a number of further international and domestic protection regimes have grappled with environmental protection and human rights. The case studies investigated for this book demonstrate that the environmental minimum has practical utility in a range of legal and economic contexts.

Chapter 8 considers the African Charter on Human and Peoples' Rights (ACHPR),[64] the American Convention on Human Rights (ACHR),[65] and the UN Covenants on Civil and Political Rights (ICCPR),[66] as well as Economic, Social and Cultural Rights (ICESCR).[67] The position in these regional human rights protection regimes is generally less encouraging than under the ECHR. In many respects, too few environmental cases have been decided to draw firm conclusions, but what has surfaced is far from a doctrinal position unequivocally supportive of the environmental minimum.

Most support comes from the ACHPR, where decisions based on the enforceable right to environment under Article 24 and related environmental rulings approach a doctrinal position in line with core claims of the environmental minimum. Less support comes from the Inter-American system. While there is a provision that specifically provides a right to a healthy environment, its practical impact is significantly reduced because it cannot be claimed through individual petitions. Finally, the doctrinal position taken under the UN Covenants is undergoing a shift. Historically, the HRC has failed to unequivocally state that environmental harm presents a relevant threat to

64 African Charter on Human and Peoples' Rights (Banjul Charter) (adopted 27 June 1981, in force 21 October 1986) 1520 UNTS 217 (ACHPR).

65 American Convention on Human Rights (Pact of San José) (adopted 22 November 1969, in force 18 July 1978), 1144 UNTS 143 (ACHR).

66 International Covenant on Civil and Political Rights (adopted 16 December 1966, in force 23 March 1976) 999 UNTS 171 (ICCPR).

67 International Covenant on Economic, Social and Cultural Rights (adopted 16 December 1966, in force 3 January 1976) 993 UNTS 3 (ICESCR).

human rights. General Comment No 36 foreshadowed a change, which was confirmed in the case of *Cáceres*, even though in this case a domestic court ruling finding a violation of legal standards was ultimately decisive to the outcome. Nonetheless, the case represents an encouraging development that hopefully signals a broader embracing of environmental human rights. This book contends that each of the examined regimes would overall benefit from the environmental minimum framework and the practical guidelines it offers.

Many domestic constitutions also contain environmental provisions, either in the form of constitutional mission statements or individual environmental rights that have attracted substantial scholarly attention. In one sense, this is a laudable development. The environmental minimum only requires basic protections from environmental harm derived from general commitments to human rights: if states go beyond this baseline through constitutional provisions, whether in the form of mission statements or codified environmental rights, then all the better. However, scholars in the vibrant field of environmental constitutionalism have almost exclusively focused on descriptive accounts that offer ever more sophisticated taxonomies of these 'green' provisions and rights in domestic constitutions and international protection regimes.[68] Comparatively little attention has been paid to the legal impact that these provisions have, or indeed ought to have, on the doctrinal approach of courts. The constitutional provisions remain abstract, and apex courts struggle to develop a clear and principled approach to their interpretation in specific environmental cases. This is precisely where the framework of the environmental minimum can provide valuable guidance. Two case studies, namely of South Africa and Germany, which implemented a constitutional environmental right and a constitutional mission statement, respectively, demonstrate that the environmental minimum can add something of value even to two comparatively sophisticated legal systems.

In the case of South Africa, the environmental minimum can provide useful guidelines for the doctrine in section 24 of the Constitution, especially with respect to the amount of deference afforded to ordinary legislation. With respect to Germany, there is similarly a concern of undue deference. However, it relates more to regulatory deference on positive obligations that arise from Articles 2 (2) GG (right to life and health) and 20a GG

[68] David R. Boyd, *The Environmental Rights Revolution – A Global Study of Constitutions, Human Rights and the Environment* (University of British Columbia Press 2012); J. R. May and E. Daly, *Global Environmental Constitutionalism* (Cambridge University Press 2015); Louis Kotzé, *Global Environmental Constitutionalism in the Anthropocene* (Hart Publishing 2016); Joshua C. Gellers, *The Global Emergence of Constitutional Environmental Rights* (Routledge 2017).

(environmental mission statement). Regulators are afforded broad discretion which permits taking a wide range of competing interests into account, especially when deciding whether and how to regulate environmental harm.

1.3.5 *Contributions to Environmental Regulation*

The environmental minimum chiefly ensures the regulation of environmental threats and consistent enforcement. This is evident in the contributions of the environmental minimum to the field of environmental regulation, where it addresses three pathologies that have historically plagued this area of law: *political lag*, the idea that environmental regulation often lags behind the best available scientific evidence due to a lack of awareness in political circles, the broader public, and among regulatory bodies; *industry resistance*, concerted efforts especially by economic stakeholders to dismiss, obfuscate, and discredit scientific evidence of environmental harm, as well as resist regulatory and enforcement actions that threaten established business models; and *regulatory inertia*, the idea that, as a result of the combined effects of political lag and industry resistance, regulators will not adopt or consistently enforce regulations to prevent environmental harm even in the face of compelling evidence. These tendencies are countered and alleviated through the environmental minimum. It requires that regulators establish and consistently enforce legal standards, and in their absence applies established and emerging standards to evaluate the regulatory response to credible environmental harm.

The environmental minimum can further specify obligations under Sustainable Development Goals (SDGs), which are phrased in broad terms and include targets that both require specification and reconciliation before they can be reasonably implemented. Human rights and their mechanisms of balancing competing interests offer an attractive solution to some of these conflicts.

This argument extends to suggested transboundary expansions of human rights in environmental contexts, specifically with respect to climate change. Prominent scholars argue an expansion of the scope of human rights can overcome persistent difficulties with the causal responsibility for environmental harm.[69] However, it appears that this suggestion is informed by a perspective on causality that is at odds with established human rights doctrine, and

[69] Dupuy and Viñuales, *International Environmental Law*, 331; Jorge E. Viñuales, 'A Human Rights Approach to Extraterritorial Environmental Protection? An Assessment' in Nihal Bhuta (ed.), *The Frontiers of Human Rights: Extraterritoriality and Its Challenges* (Oxford University Press 2016) 215.

crucially, the advocated expansion seems unnecessary to achieve meaningful improvements: it obscures rather than illuminates problems encountered in international environmental law and specifically in the context of climate change.

Finally, the environmental minimum can further contribute more generally towards overcoming deficits in compliance and effective enforcement of international environmental law.[70] Many treaties do not provide their own enforcement mechanisms, and even where these exist, compliance often proves difficult to achieve. The environmental minimum can conceivably alleviate some of these problems by providing an additional avenue of enforcement in the guise of human rights law. As we have seen, obligations under treaties constitute legal standards which become directly enforceable under established human rights regimes, provided the environmental harm amounts to a specific risk.

1.4 METHODOLOGY

From a methodological perspective, the chapters draw on several approaches to tackle a topic of considerable complexity.

The normative argument is based on current legal philosophical thinking on the core characteristics of human rights, specifically the notion of environmental human rights. Considerable effort is devoted to exploring the underlying assumptions and implications of insights from environmental ethics, as well as to the nature and scope of threats that arise from environmental harm. The success of the normative argument relies on the environmental minimum's adherence to core characteristics of human rights, its consistency with the specific normative claims of environmental human rights, and the support of and consistency with other fundamental legal principles.

The practical argument relies on the extent to which the environmental minimum accounts for the doctrinal positions expressed by courts and provides a coherent framework that can be implemented on the basis of existing protection regimes. The argument thus has strong doctrinal elements, informed by a comprehensive data analysis of case law. This methodological

[70] Massimiliano Montini, 'The Double Failure of Environmental Regulation and Deregulation and the Need for Ecological Law' [2016] C-EENRG Working Papers 1, 8; Daniel Bodansky, *The Art and Craft of International Environmental Law* (Harvard University Press 2010) 226; Jody Freeman and Charles D. Kolstad, 'Prescriptive Environmental Regulations versus Market-Based Incentives' in Jody Freeman and Charles D. Kolstad (eds.), *Moving to Markets in Environmental Regulation: Lessons from Twenty Years of Experience* (Oxford University Press 2006) 3.

approach remedies one of the central oversights in contemporary scholarship, namely the limited sample of cases considered and so alleviates the overly strong reliance on a small number of prominent 'landmark' environmental decisions. In addition to the ECtHR, the case law of the Inter-American Court of Human Rights (IACtHR), the African Court of Human and Peoples' Rights (AfCtHR), and select domestic constitutions are examined in some detail to determine what contributions the environmental minimum might make.

The data accumulated for this book is outlined in the Appendix, which describes in detail the data gathered, how it was categorised, and evaluated. What follows is an outline of the methodological approach.[71]

1.4.1 *European Convention on Human Rights*

The initial dataset was obtained through the HUDOC database provided by the ECtHR.[72] The database comprises all Court judgments and screening panel decisions from 1959 to 1998 and all judgments and admissibility decisions from November 1998 to the present. It further contains all public reports of the EComHR from 1963 onwards and some published admissibility decisions from 1955 to 1986, as well as all admissibility decisions from 1986 until the abolition of the Commission in 1998.

1.4.1.1 Approach

The database was searched using a Boolean search with the following parameters: all document collections, all language versions, and no date limitation. In order to qualify as a 'hit' the document must engage with any of the following Convention articles: Article 2, Article 3, Article 6, Article 8, Article 13, Article 1 AP-I. Additionally, the text of the decision must contain at least one of the following keywords: 'climate change', 'global warming', 'pollution', 'environment', 'nuclear', or 'ecosystem', and their respective French equivalents ('*changement climatique*', '*réchauffement climatique*', '*pollution*', '*environnement*', '*nucléaire*', or '*écosystème*'). Searches were conducted cumulatively, as opposed to *en bloc*, due to limitations of the search engine in accepting multiple complete phrases: therefore, some cases appeared in

[71] The complete dataset is available from the author on request (contact email: stefan.theil@law.ox.ac.uk) and will be made more widely available on the author's ResearchGate profile (www.researchgate.net/profile/Stefan_Theil) in the near future. The dataset may be freely used for research and teaching purposes subject to appropriate attribution and referencing of the source material.

[72] Available at: https://hudoc.echr.coe.int/.

multiple search results. For instance, while the search engine accepts searching for the term 'climate change' alone, it cannot search multiple phrases (e.g. 'climate change' and 'global warming') concurrently and instead splits the latter phrase into 'global' and 'warming', respectively, thereby vastly inflating the number of false positives, entirely unrelated to global warming.

These individual searches returned a combined total of 3,091 hits, which were supplemented through freedom of information requests to the secretariat of the ECtHR to obtain the full text of a number of known environmental decisions not otherwise publicly available.[73] The decisions were then manually examined for the presence of an environmental issue within the scope of this book: a case was deemed to be environmentally themed if either the applicants alleged a violation of the ECHR arising from environmental harm, or if a state relied on broadly defined environmental motives as a justification for an infringement of the ECHR, regardless of whether this justification was ultimately accepted.[74]

This comprehensive analysis of the case law reveals that the Convention entities have engaged with a body of approximately 371 environmentally themed cases.[75] The two types of environmental cases are reflected in the formal categorisation, which divides the cases according to whether the ECtHR is asked to rule on environmental questions

- as a substantive part of an existing Convention right claimed by an applicant (protection function, or category one), or
- as a justification for the infringement of a Convention right (limitation function, or category two).[76]

While procedural protections play an important part in the environmental case law, they do not have primary relevance to the argument in favour of the environmental minimum, which is preoccupied with establishing basic substantive protections of the environment. Hence, procedurally focused cases are included and counted towards the environmentally themed cases but are not considered as a separate category. For instance, no special attention is

[73] These decisions are on file with the author.

[74] For details see the Appendix.

[75] The cases gathered for this investigation are too voluminous to lend themselves to ordinary citation in footnotes. An overview of cases by jurisdiction, complete with citation, case history, and analytical summaries can be found in the dataset to this book.

[76] This categorisation has been adopted from Maguelonne Dèjeant-Pons, 'Le Droit de L'homme à L'environnement et la Convention Européenne de Sauvergarde des Droits L'homme et des Libertés Fondamentales' in Gerard Cohen-Jonathan, Jean-Francois Flauss, and Pierre Lambert (eds.), *Liber Amicorum Marc-Andre Eissen* (Bruylant 1995) 79–80.

given to the extent to which the Court has adopted and enforced the principles of the Aarhus Convention.

If the same case was considered by different Convention entities over time (e.g. Commission, Court, Plenary, Grand Chamber) only the final ruling was recorded. Upon confirmation of an environmental issue in a case, 23 data points were collected. It was noted whether the rights invoked were found to be violated, infringed but justified, or whether applications were dismissed, settled, dismissed for non-exhaustion of domestic remedies, or struck out of the list of pending cases. The meaning of 'dismissed' was expanded to include the use of any of the following phrases in Commission and Court decisions: 'not necessary to examine', 'no need to examine', 'rejected', 'not applicable', and 'gives rise to no separate issue.' The use of the term 'no violation' in decisions is more context sensitive, as sometimes it indicates that an infringement of a right was justified, and at other times that there has been no infringement at all: this was determined through individual examination.

The Parent List includes all cases available through the search engine on 31 December 2019. However, the dataset must be treated with a degree of caution when assessing cases originating from less than eight years ago. As Chapter 4 explores in more detail, the Court requires at least four and up to eight years to come to a final decision on roughly 80 per cent of environmental applications filed in any given year. This provides a reasonable confidence that a majority of cases filed in a given year are resolved within six years. Accordingly, the book has opted for 1950–2014 as the general reference period unless otherwise stated to avoid understating the total number of cases.

The Appendix provides more detailed information on how the data, results, and figures were compiled and is not exhaustively referenced in the book, save for the following brief description of two categories of ostensibly environmental cases that were excluded from the dataset: (1) the majority of cases regarding prison conditions, as their environmental nature is often questionable; and (2) a collection of miscellaneous cases that are not properly characterised as environmental cases.

1.4.1.2 Excluded Cases

The categorisation as environmental in nature is questionable with cases relating to conditions in prison. The ECtHR often finds violations in these cases, at times amounting to torture and inhumane and degrading treatment in the sense of Article 3 ECHR. On the one hand, inmates are often clearly subjected to intolerable conditions in their direct living environment, that result from the action (or inaction) of those imprisoning them: poor basic maintenance, lack of pest control, and failures to limit the number of occupants per cell are common. Typically, the violations arise from negligence that degrades the cells to unacceptable living conditions. Notably, most cases do not involve direct physical or psychological torture, or degrading treatment through individual guards, but rather broader institutional failures. On the other hand, the prisoners are inhabitants of a highly artificial, purpose-built environment, constructed and designed with a limited remaining natural environment. The book has opted for a narrow definition that excludes primarily artificial environments, but nonetheless acknowledges that they may experience relevant degradation and pollution.

Accordingly, the book attempts to seize the often-elusive middle ground by deciding on a case-by-case basis. If the environmental harm encountered by the applicants is considerable and would constitute a concern outside of a prison, then the case is recorded as an environmental case. However, if the harm encountered is specific to the functioning of the prison system, such that it primarily occurs within the confines of this largely artificial and highly controlled environment, then the case is not considered environmental. The following two examples illustrate this distinction.

In the case of *Tudor* v. *Romania*,[77] the applicant had complained of electromagnetic radiation originating from a cell phone antenna located on the roof of the prison. Prison guards had contractually secured a special 'hazard bonus' in light of this radiation. Leaving to one side whether the complaint can rely on a reasonable hypothesis, this type of environmental harm may be encountered anywhere where a similar cell phone tower is placed in the vicinity of an applicant's residence. The concerns raised by the applicant are hence not specific to residing in the artificial environment of a prison, and the case was accordingly counted as an environmental case.[78] Conversely, consider the case of *Khudoyorov* v. *Russia*.[79] The applicant challenged the conditions of his detainment, especially the length of his detention and the sanitary conditions in the cell. The issues raised are specific to being detained in a prison environment, as they are difficult to conceive of outside of the context of forced detention. The fact that detention in an

[77] *Tudor* v. *Romania* App no 42820/09 (ECtHR, 3 June 2014).

[78] For further examples, *see Aparicio Benito* v. *Spain* App no 36150/03 (ECtHR, 13 November 2006) [second-hand smoking in cell]; *Branduse* v. *Romania* App no 6586/03 (ECtHR, 7 April 2009) [illegal garbage dump site in vicinity to prison].

[79] *Khudoyorov* v. *Russia* App no 6847/02 (ECtHR, 8 November 2005).

institutionally deficient prison system is a necessary condition for the environmental harm to arise renders it not environmental for the purposes of the book. It should be noted that with any decisions on a case-by-case basis, there is an element of uncertainty which permits a reasonable argument to be made on both sides. When in doubt, the book has defaulted to excluding prison cases in order to preserve the overall quality and integrity of the dataset.

Following the categorisation outlined above, some cases have further been in my view erroneously counted as environmental, as neither applicants nor the respondent government makes an environmental claim. For instance, in the case of *Piermont* v. *France* the applicant was expelled from French Polynesia for taking part in protests against nuclear weapons testing.[80] While this case has an environmental dimension with respect to the object of protest, the legal issue before the ECtHR is concerned with lawful limitations on freedom of expression (Article 10 ECHR). There are no further indications that the applicant invoked any procedural safeguards directed towards participation, information, or standing with respect to an environmental harm. Indeed, the fact that the applicant was protesting specifically against nuclear testing is irrelevant to the scope of the right under Article 10 ECHR. The applicant does not make any environmental claim, nor does the state seek to justify its actions by reference to environmental motives.

1.4.2 *African Charter on Human and Peoples' Rights*

The initial datasets for the African Charter on Human and Peoples' Rights were obtained through the search engine provided by the Institute for Human Rights and Development (IHRDA).[81] The following search parameters and filters were employed: all document collections; all language versions; no date limitation; keywords ('African Commission on Human and Peoples' Rights'; 'African Court on Human and Peoples' Rights'; 'Economic, social and cultural rights'; 'Indigenous peoples'; 'Natural resources'; 'Right to a healthy environment'; 'Right to food'; 'Right to health'; 'Right to life'; 'None'). This returned 61 hits. If the same case was considered by different entities over time (Commission and Court) only the final ruling was considered. The categorisation and data extracted mirrors that of the ECHR. The Parent List includes all cases available through the search engine on 31 December 2019.

80 *Piermont* v. *France* App no 15773/89 (ECtHR, 27 April 1995).
81 Available at: http://caselaw.ihrda.org.

1.4.3 *Inter-American Human Rights*

The initial datasets on the American Declaration of the Rights and Duties of Man and the American Convention on Human Rights were obtained through the search engine *Buscador Jurídico de Derechos Humanos*.[82] The following search parameters and filters were employed in that search: all document collections; all language versions; no date limitation; keywords (*'medio ambiente'*, *'cambio climático'*, *'contaminación'*, *'ecosistema'*, *'energía nuclear'*). This returned 34 hits. If the same case was considered by different entities over time (Commission and Court) only the final ruling was counted. The categorisation and data extracted mirrors that of the ECHR. The Parent List includes all cases available through the search engine on 31 December 2019.

1.4.4 *International Covenants*

Initial datasets for the International Covenants were obtained through the search engine OHCHR Jurisprudence.[83] The following search parameters and filters were employed in that search: no date limitation; Body: CCPR; CESCR; Articles: 1, 6, 14, 17, 26, 27 of CCPR; no filtering by Article for the CESCR; keywords ('climate change', 'global warming', 'pollution', 'environment', 'nuclear', or 'ecosystem'). Due to technical limitations, the searches for the CCPR and CESCR case law had to be conducted separately. The search for CCPR cases returned 125 hits and the search for CESCR cases returned 4 hits. The categorisation and data extracted mirrors that of the ECHR. The Parent List includes all cases available through the search engine on 31 December 2019.

1.4.5 *Conclusion*

The complexities of these different methodological approaches should not distract from the overall objective: developing the environmental minimum and arguing that it constitutes a normatively appealing and practically viable framework that can address specific threats that environmental harm poses to the enjoyment of human rights. Part I of the book develops and specifies the

[82] Available at: www.bjdh.org.mx/interamericano/busqueda.

[83] Available at: http://juris.ohchr.org/.

concept and framework of the environmental minimum. Part II then examines how this normative framework fits with the existing case law from the European Court of Human Rights, while Parts III and IV examine the contributions of the environmental minimum beyond the ECHR and to international and domestic environmental regulation, respectively.

PART I

Introducing the Environmental Minimum

Part I introduces and defends the environmental minimum as a framework for conceptualising the relationship between environmental harm and human rights. The framework distinguishes between human rights, understood as an expression of idealised values and norms derived from considerations of political morality, and human rights *law*, the domestic and international protection regimes that entail practical limitations and trade-offs. There are no bright red lines between human rights and human rights law, as they frequently interact and influence one another, but the distinction is useful to explain some of the compromises the environmental minimum proposes.

The framework does not purport to facilitate the comprehensive protection of the environment in human rights law that might be justified as a matter of political morality. It does not demand an optimal level of environmental protection, nor does it claim to be the only normatively justifiable interpretation of environmental human rights. Nonetheless, the environmental minimum operationalises the abstract normative claims of environmental human rights in the context of existing human rights protection regimes. Specifically, it takes account of and incrementally develops the environmental case law of international and domestic courts.

Chapter 2 commences by elaborating the moral case in support of environmental human rights, based on fundamental commitments to human dignity, autonomy, participation, and the interests of future generations. Human rights and the fundamental commitments they serve are threatened through environmental harm. As a matter of political morality, a commitment to human rights must therefore entail basic environmental protection. The chapter describes some of the normative claims that environmental human rights make and explains how they at least require safeguards for human dignity and autonomy. Crucially, the moral case for environmental human rights is bolstered through recognition at the United Nations, codification in various domestic and international legal instruments, as well as practical adjudication before a number of domestic and international courts.

Chapter 3 defends the environmental minimum as a framework for the protections that environmental human rights require and the correlative duties they impose. It develops the guiding principles of the framework, modelled on the characteristics of human rights and specified through the lens of environmental protection. While human rights are crucial to informing the environmental minimum, their claims are tempered through the practical limitations of human rights law: most notably, the framework concedes that it can only be engaged if rightsholders successfully link environmental harm to a recognised human right on the terms of an existing protection regime. The core principles of the environmental minimum are: the specific risk principle, which acts as the trigger, requiring a specific risk to a recognised right arising from environmental harm; the minimum standards, which specify the baseline standard of review that courts ought to employ, based on legal norms, as well as established and emerging scientific evidence; and finally the high-priority characteristic, which guides the balancing of environmental protection with competing interests.

2

Environmental Human Rights

This chapter argues that a commitment to human rights necessarily presupposes the protection of an environmental minimum as a matter of political morality. This is because egregious forms of environmental harm critically undermine the fundamental values that underpin human rights, chiefly human dignity and autonomy. Human rights must therefore contain a subcategory of protections which we can conceptualise as environmental human rights. The human interests that environmental human rights protect are the environmental conditions necessary for the preservation and flourishing of human life, namely clean water, food, air, and soil within a functioning ecosystem that includes diverse species of plants and wildlife.

Those who challenge these rights as vague overlook the significant room for agreement in the pursuit of diverging notions of a 'sound' environment. They miss the fact that complete agreement on comprehensive protection is not necessary to enforce substantive environmental human rights that provide significant safeguards for the natural environment. For their part, concerns over potential conflicts between the demands of environmental protection with human rights are overstated. The balancing of competing interests is a pervasive and well-established feature of human rights law. Crucially perhaps, contemporary environmental regulations are already being challenged on the basis of human rights, for instance to property. Therefore, introducing environmental human rights into protection regimes will not diminish but arguably enhance the legal protection of the environment by giving it a level of priority on par with other human rights.

2.1 ENVIRONMENTAL HARM AS A THREAT TO HUMAN RIGHTS

The core justification for the existence of environmental human rights lies in the capacity for serious environmental harm to threaten the fundamental values that ultimately underpin all human rights: human dignity, autonomy, participation, and the interests of future generations.[1] While simplifying nuanced arguments to an extent, the following overview describes how all these justifications for human rights support the notion of environmental human rights, albeit without necessarily adding a unique perspective.

- Human dignity (well-being, natural rights, or natural law)[2]: the link between human dignity and access to clean water and air, safe food, and an environment free of toxins that seriously endanger health is well established and recognised, as is the intimate connection to any notion of well-being. Indeed, environmental protection appears as a necessary condition to sustaining human life as such and hence a prerequisite for a dignified life and the enjoyment of any human right.[3]
- Autonomy (normative agency, liberty):[4] autonomy-based justifications require a similar bundle of environmental protections as those justified through human dignity. They serve as necessary conditions to permit individuals to make meaningful decisions and retain a level of independent control of their lives.[5] Certainly, autonomy can also be invoked to justify actions that cause considerable environmental harm, for instance in the pursuit of business opportunities. However, at the very least, such autonomy would be limited through the autonomy of others, namely where actions seriously damage the environment and ecosystems upon which human life, and hence any meaningful notion of autonomy, ultimately relies.

1 Nickel, 'The Human Right to a Safe Environment', 290; Hayward, *Constitutional Environmental Rights*, 48–9; the outline of the following justifications is based on King, *Judging Social Rights*, 22–7.

2 In general, see Ronald Dworkin, *Justice for Hedgehogs* (Harvard University Press 2011) 332–9; Jack Donnelly, 'Human Rights as Natural Rights' (1982) 4 *Human Rights Quarterly* 391.

3 For a human dignity-based argument for environmental protection, see Ekkehard Habel, 'Menschenwürde und natürliche Lebensgrundlagen' (1995) 17 *Natur und Recht* 165; Erin Daly and James May, 'Bridging Constitutional Dignity and Environmental Rights Jurisprudence' (2016) 7 *Journal of Human Rights and the Environment* 16.

4 On autonomy, see John Rawls, *Political Liberalism* (Columbia University Press 1993); on normative agency, see Griffin, *On Human Rights*, chapters 2 and 10; more generally, see Sen, 'Elements of a Theory of Human Rights', 345–8; on the concept of non-domination, see Philip Pettit, *Republicanism: A Theory of Freedom and Government* (Oxford University Press 1997) 158–63.

5 For an autonomy-based argument in favour of environmental protection, see Felix Ekardt, 'Umweltverfassung und Schutzpflichten' (2013) 17 *Neue Zeitschrift für Verwaltungsrecht* 1105, 1107.

- Participation (civic republicanism, deliberative democracy):[6] meaningful political participation, such as protesting, contributing to political debate, standing for elected office and voting likewise requires certain basic environmental safeguards. Prevalent environmental harm can, for instance, exclude vulnerable groups and, in extreme cases, severely hinder the population at large from meaningful democratic participation and expression. There is some doubt whether participation provides the strongest justification for environmental human rights because it can be conceivably safeguarded even with only a semblance of a healthy environment. For now, it suffices to say that the link between democratic participation and a minimally sound environment is difficult to deny; at least where environmental harm is so severe as to deny a reasonably healthy life, meaningful democratic participation is necessarily impacted.
- Future generations: in the context of the environment, one additionally encounters concepts of intergenerational justice, which hold particular significance in the field of environmental ethics and contemporary debates on climate change.[7] Essentially, supporters rely on accounts of the rights and obligations owed towards future generations to justify environmental interventions.[8] In light of overlaps with the requirements of human dignity and autonomy, as well as the questionable legal status of as yet unborn generations as rightsholders, there is some doubt as to whether future generations provide a strong basis for environmental protection through human rights. The general idea that

6 Knud Haakonssen, 'Republicanism' in Robert E. Goodin, Philip Pettit, and Thomas Pogge (eds.), *A Companion to Contemporary Political Philosophy* (2nd ed., Wiley-Blackwell 2012), 729; Jürgen Habermas, *Faktizität und Geltung: Beiträge zur Diskurstheorie des Rechts und des Demokratischen Rechtsstaats* (2nd ed., Suhrkamp 1992).

7 For more recent examples, see Holmes Rolston, *A New Environmental Ethics: The Next Millennium for Life on Earth* (Routledge 2012); Joseph Mazor, 'Liberal Justice, Future People, and Natural Resource Conservation' (2010) 38 *Philosophy & Public Affairs* 380; Edward A. Page, *Climate Change, Justice and Future Generations* (Edward Elgar Publishing 2006). For a recent legal argument in favour of the rights of future generations, *see* Michael Kleiber, *Der Grundrechtliche Schutz Künftiger Generationen* (Mohr Siebeck 2014); as well as the early statements in the *Report of the World Commission on Environment and Development – Our Common Future* (1987) 52, para [55].

8 Edith Brown Weiss, 'Our Rights and Obligations to Future Generations for the Environment' (1990) 84 *American Journal of International Law* 198; Kristian Skagen Ekeli, 'Green Constitutionalism: The Constitutional Protection of Future Generations' (2007) 20 *Ratio Juris* 378; Henry Shue, 'Changing Images of Climate Change: Human Rights and Future Generations' (2014) 5 *Journal of Human Rights and the Environment* 50.

> future generations have a moral interest in environmental protection and sustainable development should nonetheless not provoke much controversy.[9]

Notwithstanding the differences between these outlined justifications, all of them provide sound moral reasons for upholding and protecting environmental human rights.

The moral imperative of protecting humans from environmental harm has also become the generally accepted position at the HRC[10] and is supported by John Knox, the former independent expert and special rapporteur on the issue of human rights obligations relating to the enjoyment of a safe, clean, healthy, and sustainable environment. Knox's final report concludes that '[a] safe, clean, healthy and sustainable environment is necessary for the full enjoyment of a vast range of human rights, including the rights to health, food, water and development.'[11] As this view is shared by virtually every source reviewed in his report, Knox concludes that states have '(. . .) obligations under human rights law to protect against such harm.'[12] From this foundation, he develops 16 framework principles that outline the specific environmental protection obligations of states, which have been expressly endorsed by his successor as special rapporteur, David Boyd.[13] Regrettably, the report declines to acknowledge that states are already bound under existing human rights protection regimes to provide a minimum level of environmental protection. Instead, the framework principles describe themselves only as 'best practices' that states will 'move to adopt as expeditiously as possible.'[14] Boyd goes somewhat further in finding that states must 'dedicate the maximum available resources to comply with their human rights obligations in addressing environmental challenges', but neither report gives much indication how this might be practically determined and legally evaluated: especially in light of competing

9 For instance as reflected in domestic constitutions, see Article 20a German Basic Law, and international agreements, see *Rio Declaration on Environment and Development*, UNCED Doc A/CONF/151/5/Rev. 1 (United Nations Conference on Environment and Development, 1992), Principle 3.

10 Human Rights and the Environment, UN A/HRC/25/L.31.

11 John H. Knox, Final Report of the Special Rapporteur on the Issue of Human Rights Obligations Relating to the Enjoyment of a Safe, Clean, Healthy and Sustainable Environment (UN A/HRC/37/59, 2018) para 2.

12 Ibid., para 3.

13 David R. Boyd, *Right to a Healthy Environment: Good Practices* (UN A/HRC/43/53, 2020) para 8.

14 Knox, Final Report of the Special Rapporteur, para 9.

and arguably equally pressing demands on scarce state resources, for instance in eradicating absolute poverty or providing sufficient healthcare.

The framework principles remain vague and generally content themselves with sketching aspirational goals. With respect to framework principles one and two, the explanatory notes for instance state that

> While it may not always be possible to prevent all environmental harm that interferes with the full enjoyment of human rights, States should undertake due diligence to prevent such harm and reduce it to the extent possible, and provide for remedies for any remaining harm.[15]

While it is difficult to disagree with these statements, much of the report notably avoids discussing the delicate balancing of interests, the scientific uncertainty and enforcement difficulties that are at the heart of human rights adjudication and environmental regulation. The guidance on procedural environmental rights is somewhat detailed but exhausts itself in effectively restating many well-known principles adopted, for instance, under the Aarhus Convention.[16] Boyd has since gone into some detail and gathered evidence on these best practices from states. However, in its current form the report reads as little more than a loose collection of legislative and regulatory initiatives from a limited number of states, without much consideration of their domestic legal context.[17] In this form, the framework principles and the best practices are malleable to the interests and peculiarities of states and fail to provide a sufficient basis for a principled legal evaluation of environmental regulation that could be effectively enforced. This is a recurring theme and shortcoming particularly in international environmental law that the book revisits in its discussion of enforcement deficits and especially in the context of the Sustainable Development Goals in Chapter 11.

As we shall see, the ICCPR potentially, and regional and domestic human rights protection regimes certainly require that states provide environmental safeguards that stretch far beyond the progressive realisation suggested by the special rapporteurs. More importantly, the legal reality of extensive court engagement with environmental cases is beyond credible empirical doubt, while paradoxically receiving only passing mention in scholarly and environmental policy debates.

[15] Ibid., 8.

[16] Aarhus Convention. See, for instance, the explanatory notes to framework principles 3, 7, 8, 9, and 10.

[17] Boyd, *Right to a Healthy Environment*.

Courts routinely engage with environmental harm under a range of human rights protection regimes, often addressing environmental harms that are the subject of intense policy debates and extensive media reporting. This is because environmental protection is often legally mandated, either through specifically tailored environmental rights or through a 'greener' interpretation of existing human rights and mission statements. Prominent examples include the Constitution of South Africa, the Constitution of India, the German Basic Law, and countless other constitutions,[18] as well as leading regional human rights protection regimes such as the ECHR, the ACHPR, and the ACHR. Crucially, the framework principles developed by the rapporteurs cannot provide practical guidance to these courts when they are confronted with environmental claims under human rights. Closing this normative gap with a practical framework is the core task this book has set itself. While many of the caveats and limitations of the reports are understandable given the political mandate of the special rapporteur and their International Law focus, the environmental minimum is free from such constraints.

In Chapter 4 we integrate some of the helpful substantive elements expressed in the reports, especially framework principles 11 and 12, into the environmental minimum framework.

For now, the crucial takeaway is that the general moral notion of environmental human rights finds widespread support. They address an empirically undeniable threat to human rights, but there are difficulties in translating their abstract and general commitments into practical measures that benefit the environment. Therefore, the following section turns to illuminate and defend specific normative claims of environmental human rights and explains how they relate and ultimately support the environmental minimum framework.

2.2 THE NORMATIVE CLAIMS OF ENVIRONMENTAL HUMAN RIGHTS

The account of human rights adopted from James Nickel told us much about the core characteristics that render certain rights human rights. However, the definition cannot provide us with the specific demands of environmental human rights: potentially any number of normative claims meet the characteristics set out by Nickel and yet bear very little relation to environmental human rights. Therefore, we require an account that allows us to develop normative claims that are specific to environmental human rights, not just human rights.

[18] For a comprehensive overview, *see* May and Daly, *Global Environmental Constitutionalism*.

Based on the adaptation of Joseph Raz's interest theory of rights by Jeff King, we may formulate guidelines that help define some of the normative claims that are unique to environmental human rights.[19] In order to define relevant normative claims, we must set out

- the human interest that environmental human rights protect and why this interest is of a fundamental character;
- the state of affairs that would satisfy this interest; and
- the duties owed by rights-addressees to rightsholders in respect of those interests.[20]

As previously outlined, the human interest that environmental human rights seek to protect are the environmental conditions necessary for the preservation and flourishing of human life. This interest is of a fundamental character because it flows from the values that human rights are built upon. But what state of affairs would satisfy these fundamental values and the interests of environmental human rights? Certainly, they would demand avoiding and alleviating the most egregious forms of environmental harm, particularly those that threaten human dignity and autonomy. There are some lingering doubts as to the normative contribution that the interests of future generations, participation, and the rights of nature might make to environmental human rights, which merit separate consideration.

2.2.1 *Human Dignity and Autonomy*

This book suggests that the interests of environmental human rights are satisfied when two thresholds are met that anchor and shape their normative claims. The human dignity threshold requires states to provide and enforce regulations that protect humans from serious threats to their health and lives, including from environmental harm. Basic protections are necessary under environmental human rights for a dignified life, including clean air, water, and soil, as well as for the ecosystems that sustain and provide environmental necessities. The details depend on the specific context but would, for instance, involve regulations that ensure access to safe drinking water, breathable air, and clean soil suitable for sustainable food production. I take it as uncontroversial that environmental harm that seriously threatens the health and life of individuals; undermines their access to reasonably clean air, water, and soil; or

[19] Joseph Raz, *The Morality of Freedom* (Oxford University Press 1986) 166.
[20] King, *Judging Social Rights*, 28.

threatens ecosystems that ultimately support all human life, cannot satisfy any meaningful commitment to human dignity.

The autonomy threshold goes further: it requires that states provide and enforce regulations that safeguard an environment which permits for an autonomous life, that is to say '(...) a vision of people controlling, to some degree, their own destiny, fashioning it through successive decisions throughout their lives.'[21] Meaningful life decisions also imply an absence of undue influence from severe environmental harm. Details are again dependent on the specific circumstances. For instance, states may have to ensure basic protections of the environment in a uniform manner across their territory, so that people can make autonomous decisions on their place of residence and profession. This autonomy threshold is more difficult to justify because it moves beyond the environmental requirements for sustaining human life and well-being: it raises the possible objection that autonomy considerations exceed the minimum standards that are crucial characteristics of human rights.

The strongest line of defence to this objection comes from the insight that any notion of autonomy requires more than mere survival. This idea is best explained through a thought experiment adapted from Joseph Raz.[22] We can imagine that all the environmental requirements of the human dignity threshold are met in the context of a dystopian future, which is distinctly lacking in any meaningful autonomy: suppose that the air, water, and soil on Earth have been fatally contaminated and the broader ecosystem has all but collapsed. Human life is sustainable only in small isolation chambers not unlike modern spacecraft, which use feeding tubes and sophisticated air filtration to provide basic environmental resources. The chambers thus ensure a physically healthy and sustainable environment for humans, but one which jeopardises their mental health and which they cannot ever leave without endangering their health and risking death. A life in such an isolation chamber would not satisfy even a very basic notion of autonomy because individuals would effectively be prisoners of this artificial environment. While the example may seem farfetched, it in fact comes close to the living conditions of millions of people in major cities such as Beijing during times of high air pollution.[23] Leaving

[21] Raz, *The Morality of Freedom*, 369.

[22] Ibid., 373.

[23] Defined as 'the presence in the air of any air pollutant that reduces air quality enough to threaten the health and welfare or people, plants, and animals, to adversely affect materials and structures, and/or to interfere with the enjoyment of life and property', see 'Air Pollution', in Park (ed.), *A Dictionary of Environment and Conservation* and Wainwright, 'Inside Beijing's Airpocalypse'.

air-filtered indoor environments in these cities has immediate and detrimental impacts on health that can lead to chronic illnesses and premature death.

However, even less dramatic examples demonstrate that environmental harm can fail to satisfy the autonomy while nominally meeting the dignity threshold. Humans are routinely forced to choose between economic deprivation and exposure to severe environmental hazards: think only of the workers employed in the asbestos and coal mining industries who endured frequent accidents, reduced lifespans, and chronic illnesses to earn their wages.[24] Even today, for many individuals their 'autonomy' can boil down to a choice between economically viable employment in major cities such as London, a city with substantial air pollution that significantly shortens lifespans, or (relative) economic deprivation and underemployment in more rural areas, albeit with significantly less exposure to environmental harm.[25] Plainly put, human rights, and by extension, environmental human rights safeguard more than the possibility for survival.

2.2.2 *Future Generations and Participation*

The interests in participation and of future generations appear as less compelling sources for normative claims of environmental human rights. I do not take a view on whether these interests might offer preferable justifications for other reasons but would contend that it is questionable whether they make normative claims beyond those based on human dignity and autonomy just outlined.

The interests of future generations are often the subject of discussion in environmental law.[26] It is arguable that these interests would require basic projections into the future and giving weight to environmental concerns in balancing with competing considerations. However, under the definition of human rights adopted here, it is questionable whether unborn generations can

[24] Geoffrey Tweedale, *Magic Mineral to Killer Dust: Turner & Newall and the Asbestos Hazard* (Oxford University Press 2000); A. M. Donoghue, 'Occupational Health Hazards in Mining: An Overview' (2004) 54 *Occupational Medicine* 283; Michael Hendryx and Melissa M. Ahern, 'Relations between Health Indicators and Residential Proximity to Coal Mining in West Virginia' (2008) 98 *American Journal of Public Health* 669.

[25] Gowers and others *Estimating Local Mortality Burdens*, 15.

[26] Shue, 'Changing Images of Climate Change'; Ekeli, 'Green Constitutionalism'; Wilfred Beckermann, 'Sustainable Development and Our Obligations to Future Generations' in Andrew Dobson (ed.), *Fairness and Futurity: Essays on Environmental Sustainability and Social Justice* (Oxford University Press 1999); Peter Laslett and James S. Fishkin, *Justice between Age Groups and Generations* (Yale University Press 1992); Weiss, 'Our Rights and Obligations to Future Generations'.

be rightsholders. Rightsholders likely have to be in existence in order to impose obligations on rights-addressees. Be that as it may, it appears doubtful that future generations and their interests would add much to the clean air, water, soil, and ecosystem requirements of the human dignity and autonomy thresholds: environmental harm that jeopardises the well-being and autonomy of living humans impacts any interests that could be invoked on behalf of future generations. There is a possible exception for gradual environmental harm that does not pose an immediate threat to living humans but will materialise greater harm in the future. A future generations threshold could require a general commitment to the ideas embodied in the precautionary principle, that is to say, a willingness to act on the basis of unclear scientific evidence to prevent potential future harm. In this limited sense, environmental human rights and the environmental minimum incorporate normative claims based on the interests of future generations.[27]

Interests in democratic participation are generally more appropriately protected through civil and political, as well as social rights.[28] Although the natural environment is of fundamental importance to human life, this should not tempt us to include every aspect of human life in the notion of environmental human rights. Environmental human rights cannot offer comprehensive protection of every human interest tangentially impacted by environmental harm. For instance, an individual who suffers from asthma might argue that an absence of air pollution regulation impacts not only their health but also their ability to campaign for elected office, because it impedes public speaking and election rallies. Air pollution can undoubtedly have such detrimental impacts beyond general health and autonomy. However, civil and political human rights are more closely related to campaigning for elected office and can more readily uphold the fundamental interests in political participation. Otherwise, we might risk turning environmental human rights into a comprehensive statement of an ideal environment for the existence of all human life and the exercise of all human rights. Apart from the conceptual difficulties this raises, such a comprehensive protection is precisely what Nickel's definition seeks to exclude from the notion of human rights: they would no longer represent a minimum standard for a discrete human interest that enjoys high priority. To those who advocate the rights of nature, this minimalism of environmental human rights appears as a key conceptual weakness.

[27] See the discussion in Chapter 3.

[28] King, *Judging Social Rights*, 32–5.

2.2.3 *Rights of Nature*

A conceptual weakness of environmental human rights might be seen precisely in their anthropocentric focus and the minimalism advocated in this book. Such objections are often nested within a broader debate in environmental ethics between those that argue in favour of the intrinsic value of the environment,[29] rejecting the historic anthropocentric approach to environmental protection, and those who justify the value of the environment at least to some degree with its utility to humans.[30] The latter view sees humans as either the sole bearers, or at least the bearers of significantly higher intrinsic value than the environment.[31]

From a position that assigns intrinsic value, some scholars call for ambitious and comprehensive protection of the environment for its own sake on a global scale. They allege that the anthropocentric focus underpinning human rights wrongly conceives of humans as distinct from their environment. The seminal work of Arne Naess on deep ecology has been particularly influential on this point.[32] In essence, Naess argues that all living things are alike in intrinsic value, and independent from their utility, all organisms are interconnected and constituted in their relationship to other living things.[33] Legal scholarship has been slow to receive these ideas, but several strands have emerged that adopt views broadly along these lines.

Some legal scholars reject human rights as ill-suited to providing the necessary protection of the natural environment and climate,[34] or as unduly dismissive of the rights of the broader ecosystem, for instance the protection of (sentient) animals and the legal personhood of nature.[35] The normative claims of environmental human rights indeed appear limited when compared to a

29 Dale Jamieson, *Morality's Progress: Essays on Humans, Other Animals, and the Rest of Nature* (Oxford University Press 2002); John O'Neill, 'The Varieties of Intrinsic Value' (1992) 75 *Monist* 119.

30 Passmore, *Man's Responsibility for Nature*, 177–81; Murray Bookchin, *The Ecology of Freedom – The Emergence and Dissolution of Hierarchy* (Cheshire Books 1982) 315–18.

31 Andrew Brennan and Lo Yeuk-Sze, 'Environmental Ethics' in Edward N. Zalta (ed.), *The Stanford Encyclopedia of Philosophy* (Stanford University 2015) 1.

32 Naess, 'The Shallow and the Deep, Long-Range Ecology Movement'; Andrew Light, 'Callicott and Naess on Pluralism' (1996) 39 *Inquiry* 273.

33 Naess, 'The Shallow and the Deep, Long-Range Ecology Movement', 95–6; Brennan and Yeuk-Sze, 'Environmental Ethics', 7.

34 Redgwell, 'Life, The Universe and Everything', 71; Francioni, 'International Human Rights in an Environmental Horizon'; Grear, 'Towards New Legal Futures?'; Grant, 'Re-imagining Adjudication'; Dupuy and Viñuales, *International Environmental Law*, 300.

35 Stone, 'Should Trees Have Standing?'; Goodkin, 'The Evolution of Animal Rights'; Singer, *Animal Liberation*; Regan, *The Case for Animal Rights*; Regan and Singer, *Animal Rights and*

broader protection of biological diversity and ecological processes not immediately beneficial or useful to humans.[36] The core problem, argues Anna Grear, is that liberal legal theories unduly separate a 'hyper-rational' human from 'an objectified, exploitable nature.'[37] In many respects, the emerging debate on the implications of the Anthropocene for legal studies reflects the same desire for a reconceptualisation, beyond the prevailing anthropocentric focus.[38] This raises the question of whether environmental human rights are sufficiently ambitious, given that they are expressly limited by human needs.

Expanding the protected interests beyond human needs would conflict with the understanding of human rights adopted here, and their necessarily anthropocentric perspective.[39] The minimum standards would be replaced with a very specific idea of ecological justice that does not necessarily follow from moral commitments to human rights: the concept that all life on Earth has intrinsic value, and is, at least as a starting point, worthy of protection from all (human) interference. More importantly, it is not at all clear whether the scepticism towards anthropocentric human rights is warranted. Drawing on the debate in environmental ethics, there is a strong argument that anthropocentric environmentalism, conceived in broader terms, can accommodate many of the concerns voiced by critics. The core argument is that a retention of the anthropocentric focus provides a more practical and effective legal

Human Obligations; Giagnocavo and Goldstein, 'Law Reform or World Re-form'; Montini and Volpe, 'Rethinking Regulation', 24; Hsiao, 'Whanganui River Agreement'.

36 Brian J. Preston, 'Internalizing Ecocentrism in Environmental Law' in Michelle M. Maloney and Peter Burdon (eds.), *Wild Law: In Practice* (Routledge 2014) 75; Douglas E. Fisher, 'Jurisprudential Challenges to the Protection of the Natural Environment' in Maloney and Burdon (eds.), *Wild Law*, 95; Jamieson, *Morality's Progress*, chapters 14 and 16; O'Neill, 'The Varieties of Intrinsic Value', 119.

37 Anna Grear, 'The Vulnerable Living Order: Human Rights and the Environment in a Critical and Philosophical Perspective' (2011) 2 *Journal of Human Rights and the Environment* 23; for a similar argument against the backdrop of reform efforts in Ecuador, see Erin Fitz-Henry, 'Decolonizing Personhood' in Michelle M. Maloney and Peter Burdon (eds.), *Wild Law: In Practice* (Routledge 2014) 133.

38 A new geological era recognising the lasting impact of man on Earth, see Paul J. Crutzen and Eugene F. Stoermer, 'The Anthropocene' (2000) 41 *Global Change Newsletter* 17; on possible legal implications, see Jorge E. Viñuales, 'Law and the Anthropocence' [2016] C-EENRG Working Papers 1; Kotzé, *Global Environmental Constitutionalism in the Anthropocene* 6–7.

39 This is inevitable under a human rights–based approach, see Hayward, *Constitutional Environmental Rights*, 32; Alan Boyle, 'The Role of International Human Rights Law in the Protection of the Environment' in Michael R. Anderson and Alan E. Boyle (eds.), *Human Rights Approaches to Environmental Protection* (Clarendon 1996); for a critique, see Redgwell, 'Life, The Universe and Everything', 72; Ulrich Beyerlin, 'Umweltschutz und Menschenrechte' (2005) 65 *Zeitschrift für ausländisches öffentliches Recht und Völkerrecht* 525, 540.

framework for tangible improvements, as it is more consistent, and hence fits more easily into the undeniably anthropocentric status quo.[40] Moreover, it can provide improvements without risking past achievements in environmental law that would inevitably be drawn into question following a major paradigm shift.

To a significant extent, both an anthropocentric approach, and one centred on the intrinsic value of the natural environment, make overlapping normative claims: the differences in perspectives are not necessarily irreconcilable.[41] Both accept that there is no fundamental separation between humans and their ecological surroundings, in the sense that we could imagine a non-trivial environmental harm entirely harmless to humans.[42] Almost any environmental harm will at some point impact humans, even if the effects are not immediate, mitigated through degrees of separation, and unfold over time. This is a conclusion broadly shared by those who advocate for anthropocentric environmentalism,[43] as well as those who assume an intrinsic value of the natural environment.[44] Both accept that all environmental harm inevitably impacts on a complex system of interconnection between organisms, of which humans form an integral part. This basic agreement provides the foundation for some level of reconciliation and jointly endorsed normative claims. Dinah Shelton has argued this view, concluding that both human rights and ecological protection ultimately seek to 'achieve the highest quality of sustainable life for humanity within existing natural conditions.'[45]

Basic agreement should however not be mistaken for complete complementarity. It is, for instance, difficult to imagine how a wide range of human

40 de Shalit, *Why Does Posterity Matter*, 7–8.

41 Hayward, *Constitutional Environmental Rights*, 34–5; de Shalit, *Why Does Posterity Matter*, 7–8; Light, 'Compatibilism in Political Ecology', 161; Norton, *Toward Unity among Environmentalists*, chapter 10.

42 Sandra B. Rosenthal and Rogene A. Buchholz, 'How Pragmatism Is an Environmental Ethic' in Eric Katz and Andrew Light (eds.), *Environmental Pragmatism* (Routledge 1996) 38; Jamieson, *Morality's Progress*, 243.

43 Tim Hayward, *Political Theory and Ecological Values* (Polity 1998) xx ; William Aiken, 'Human Rights in an Ecological Era' (1992) 1 *Environmental Values* 191, 196; Brennan and Yeuk-Sze, 'Environmental Ethics', 3; Passmore, *Man's Responsibility for Nature*, chapter 2; Bookchin, *The Ecology of Freedom*, 315–18; de Shalit, *Why Does Posterity Matter*, 7–8; also referred to as weak anthropocentrism, see Bryan G. Norton, 'Environmental Ethics and Weak Anthropocentrism' (1984) 6 *Environmental Ethics* 131.

44 Naess, 'The Shallow and the Deep, Long-Range Ecology Movement'; O'Neill, 'The Varieties of Intrinsic Value'.

45 Dinah Shelton, 'Environmental Rights' in Philip Alston (ed.), *Peoples' Rights* (Oxford University Press 2001) 190, citing Dinah Shelton, 'Human Rights, Environmental Rghts, and the Right to Environment' (1991) 28 *Stanford Journal of International Law* 103.

rights can be enjoyed without a certain level of economic development, which in turn inevitably causes some harm to the natural environment. Rather than complete reconciliation, we ought to determine the degree of environmental harm that is acceptable in exchange for an anticipated benefit towards other interests based on human rights. The environmental minimum provides a framework to answer this question by defining the minimum requirements under any serious commitment to human rights. At times, business opportunities may take precedence over ecological convictions, but at the same time, not every benefit (economic or otherwise) justifies harming the environment that human survival ultimately relies upon. Striking a balance and building on existing policy agreement under a principled and practical framework is the ultimate promise of the environmental minimum.

2.3 SCEPTICISM

Notwithstanding the moral case and the specification of the duties developed in the previous section, there there is considerable scepticism of environmental human rights in contemporary literature. The accounts no longer deny the relevance or existence of environmental human rights, as was more fashionable in earlier decades. Instead, they focus on obstacles to practical implementation, chiefly, concerns over vagueness of environmental human rights, and undesirable trade-offs when environmental human rights conflict with other interests.

2.3.1 *Vagueness*

One important line of scepticism in contemporary scholarship is the idea that the normative claims of environmental human rights are inherently vague and therefore unsuitable to practical implementation. It is difficult to deny that such views are influential in legal scholarship. Many authors see the notion of a 'sound' (or 'safe', 'healthy', 'healthful') environment as inherently uncertain,[46] even those who generally support environmental human rights in some form.[47] It is important to note that the sceptics do not take issue with the

[46] Birnie and others, *International Law and the Environment*, 281, 302; Nickel, 'The Human Right to a Safe Environment', 285; Günther Handl, 'Human Rights and Protection of the Environment' in Asbjørn Eide, Catarina Krause, and Allan Rosas (eds.), *Economic, Social and Cultural Rights – A Textbook* (2nd ed., Martinus Nijhoff 2001) 313–15; Beyerlin and Marauhn, *International Environmental Law*, 392.

[47] Karen E. MacDonald, 'A Right to a Healthful Environment – Humans and Habitats' [2008] *European Energy and Environmental Law Review* 213, 214; Sueli Giorgetta, 'The Right to a

suggestion that human rights are threatened through environmental harm. Their concerns centre on our ability to formulate environmental human rights with sufficient clarity to allow them to become practically enforceable in human rights protection regimes.[48] Alan Boyle epitomises these concerns when he channels John Rawls' famous statement on differences of opinion on policy matters: 'What constitutes a decent environment is a value judgment, on which reasonable people will differ.'[49] Although Boyle is prepared to entertain the notion of a human rights–based approach to environmental protection, his core counterargument is that environmental human rights are aspirational rights, those that states must only progressively achieve in line with available resources. Hence, he believes we should focus on procedural guarantees over substantive protections of the environment.[50]

When environmental human rights are presented as aspirational principles, they are often modelled on the International Covenant for Economic, Social and Cultural Rights (ICESCR),[51] which requires realisation only in step with available resources, subject to few firm commitments and largely in absence of enforcement beyond periodic monitoring.[52] Alan Boyle concludes that this approach simply 'makes more sense' over an enforceable human rights approach.[53] This position was also notably endorsed by John Knox in his final report to the UN Human Rights Council.[54] An important variant of this

Healthy Environment, Human Rights and Sustainable Development' (2002) 2 *International Environmental Agreements* 171, 187; Prudence E. Taylor, 'From Environmental to Ecological Human Rights: A New Dynamic in International Law' (1997) 10 *Georgetown International Environmental Law Review* 309, 360–1; Tomuschat, *Human Rights*, 57; J. G. Merrills, 'Environmental Rights' in Daniel Bodansky, Jutta Brunnée, and Ellen Hey (eds.), *The Oxford Handbook of International Environmental Law* (Oxford University Press 2007) 673; Francioni, 'International Human Rights in an Environmental Horizon', 54; Ernst Brandl and Hartwin Bungert, 'Constitutional Entrenchment of Environmental Protection: A Comparative Aanalysis of Experiences Abroad' (1992) 16 *Harvard Environmental Law Review* 1, 88–90.

48 Birnie and others, *International Law and the Environment*, 281, 302; Nickel, 'The Human Right to a Safe Environment', 285; Handl, 'Human Rights and Protection of the Environment', 313–15; Beyerlin and Marauhn, *International Environmental Law*, 392; Francioni, 'International Human Rights in an Environmental Horizon', 54.

49 Boyle, 'Human Rights and the Environment', 219; Rawls, *Political Liberalism*, 8.

50 Boyle, 'The Role of International Human Rights Law', 43.

51 International Covenant on Economic, Social and Cultural Rights.

52 Birnie and others, *International Law and the Environment*, 281; Gudmundur Alfredsson and Alexander Ovsiouk, 'Human Rights and the Environment' (1991) 60 *Nordic Journal of International Law* 19, 25; Weston and Bollier, 'Toward a Recalibrated Human Right to a Clean and Healthy Environment'; Harrison, 'Reflections on the Role of International Courts and Tribunals in the Settlement of Environmental Disputes', 503.

53 Boyle, 'Human Rights and the Environment' 238–9.

54 Knox, *Final Report of the Special Rapporteur on the Issue of Human Rights Obligations.*

argument advocates for procedural guarantees as allegedly easier to implement within existing human rights protection regimes. Advocates suggest that procedural rights can avoid the uncertainty and controversy surrounding the normative claims of substantive environmental rights.[55] The underlying assumption appears to be that aspirational and procedural environmental rights do not make vague normative claims, or at least do not raise the same level of concern as substantive rights.

It is of course reasonable to advocate for the introduction of procedural and aspirational rights in addition to substantive environmental rights. All three can secure meaningful improvements and enhanced public participation in environmental decision making is certainly desirable. What is more problematic is to reject substantive environmental rights on the basis of allegedly vague and contentious normative claims, while arguing that procedural environmental rights somehow do not present the same pitfalls.[56] As Tim Hayward points out in the context of constitutionalising procedural environmental rights:

> It cannot consistently be maintained both that the procedural rights in question are distinctively environmental and that they are entirely neutral with regard to the potential outcomes of their exercise. (...) If explicit acknowledgement of this substantive environmental right [a right of every person to live in an environment adequate to his or her health and well-being] is accepted as the reason why there should be constitutional provision for procedural environmental rights, then the argument that procedural rights only should be constitutionalized, because the substantive environmental right somehow illicitly skews outcomes, loses not only normative force, but even conceptual coherence.[57]

In other words: if procedural environmental rights are to offer anything beyond neutral and general fairness guarantees that are universally applicable, a view evidently taken by some scholars, then this necessarily implies a normative agenda that somehow, however covertly, favours environmental

55 Handl, 'Human Rights and Protection of the Environment', 318; Sionaidh Douglas-Scott, 'Environmental Rights in the European Union – Participatory Democracy or Democratic Deficit?' in Michael R. Anderson and Alan E. Boyle (eds.), *Human Rights Approaches to Environmental Protection* (Clarendon 1996) 113; Marc Pallemaerts, 'The Human Right to a Healthy Environment as a Substantive Right' in Maguelonne Déjeant-Pons and Marc Pallemaerts (eds.), *Human Rights and the Environment* (Council of Europe 2002) 15.

56 Handl, 'Human Rights and Protection of the Environment', 318; Douglas-Scott, 'Environmental Rights in the European Union' 113; Pallemaerts, 'The Human Right to a Healthy Environment as a Substantive Right', 15.

57 Hayward, *Constitutional Environmental Rights*, 87–8, footnotes omitted.

interests.[58] This in turn means that procedural environmental rights must make some normative claims, and hence the same critique that scholars level against substantive environmental rights would in principle apply. As it turns out, vagueness and enforcement are also salient concerns for aspirational and procedural guarantees. Empirically, the data gathered for this book demonstrates that a plethora of legal questions are raised under the procedurally focused Articles of the ECHR in environmental cases. Articles 6 and 13 ECHR were invoked 138 and 77 times in environmental cases, respectively, and jointly surpass any other Article of the Convention.[59] Procedural claims for environmental harm, it would therefore seem, are no less contested, vague or any easier to enforce than substantive environmental protections available under the ECHR.

However, one might also raise a basic objection that addresses the vagueness argument more directly: what is it about the protection of the environment that should occasion us to conceive of the relationship to human rights as merely aspirational or procedural, as opposed to formulating enforceable claims that challenge an endemic threat to human well-being? Are the demands of human well-being and autonomy currently adequately protected through procedural and aspirational demands? Persistent environmental harms demonstrate that they are not sufficient on their own. The often-overlooked reality is that while scholars, environmentalists, human rights activists, policymakers, and the public might not always agree on what constitutes an ideal environment or the best means to achieve it, it remains entirely feasible to formulate a minimum standard: one that the environmental minimum seeks to express, and which is based on only a handful of broadly shared normative assumptions. Without these basic environmental safeguards, no vision of an ideal environment can be realised, regardless of whether one emphasises human needs or the intrinsic value of the natural environment, or prefers an untouched Amazon rainforest to an English countryside historically shaped by agriculture.

This room for agreement is all too often overlooked in the pursuit of a complete and universally applicable notion of a 'sound' environment. It obscures the fact that a comprehensive protection regime for the natural

[58] Jonas Ebbesson, 'Public Participation' in Daniel Bodansky, Jutta Brunnée, and Ellen Hey (eds.), *The Oxford Handbook of International Environmental Law* (Oxford University Press 2007) 687–8 Birnie and others, *International Law and the Environment*, 288; a view also notably expressed in the preamble to the Aarhus Convention: 'In order to contribute to the protection of the right of every person of present and future generations to live in an environment adequate to his or her health and well-being . . .'.

[59] See the Results Table of the dataset gathered for this book.

environment is not necessary to enforce substantive environmental human rights that provide significant safeguards.[60] This is especially true in law, a field that is well accustomed to working with terms, provisions, legislation, and treaties that require further specification through legal interpretation, particularly when abstract rules are applied to practical cases. Terms such as 'fair', 'equitable', 'necessary', and 'sufficient' have been specified in a plethora of legal systems for various purposes, and there is no reason to suppose the same cannot be achieved for environmental human rights.[61] Likewise, the environmental minimum need not provide a complete and self-executing set of rules, where a framework and guidelines for adjudication can achieve tangible improvements.

2.3.2 *Conflicts*

Some critics are less concerned with the vagueness of environmental human rights, but suggest that they threaten established human rights through their potential for conflicting normative claims.[62] They perceive a risk that introducing rights-based protections of the environment undermines both human rights and environmental protection: human rights are undermined because environmental harm introduces new interests into protection regimes resulting in conflicts that ultimately limit and qualify available protections (the dilution argument); and environmental protection is undermined because human rights protection regimes could be used to blunt existing environmental regulations (the cynical rightsholder argument).

60 John Lee, 'The Underlying Legal Theory to Support a Well-Defined Human Right to a Healthy Environment as a Principle of Customary International Law' (2000) 25 *Columbia Journal of Environmental Law* 283, 338.

61 Boyle, 'The Role of International Human Rights Law', 51.

62 A focus on conflicts between rights can be found in Miller, *Environmental Rights*, xi; Chris Miller, 'Case Law Analysis: The European Convention on Human Rights: Another Weapon in the Environmentalist's Armoury' (1999) 11 *Journal of Environmental Law* 157, 176; Karen Morrow, 'After the Honeymoon: The Uneasy Marriage of Human Rights and the Environment under the European Convention on Human Rights and in UK Law under the Human Rights Act 1998' (2013) 43 *Revue Générale de Droit* 317, 323; for an emphasis on conflicts with environmental protection and regulation, see Dupuy and Viñuales, *International Environmental Law*, 297–8 and 331–5; Dinah Shelton, 'Resolving Conflicts between Human Rights and Environmental Protection: Is There a Hierarchy?' in Erika De Wet and Jure Vidmar (eds.), *Hierarchy in International Law: The Place of Human Rights* (Oxford University Press 2012) 235.

2.3.2.1 Diluting Human Rights Protections

The starting point of critics is certainly sound: environmental human rights naturally conflict with other human rights, and this potential for conflict could be exacerbated when environmental rights are formally recognised (either through green interpretation or express amendment) in human rights protection regimes. However, the theory and practice of human rights demonstrate that conflicts are inevitable, and adjudication likewise gives hope that these conflicts can be resolved in the context of specific cases.

Contradictory normative claims are made by a host of human rights, and balancing these interests is a pervasive feature of any human rights protection regime, frequently through proportionality assessments:[63] for instance, freedom of the press is balanced with a right to privacy,[64] freedom of expression with the harmful effects of hate speech,[65] equality is balanced with positive discrimination,[66] and data protection with national security interests.[67] The same pragmatic approach is applied in a wide variety of jurisdictions that have introduced constitutional environmental rights.[68] If the possibility for contradictory claims and conflicts were a decisive concern, then we would have to eliminate a host of existing human rights and many foundational principles in almost every area of the law.[69]

There is also a pragmatic reason not to avoid conflict, particularly around environmental human rights: it would be to close the stable door after the

63 For instance in Germany under the Basic Law, see Ulrich Vosgerau, 'Zur Kollision von Grundrechtsfunktionen: Ein zentrales Problem der Grundrechtsdogmatik' (2008) 133 *Archiv des öffentlichen Rechts* 346; Hans D. Jarass and Bodo Pieroth, *Grundgesetz für die Bundesrepublik Deutschland – Kommentar* (14th ed., C. H. Beck 2016) 554 [13]; in England and Wales, *see* Mark Elliott and Robert Thomas, *Public Law* (3rd ed., Oxford University Press 2017) 609–10; in the context of the ECHR, *see* Pätzold, 'Artikel 8 EMRK' 277 [97]; in the context of the ACHPR, *see* Frans Viljoen, *International Human Rights Law in Africa* (2nd ed., Oxford University Press 2012) 330.

64 *Von Hannover* v. *Germany* App no 59320/00 (ECtHR, 24 June 2004); *Von Hannover* v. *Germany (No. 2)* App nos 40660/08; 60641/08 (Grand Chamber, 7 February 2012).

65 *Jersild* v. *Denmark* App no 15890/89 (Grand Chamber, 23 September 1994); *Gündüz* v. *Turkey* App no 35071/97 (ECtHR, 4 December 2003); *Norwood* v. *United Kingdom* App no 23131/03 (ECtHR, 16 November 2004); *Vejdeland and others* v. *Sweden* App no 1813/07 (ECtHR, 9 February 2012).

66 Case C-450/93 *Kalanke* v. *Freie Hansestadt Bremen* [1995] ECLI:EU:C:1995:322; Case C-409/95 *Hellmut Marschall* v. *Land Nordrhein-Westfalen* [1997] ECLI:EU:C:1997:533.

67 Case C-230/14 *Weltimmo sro* v. *Nemzeti Adatvédelmi és Információszabadság Hatóság* [2015] ECLI:EU:C:2015:639; Case C-362/14 *Maximillian Schrems* v. *Data Protection Commissioner* [2015] ECLI:EU:C:2015:650.

68 May and Daly, *Global Environmental Constitutionalism*, 146–8.

69 King, *Judging Social Rights* 5.

horse has already bolted. As much as critics may have concerns over the introduction of environmental interests into human rights protection regimes, they are already widely recognised and adjudicated under various international, regional, and domestic protection regimes. Fundamental opposition to these developments is futile, especially considering pervasive environmental harm and the serious threat that it poses to the enjoyment of human rights. Scholarship ought therefore to focus on more constructive endeavours, such as addressing and channelling conflicts into a coherent framework that alleviates problems and provides a mechanism for their resolution. As we shall see, one such framework is the environmental minimum.

2.3.2.2 Diluting Environmental Protection

A more nuanced objection challenges precisely this balancing of interests because of the repercussions it might have for environmental protection itself. Specifically, sceptics allege that human rights-based challenges risk diluting environmental protection over time. Leaving environmental interests separate from the realm of human rights, so goes the argument, insulates them from challenges and permits for more far-reaching and radical protections of the environment. While this view has certain overlaps with those who advocate for entirely separate rights of nature, it focuses on the specific idea that the negative consequences for environmental protection from balancing and accommodating human rights are underestimated.[70]

Suppose a 'cynical' rightsholder, a claimant with vested interests in business that causes environmental harm, challenges an environmental regulation based on a right to property and freedom to conduct business. Even if environmental human rights might improve the environmental situation in some areas, the overall protection across the board might be weakened, as regulations are successfully challenged and thus overall gravitate towards a lowest common denominator that lags behind currently available safeguards. There are several possible rejoinders to this line of argument.

First, the argument falsely assumes that current environmental regulatory regimes, whether at a national or international level, stand apart and are somehow insulated from rights-based challenges. The truth is that the 'cynical' rightsholder is already challenging environmental regulations, often based on rights to property and the freedom to conduct business.[71] The recognition of

[70] Dupuy and Viñuales, *International Environmental Law*, 297–8 and 331–5; Shelton, 'Resolving Conflicts between Human Rights and Environmental Protection', 235.

[71] See, for instance, the dataset gathered for this book.

environmental human rights would arguably make protection regimes less vulnerable to such challenges. Environmental regulations are more likely to be accepted as justified limitations on rights when the underlying interests and policy objectives are likewise recognised and protected under human rights law.[72]

This is especially relevant because 'cynical' rightsholders often rely on a limited set of rights that are open to reasonable justifications through competing interests of a comparable priority. For instance, although the right to property features prominently in the ECHR case law, it often yields only limited practical safeguards because the ECtHR routinely accepts the interests of environmental protection as sufficient to justify infringements.[73] It is therefore not surprising that industry resistance to environmental regulation often takes more subtle forms, for instance lobbying of legislatures and challenging the validity of scientific evidence, a phenomenon explored in more detail in Chapter 10. At the very least, introducing environmental human rights would not appear to give 'cynical' rightsholders more tools to challenge environmental regulations than they already wield.

Second, it is worth pointing out that many regional protection regimes expressly view themselves as setting a minimum standard of protection, permitting states to go further if they wish. For instance, Article 53 ECHR states that the provisions cannot be relied upon to limit the level of protection available under domestic and international rights.[74] This reinforcing function of the ECHR is also generally accepted by domestic courts, for instance the German Constitutional Court.[75] To the extent therefore that environmental regulations go beyond what is required under the ECHR, human rights law does not stand in the way of higher protection standards, provided the measures are proportionate.

Naturally, the matter becomes more complicated in cases where human rights are invoked on both sides of an issue. These are so-called multipolar human rights relationships, as two or more rightsholders make conflicting

[72] Redgwell, 'Life, The Universe and Everything', 87; Hayward, *Constitutional Environmental Rights*, 34; Brandl and Bungert, 'Constitutional Entrenchment of Environmental Protection', 87.

[73] See the plethora of cases in the category two dataset gathered for this book.

[74] *United Communist Party of Turkey and others* v. *Turkey* App no 19392/92 (Grand Chamber, 30 January 1998) [28] [referencing the former Article 60, now Article 53].

[75] This is settled case law, *see* Federal Constitutional Court, 2 BvR 589/79; 2 BvR 750/81; 2 BvR 284/85, 26 March 1987, BVerfGE 74, 358 (370); Federal Constitutional Court, 2 BvR 1481/04, 14 October 2004, BVerfGE 111, 307 (317); Federal Constitutional Court, 1 BvR 1602/07; 1 BvR 1606/07; 1 BvR 1626/07, 26 February 2008, BVerfGE 120, 180 (200).

demands on the state.[76] This is a typical constellation that arises in environmental cases, as the environment is rarely degraded through direct state action but instead through private actors. In such cases, regimes like the ECHR naturally cannot be applied selectively to benefit only the interests of one party.[77] Rather, the Convention must stand outside the domestic protection regime and examine the balance struck by the state between competing rights as a whole.[78] The Court generally does not second guess state balancing, and therefore, if a state opts for a higher environmental protection standard, it is difficult to see how environmental human rights would invariably stand in its way. At least it is not clear that under the environmental minimum a regression on established environmental regulations and practices is probable.

Third, it is useful to recall that environmental human rights do not demand perfect environmental regulations. In particular, as we shall see in Chapter 3, the environmental minimum framework primarily enforces existing regulations and combats regulatory inertia: it intentionally avoids the conceptual uncertainty that a comprehensive substantive environmental right might bring to the debate. Provided a good faith regulation exists and is consistently enforced, the environmental minimum is irrelevant, for instance, to a debate surrounding a conflict between freedom of religion, invoked to seek the kosher slaughter of animals, and animal welfare regulations.[79] Only where regulations fall short of generally accepted minimum standards does the framework place demands on regulators.

2.4 SUMMARY

This chapter presented the argument for the recognition and the specific normative claims of environmental human rights. It has become a widely

[76] Wolfgang Hoffmann-Riem, 'Kontrolldichte und Kontrollfragen bei nationalen und europäischen Schutz von Freiheitsrechten in mehrpoligen Rechtsverhältnissen' [2006] *Europäische Grundrechtezeitschrift* 492; Tobias Thienel, 'Artikel 53 EMRK' in Ulrich Karpenstein and Franz C. Meyer (eds.), *Europäische Menschenrechtskonvention – Kommentar* (2nd ed., C. H. Beck 2015) 704 [5]; *Anayo* v. *Germany* App no 20578/07 (ECtHR, 21 December 2010) [70] [competing Article 8 ECHR claims of a child, biological mother, legal father, biological father, and their respective biological children].

[77] Thienel, 'Artikel 53 EMRK', 703–4 [5].

[78] Ibid., 705 [8]; Oliver Klein, 'Straßburger Wolken am Karlsruher Himmel – Zum geänderten Verhältnis zwischen Bundesverfassungsgericht und Europäischen Gerichtshof für Menschenrechte seit 1998' [2010] *Neue Zeitschrift für Verwaltungsrecht* 221, 223; Markus Ludwigs, 'Kooperativer Grundrechtsschutz zwischen EuGH, BVerfG und EGMR' [2014] *Europäische Grundrechtezeitschrift* 273, 282.

[79] See, for instance, *Cha'are Shalom Ve Tsedek* v. *France* App no 27417/95 (Grand Chamber, 27 June 2000).

accepted position and a clearly established empirical fact that environmental harm of sufficient severity ultimately threatens the fundamental moral justifications for all human rights, among them, human dignity, autonomy, participation, and the interests of future generations. Thus, the chapter argued that human rights must contain a sub-category of environmental human rights that provides corresponding protections.

In developing the normative claims of environmental human rights, it is important to recall the general theme of this book: minimalism and a commitment to incremental improvements that shift the status quo towards greater environmental protection. While environmental human rights admittedly fall short of what perhaps ought to be protected under a complete ecological adjustment of society, there are significant normative overlaps between competing visions of a sound environment. It is this space for cooperation and compromise that the environmental minimum leverages. Progress in line with established legal categories and principles is useful, even if it ultimately fails to meet grander ambitions, especially in a political environment where concessions to the practical constraints of existing human rights protection regimes are required. Ultimately, the book defends a framework that fits well with the approach of courts and tribunals. This necessarily entails that not all the normative claims made by environmental human rights are represented through the environmental minimum: as we shall see, there are undeniable trade-offs and compromises that some will find objectionable. Nonetheless, there is value in a framework that systematises and normatively develops the notion of environmental harm and translates it into the context of existing human rights protection regimes.

Finally, objections based on vagueness and conflicts between rights are ultimately unconvincing. The law in particular is a field where vague concepts are routinely and successfully clarified in the context of specific cases through courts. Procedural and aspirational rights only seemingly avoid the difficult balancing with competing interests and the need for interpretive specification. Undoubtedly, the introduction of environmental protection into human rights regimes has and will continue to produce conflicts. Human rights have well-established mechanisms to resolve conflicts and, on balance, it appears more helpful to elevate environmental protection to the same level of priority as other human rights, rather than artificially separate it from protection regimes. Environmental protection is already widely recognised and adjudicated under a plethora of existing human rights protection regimes. Given that no amount of conceptual unease is likely to alter this fact, scholarship is more productively focused on engaging and providing practical frameworks that can guide this process. Even if the environmental minimum should ultimately not provide an optimal, or perhaps even a satisfactory approach to reconciling human rights with environmental protection, the framework hopefully enables future attempts to 'fail better.'

With the normative claims of environmental human rights specified and qualified, we can now turn to developing and defending the specific operational principles of the environmental minimum in line with these characteristics.

3

The Environmental Minimum

This chapter introduces and defends the environmental minimum as a framework for the protections (and correlative duties) that environmental human rights require. The guiding principles of the framework are modelled on the characteristics adopted from James Nickel and developed through the lens of environmental protection. The framework must therefore make specific claims, respect the inherent minimalism of human rights, and adhere to the high-priority principle. However, as we have seen, adherence to these characteristics does not answer the question as to what specific normative demands satisfy environmental human rights. Following Raz's interest theory of rights, the framework must clarify what state of affairs would satisfy the fundamental interests in environmental protection based on the values of human dignity and autonomy that we identified as core moral justifications.

At the same time, it is important not to overstate the case. The environmental minimum has the ambition of being a practically viable framework in the context of existing human rights protection regimes. Therefore, the framework only becomes operational if individuals affected by environmental harm succeed in establishing a specific risk to a human right guaranteed in the invoked protection regime. The environmental minimum therefore does not purport to introduce novel standalone rights. The goal is to conceptualise the environmental dimensions of existing rights in a principled manner that is normatively aligned with the claims of environmental human rights. While this may appear as insufficiently ambitious to some, it is indispensable to securing tangible improvements under the protection regimes examined in this book. The principles of the environmental minimum developed on this basis are as follows:

- The **specific risk principle** provides the conditions under which the environmental minimum applies to a given set of facts (trigger).

Notably, it requires that there must be a specific risk to a recognised human right arising from environmental harm;
- The **minimum standards** then specify the standard of review for environmental harm that constitutes a specific risk. The standards of review are determined through three different types of minimum standards: legal, established, and emerging standards. Their applicability depends on the existence and enforcement of regulations, and the level of scientific certainty and consensus;
- The **high-priority characteristic** guides the balancing of environmental interests with competing considerations. While the environment ought to be afforded a certain level of priority over other interests, it does not demand that environmental interests prevail in all cases. States are required to strike a good faith balance with competing interests but not all measures in furtherance of environmental protection are proportional to the impairments of competing interests.

3.1 SPECIFIC RISK PRINCIPLE

The specific risk principle is the trigger of the environmental minimum, defining when the minimum standards of review are applicable to a given environmental harm. The principle requires both a specific environmental harm and a resulting threat to a specific human right guaranteed under the invoked protection regime. All environmental harms are potential violations as long as applicants can establish a sufficient link to human rights through a reasonable hypothesis, based on scientific evidence. The specific risk principle is therefore best understood as an inquiry into the scope of protection offered by environmental human rights.

3.1.1 *Environmental Harm*

Environmental harm that threatens human rights can arise from a number of sources: for instance, pollution and degradation affecting air quality, water, food safety, soil, flora, and fauna as a part of functioning ecosystem, or radiation and noise that are introduced into the natural environment. A central claim of the environmental minimum is that all environmental harms must be taken seriously as potential threats to human rights. Human dignity and autonomy demand basic environmental protections, regardless of whether environmental harm has been conventionally acknowledged as significant under human rights law. Helpfully, the empirical evidence demonstrates that environmental harm is in principle widely and generally accepted as relevant across all examined human rights protection regimes.[1]

3.1.2 *Recognised Human Right*

The need for a threat to a recognised human right forms an integral part of the trigger of the environmental minimum. At first glance, this might appear at odds with the normative demands of environmental human rights. After all, Chapter 2 argued at some length that environmental protection is crucial to any commitment to human rights. The normative claims of environmental human rights hold regardless of express recognition in human rights law or whether a specific environmental harm can be linked to a recognised human right. As normatively appealing as this line of argument is, it is somewhat removed from the legal reality and practice of the examined human rights protection regimes. If the environmental minimum is to be practically viable, it needs to strike a compromise between the sound normative position and existing practical and institutional constraints.

The increased recognition of environmental threats has undoubtedly steered the doctrinal position of courts towards the normative claims of environmental human rights, especially under so-called evolutionary interpretative approaches.[2] Indeed, it is difficult to make sense of the current doctrinal position of many courts without assuming that they are guided at least to some extent by the idea of environmental human rights: the safeguards provided under the ECHR for instance are difficult to explain otherwise, as the Convention does not acknowledge a separate right to environmental protection, or even so much as mentions the natural environment.

That being said, it would be naïve to assume that the general normative influence of environmental human rights is sufficient to develop a principled and consistent doctrinal position on environmental harm. Courts have adapted their case law to address some aspects of environmental pollution and degradation, but these adaptations have their limitations. Most notably,

1 See the dataset gathered for this book; the position of the ECtHR will be examined in more detail in Part II of the book, the view taken under other international and regional systems, as well as South Africa and Germany in Part III.

2 Ronald Dworkin, *Freedom's Law: The Moral Reading of the American Constitution* (Oxford University Press 1996); George Letsas, 'The ECHR as a Living Instrument: Its Meaning and Legitimacy' in Andreas Føllesdal, Birgit Peters, and Geir Ulfstein (eds.), *Constituting Europe: the European Court of Human Rights in a National, European and Global Context* (Cambridge University Press 2013); Eirik Bjørge, *The Evolutionary Interpretation of Treaties* (Oxford University Press 2014).

courts operating under regional protection regimes based on international treaties cannot conjure up standalone rights to environmental protection: if the environmental minimum is to be practically viable, then it must acknowledge and respect this limitation as well as the incremental development implied by evolutive treaty interpretation.

At the same time, it would be a mistake to attach onerous requirements to what counts as a recognised right. The standard set certainly does not require that the protection regime contains a comprehensive and independent right to a healthy environment. The book argues that a greener interpretation of an existing right can provide a sufficient basis for the key safeguards the environmental minimum envisions. This position goes against a common historical argument levied against novel human rights in general, and environmental rights in particular, namely that prior recognition through states is of crucial importance.[3] The most prominent exponent of this argument is Günther Handl's 'mildly revisionist' article on environmental protection and human rights. It should be noted that his view is not commonly held in scholarship on environmental human rights, but its echoes can still be traced in modern accounts.[4] Therefore, it is worthwhile briefly refuting his argument here. Handl's piece can be interpreted as advancing a broad or a narrow argument, both of which however fail to convincingly challenge the environmental minimum framework.

The broad objection simply denies that, in the absence of recognition of environmental harm as a relevant threat, anything of legal relevance can be inferred from either environmental threats to existing rights, or to underlining values like human dignity and autonomy. According to this view, express recognition in some form is required before existing protection regimes can offer any meaningful protection: most directly through codification of environmental human rights. By contrast, the narrow objection only suggests that existing human rights cannot be interpreted to provide a comprehensive protection of the environment, leaving space for limited and specific safeguards based on existing rights. This narrow variant rejects the idea that, in the absence of recognition, protection regimes can offer environmental protections that go beyond what is directly linked to existing human rights commitments. In the broader sense, the argument would appear to fail even on its own terms, while in the narrow sense, the environmental minimum framework avoids the essence of the objection.

3 Martin, A *System of Rights*; Alston, 'Conjuring up New Human Rights'.

4 The book explored some of these when considering vagueness and the potential for conflicts as objections to environmental human rights in Chapter 2.

3.1.2.1 Broad Objection

On a broader interpretation, Handl operates on the assumption that a lack of codification of environmental rights equates to an absence of protection from environmental threats:

> (. . .) the existence of an established human right that is conceptually related, perhaps in some way even logically antecedent, to the claimed environmental entitlement, cannot be invoked dispositively to establish the latter's international normativity as 'derived from', or 'subsumed under', the former's. (. . .) Another illustration of this point involves the right to health. The fact that states have committed themselves internationally to recognizing a human right to the protection of health, cannot be taken to imply that individuals also have a corresponding human right to a healthy environment.[5]

However, this recognition argument begs the question: if states have committed themselves to upholding rights that (logically) entail at least some protection from environmental threats, then have states not also (implicitly) recognised such protections and accepted environmental harm as a credible threat? It is not the source of the threats, but rather the requirements and scope of recognised rights that determine the relevance of environmental harm to a given protection regime. It is difficult to imagine a principled basis on which a state on the one hand commits itself to protecting human health, but on the other denies that this implies any responsibility for environmental threats to human health.

The protection regimes examined for this book certainly do not rule out environmental safeguards where the scope of the right is impacted by state action: whether that is the case is a matter of interpretation by the authoritative judicial bodies. Various regional and international regimes have thus through a 'green' interpretation of their provisions provided environmental safeguards even in the absence of a recognised standalone right to a healthy environment. In this meaningful empirical sense, many states have already committed themselves to precisely what the broad objection disputes. This is where elements of Handl's argument resonate with sceptics of evolutive judicial interpretation of rights. Most notably, these include the proponents of

[5] Günther Handl, 'Human Rights and Protection of the Enviroment, a mildly "revisionist" view' in Canrado Trindade AA (ed.), Human Rights, Sustainable Development and the Environment (Instituto Interamericano de Derechos Humanos 1992); an argument restated in his more recent work, *see* Handl, 'Human Rights and Protection of the Environment', see also Boyle, 'Human Rights and the Environment', 43.

'originalism', who argue that judges ought to prioritise the historic understanding of provisions over all other possible interpretations.[6] However, there are strong reasons to afford greater weight in interpretation to general underlying principles over specific historical threat scenarios that may have been on the minds of drafters.

Ronald Dworkin helpfully draws a distinction between the so-called concrete and abstract intentions of drafters to illustrate this point. Concrete intentions are those based, for instance, on specific threat scenarios, and the corresponding protections, and entitlements that the drafters had in mind, given their specific legal, historical, cultural, and social contexts. Abstract intentions, on the other hand, express more general underlying principles the drafters sought to express, for instance, moral ideas about the nature and conflicts between competing notions of justice, equality, autonomy, and human dignity.[7]

Dworkin argues that the interpretation of rights, in our case the extent to which they require environmental protection, should be primarily informed by abstract intentions. Apart from the fact that it can prove difficult, potentially impossible, to accurately determine the concrete intentions of drafters, proponents of an originalist view have failed to convincingly establish why it alone ought to be decisive on matters of interpretation.[8] Originalism often appears merely as a convenient placeholder for ideologically motivated preferences, for instance, in a highly politicised and institutionally damaged US Supreme Court. Crucially perhaps, this eccentric approach to legal interpretation celebrates as gospel the values and moral judgments of past generations that systematically discriminated and suppressed the political views of women, minorities, and the economically destitute in their societies. The suggestion that amendments to laws and constitutions ought to be the primary remedy for dubious moral judgments and systemic discriminations is not on its face unreasonable. However, that view assumes the prevailing distribution of political power no longer denies equal protection and influence to marginalised groups seeking change: there is a plethora of recent evidence to cast doubt on that assertion. At the very least, pointing to legislative and constitutional amendment as remedies is not a politically neutral position that originalists claim to adopt on matters of legal interpretation.

[6] Antonin Scalia, *A Matter of Interpretation: Federal Courts and the Law* (Amy Gutmann ed., Princeton University Press 1997).

[7] Ronald Dworkin, *A Matter of Principle* (Harvard University Press 1985) 48–50.

[8] Ronald Dworkin, *Law's Empire* (Hart 1998) 364; see also Dworkin's direct response to Scalia in Scalia, *A Matter of Interpretation*, 144–9.

It is for good reason that the vast majority of apex courts worldwide, and indeed a significant majority of US Supreme Court judges have not accepted, nor practised, the historic absolutism of originalism. Its cousin, the historic approach, has its place in the cannon of legal interpretation, but should operate alongside the text, context, and object and purpose of a provision. Legal interpretation should embrace a variety of interpretive approaches without an obvious hierarchy: recognising the necessity of updating and adapting the law to meet modern threats, rather than seeking to perpetuate, of all things, the ill-defined concrete intentions of historic lawmakers. A focus on abstract intentions ultimately forms an important part of the justification of the evolving interpretation of human rights and treaties as consistent with the obligations originally accepted by states.[9]

As Chapter 2 demonstrated, denying the relevance of environmental threats in an age of endemic harm is tantamount to undermining the foundations of human rights and the commitments they represent. States that take fundamental commitments to human rights seriously cannot reasonably exclude environmental threats from consideration, and indeed there is strong empirical evidence that many states generally accept this moral position and its legal implications.[10] The broad recognition argument is thus based on the dubious premise that environmental human rights and threats from environmental harm have not been recognised, and only retains coherence when combined with an eccentric interpretative doctrine that, for good reason, lacks widespread recognition and adaptation.

3.1.2.2 Narrow Objection

In a narrow sense, Handl's argument would challenge only the idea that comprehensive environmental protection can follow from existing rights, absent state recognition. Handl contends that there is a significant difference between an environmental human right and the environmental protection available through other human rights. The former necessitates recognition and codification before it can be considered protected, while the latter offers only limited protection. This narrow objection would hence refute an independent and comprehensive right to a sound environment, derived exclusively

[9] George Letsas, *A Theory of Interpretation of the European Convention on Human Rights* (Oxford University Press 2007) 70; Bjørge, *The Evolutionary Interpretation of Treaties*, 62; Stefan Theil, 'Is the 'Living Instrument' Approach of the European Court of Human Rights Compatible with the ECHR and International Law?' (2017) 23 *European Public Law* 587.

[10] The work of May and Daly, *Global Environmental Constitutionalism*, provides a comprehensive overview of this vibrant field of environmental constitutionalism.

from human rights not primarily related to the environment, especially one that allows individuals to enforce broad environmental claims.[11] However, on that narrow reading, the recognition argument does not appear to challenge the environmental minimum at all.

The environmental minimum does not posit a right to comprehensive environmental protection divorced from threats to specific human rights currently recognised in domestic and international law instruments. Provided that the environmental harm constitutes a specific risk, for instance to health, both a (hypothetical) comprehensive environmental right and the right to health would be engaged and make overlapping normative claims.[12] At the basic level where the framework operates, there is little meaningful separation between a comprehensive environmental human right and the environmental requirements imposed through other rights that function as the normative trigger for the environmental minimum.

3.1.2.3 Addressees

While scholarship in International Law increasingly presents private actors as addressees of human rights obligations alongside states, this aspect is not of primary relevance to the argument.[13] The environmental minimum does not concern itself with the distribution of responsibility for environmental harm between private actors. Indeed, the difficulties in attributing responsibility amongst domestic private actors have led some scholars to abandon key characteristics of a human rights–based approach to environmental harm that form a central premise for this book. Richard Hiskes for instance concedes that with environmental harm, the '(. . .) responsibility is diffuse across persons, institutions, and time in that they are often the result of years of actions, decisions, or often collective neglect.'[14] He concludes that the only way to make sense of a human right to protection from such harm is as a collective responsibility.[15]

This book humbly disagrees. Given a modern regulatory state, human rights have a clear and capable addressee: states are both comprehensively bound under human rights and can bring the necessary resources to bear on

[11] Handl, 'Human Rights and Protection of the Environment', 128.
[12] Shelton, 'Human Rights and the Environment', 170.
[13] David Jason Karp, *Responsibility for Human Rights: Transnational Corporations in Imperfect States* (Cambridge University Press 2014).
[14] Richard P. Hiskes, *The Human Right to a Green Future – Environmental Rights and Intergenerational Justice* (Cambridge University Press 2009) 46.
[15] Ibid., 38.

whichever number of (private) actors are ultimately responsible for the harm. As past regulatory experiences have demonstrated, for instance with asbestos, this can involve extensive intervention, and as a last resort, regulating persistently harmful industries out of existence entirely.[16] Accordingly, the environmental minimum is designed to operate in protection regimes where states are the primary addressees, which encompasses both domestic legislative and constitutional bills of rights as well as regional and international protection regimes.[17]

The environmental minimum does not claim to apportion responsibility among private actors for environmental harm, nor does it presume to determine how these actors should be burdened with regulations or enforcement actions. As we shall see in Section 3.2, the primary duty of states is to regulate in good faith and consistently enforce regulations with available legal mechanisms. In that sense, the environmental minimum does not preclude recourse to other enforcement methods, but rather supplements existing domestic and international remedies, including private and public law litigation, such as toxic torts in the United States, or public interest litigation in Germany under the Federal Nature Protection Act.[18]

3.1.3 *Reasonable Hypothesis*

An established environmental threat must rise to the level of a specific risk to a human right. This requires environmental harm to be linked through a reasonable hypothesis to a threat towards a specific human right, based on scientific evidence.[19] A hypothesis would be reasonable if it appears possible, on the basis of the currently available evidence, that the hypothesis could be proven correct through further research. That is not to say that a reasonable hypothesis need be free from uncertainty, or even be a leading hypothesis, as long as it is supported by a sufficient body of credible scientific evidence. Whether a hypothesis meets these requirements would have to be established through expert evidence and should generally be treated as a question of fact in the process of adjudication.

16 See the adding of asbestos to Annex I of the Council Directive 76/769/EEC of 27 July 1976 on the Approximation of the Laws, Regulations, and Administrative Provisions of the Member States Relating to Restrictions on the Marketing and use of Certain Dangerous Substances and Preparations; on the ozone layer, see Chapter 11.

17 For instance, the data on the ECHR, ACHR, ACHPR, ICCPR, ICESR, and the German and South African Constitutions.

18 Carl F. Cranor, *Toxic Torts: Science, Law, and the Possibility of Justice* (2nd ed., Cambridge University Press 2016); Bundesnaturschutzgesetz – German Federal Nature Conservation Act.

19 Adapted from Wansley, 'Regulation of Emerging Risks', 431.

Admittedly, this approach brackets the vexed question on when everyday environmental risks reach the level of legally significant risks, which depends chiefly on the degree of scientific certainty, the prevailing regulatory attitude, and cultural perspectives on acceptability of risks.[20] The reasonable hypothesis is only intended as a rough filter that removes some risks from consideration as implausible given current scientific knowledge.[21]

A good example for a commonly debated, but scientifically unsubstantiated, hypothesis is the fluoridation of drinking water. Not only is the hypothesis of harmfulness of fluoride not supported by scientific evidence, there is a decade-old consensus to the contrary. Adequately limited fluoridation of drinking water is generally not harmful.[22] Indeed, scientific evidence points to the benefits to the public with respect to tooth decay, especially in children.[23] Hence, any hypothesis alleging a harm from fluoridation of drinking water would likely fail the requirements of the specific risk principle, and not engage the environmental minimum. Conversely, a hypothesis alleging a threat from the commonly used herbicide glyphosate can draw on scientific evidence that has found it to be carcinogenic.[24] Final certainty has not yet been achieved, with conflicting studies and views expressed by a number of organisations.[25] Regardless of the eventual scientific consensus position, evidence currently supports a hypothesis linking glyphosate to a threat to human rights, and hence it would meet the requirements of the specific risk principle. Once a specific risk is established, a court then applies the minimum standards to

[20] For an excellent account of the complex interactions between risk management and the modern regulatory state, see Elizabeth Fisher, *Risk Regulation and Administrative Constitutionalism* (Hart Publishing 2007), especially chapter 1.

[21] A similar position is taken under EU law on the precautionary principle, see Case T-13/99 *Pfizer Animal Health SA* v. *Council of the European Union* [2002] ECLI:EU:T:2002:209 [143]–[144]; Case T-483/11 *Sepro Europe Ltd* v. *European Commission* [2013] ECLI:EU:T:2013:407 [44] and [94].

[22] Fawell and others, *Fluoride in Drinking-Water*; *Water Fluoridation Basics*, Centers for Disease Control and Prevention (2016) www.cdc.gov/fluoridation/basics/index.htm, accessed 1 April 2021.

[23] S. O. Griffin and others, 'Effectiveness of Fluoride in Preventing Caries in Adults' (2007) 86 *Journal of Dental Research* 410; N. Young and others, 'Community Water Fluoridation and Health Outcomes in England: A Cross-Sectional Study' (2015) 43 *Community Dentistry and Oral Epidemiology* 550.

[24] L. Fritschi and others, 'Carcinogenicity of Tetrachlorvinphos, Parathion, Malathion, Diazinon, and Glyphosate' (2015) 16 *The Lancet* 490.

[25] *Pesticide Risk Assessment of the Active Substance Glyphosate* (European Food Safety Authority, 2015); *Neue Meta-Analyse zu glyphosathaltigen Pflanzenschutzmitteln ändert die Bewertung des Wirkstoffs nicht – Stellungnahme Nr 008/2019* (Bundesinstitut für Risikobewertung, 2019); *Joint FAO/WHO Meeting on Pesticide Residues, Geneva 9–13 May 2016 – Summary Report* (World Health Organization – Food and Agriculture Organization of the United Nations, 2016).

make sense of the environmental harm in a manner consistent with the normative claims of environmental human rights.

3.2 MINIMUM STANDARDS

An environmental harm that constitutes a specific risk is evaluated against the minimum standard before any determination on a violation of the invoked human right can be made. The environmental minimum holds that breaching the applicable minimum standard constitutes a violation of the human right identified at the specific risk stage, regardless of otherwise applicable exceptions and permissible limitations.

For instance, a specific risk might arise from high levels of air pollution, based on a threat to private and family life,[26] and the reasonable hypothesis linking the former to the latter.[27] While a right to private and family life is generally subject to limitations, for instance such that states can justify proportionate infringements in pursuit of legitimate aims, a state would be barred from relying on these justifications whenever the applicable minimum standard under the environmental minimum is breached. The framework thus modifies the determination of rights violations that would ordinarily apply in a given protection regime. The minimum standards for environmental harm draw on three distinct sources: legal norms, established, and emerging scientific evidence. Notably, unclear or contradictory scientific evidence typically grants states a wide margin of discretion in formulating a regulatory response. These three minimum standards and their effects admittedly seek significant modifications of existing doctrinal approaches to environmental harm that require careful justification.

3.2.1 *Legal Standards*

Legal norms form the core standard of review in the environmental minimum framework. They are comprised of domestic and international regulations with legal force that have been freely accepted by states. Examples are abound in various pieces of domestic environmental protection legislation and specific agreements under International Law, for instance protecting the ozone

26 *Budayeva and others* v. *Russia* App nos 15339/02; 21166/02; 20058/02; 11673/02; 15343/02 (ECtHR, 20 March 2008); for a detailed analysis explaining that Article 8 ECHR has taken on the role of a de facto right to health, see Chapter 4.

27 *Burden of Disease from Ambient Air Pollution for 2012*.

layer.[28] The effective enforcement of these regulations should be an unconditional necessity in any state that takes the rule of law, international treaty obligations, human rights, and environmental harm seriously: a failure to enforce legal norms invariably constitutes a violation of the invoked human right under the environmental minimum framework.

3.2.1.1 Rule of Law

Scholarly work on the rule of law abounds in legal studies and political philosophy. Scholars from a range of different legal systems have offered their perspectives, and courts often apply the domestic incarnations of the principle in constitutional and public law cases. Paul Craig helpfully draws a distinction between formal and substantive accounts of the rule of law to delineate a key area of disagreement: formal conceptions primarily concern themselves with the rules of law making and their application, while substantive accounts additionally pass more detailed judgment on the contents of law. Substantive accounts hence provide a broader basis for distinguishing what we might loosely term 'good' law, which follows the requirements of the rule of law, from 'bad' law, which does not.[29] The formal view tends to be more strongly aligned with legal positivism and its desire to separate what the law is from questions of political morality.[30]

A prominent exponent of the formal view is found in Joseph Raz's early work, which suggests that the rule of law '(...) is not to be confused with democracy, justice, equality (before the law or otherwise), human rights of any kind or respect for persons or for the dignity of man.'[31] Raz suggests these are important values and desirable in a legal system, but holds that it is crucial the rule of law remains an independent concept from political morality. Otherwise, it risks losing its utility, as any determination on adherence with the rule of law would collapse entirely into the peculiarities of ostensibly more subjective moral and political convictions. The rule of law should thus according to Raz specify the rules that ought to be followed in law making

[28] Vienna Convention for the Protection of the Ozone Layer (adopted 22 March 1985, in force 22 September 1988) 1513 UNTS 323.

[29] Paul Craig, 'Formal and Substantive Conceptions of the Rule of Law: An Analytical Framework' [1997] *Public Law* 467.

[30] Ibid., 477.

[31] Joseph Raz, *The Authority of Law: Essays on Law and Morality* (Oxford University Press 1979) 211; although his more recent restatement does cast doubt as to whether he is still committed to a strictly formal view, see Joseph Raz, 'The Law's Own Virtue' (2019) 39 Oxford Journal of Legal Studies 1.

and the application of the law, but generally avoid passing judgment on content.

Several influential authors including Ronald Dworkin and Trevor Allan broadly disagree with this formal conception for that very reason, and instead advocate the inclusion of ideas about justice and equality into the rule of law.[32] They build on elements of the formal conception by adding substantive values into any assessment of what counts as law and how it is enforced. There are certainly more complexities and nuances to the views expressed in legal scholarship than are captured by this simplified analytical framework.[33] Allan has noted, for instance, that even a formal conception of the rule of law contains some implicit substantive judgments about autonomy and the individual. Indeed, Raz formulates a number of principles that seem to go well beyond a bare formal conception of the rule of law, allegedly stripped of normative values.[34] For Allan therefore, seeking to maintain a distinction between formal and substantive concepts of the rule of law is futile, as it was for the account of Lord Bingham.[35] The difference between formal and substantive accounts of the rule of law indeed appears to come down to the degree to which normative values are (expressly or implicitly) integrated into the concept, rather than the stark choice between rejection and integration that Raz presents. There may well be value in a less normatively loaded conception of the rule of law, as indeed there is in not simply collapsing all human rights safeguards into a comprehensive right to life, but that is a more nuanced argument that requires careful elaboration.

It is unnecessary for the purposes of this book to engage more deeply with this debate: as with the environmental minimum, what matters are the overlapping points of agreement across most contemporary accounts. For present purposes, it suffices that any conception of the rule of law includes a duty on competent state entities to implement and enforce the law,[36] which of course includes enacted environmental regulations.[37] In many respects, the

32 Dworkin, *A Matter of Principle*, 11–12; T. R. S. Allan, *Law, Liberty, and Justice: The Legal Foundations of British Constitutionalism* (Oxford University Press 1995) chapter 2.

33 A notable example of a more nuanced and detailed categorisation is found in the appendix of Frank Lovett, *A Republic of Law* (Cambridge University Press 2016).

34 Raz, *The Authority of Law*, 214–19; Craig, 'Formal and Substantive Conceptions of the Rule of Law', 469.

35 Allan, *Law, Liberty, and Justice*, 39; Lord Bingham, 'The Rule of Law' (2007) 66 The Cambridge Law Journal 67.

36 Joseph Raz, *Ethics in the Public Domain: Essays in the Morality of Law and Politics* (Clarendon 1994) 358; Allan, *Law, Liberty, and Justice*, 20–1; Dworkin, *A Matter of Principle*, 11.

37 Knox, *Final Report of the Special Rapporteur*, para 34.

functional equivalent of the rule of law in International Law is the principle of *pacta sunt servanda*.

3.2.1.2 *Pacta Sunt Servanda*

Pacta sunt servanda is a principle which forms a fundamental pillar of International Law and is in many respects the corollary to the rule of law in a domestic setting. It states that agreements are binding and must be implemented in good faith.[38] The importance for the proper functioning of International Law is difficult to overstate,[39] and the principle has been codified in Article 26 VCLT.[40] As such, it is 'axiomatic and self-evident',[41] and can be considered a *Grundnorm* of International Law in the Kelsian tradition of legal positivism.[42] The implications for international legal standards of the environmental minimum are clear: states have an obligation to stand by their commitments, and hence must consistently enforce any environmental protection measures agreed in treaties, save for highly unlikely conflicts with peremptory norms of international law. Examples for legal standards in International Law include environmental directives adopted by the European Union,[43] and various treaties and conventions in international environmental law.[44]

3.2.1.3 Good Faith Requirement

Not all environmental regulations automatically count as legal standards for the purposes of the environmental minimum. The framework specifies that regulations made in bad faith are not legal standards because they manifestly fail to address specific risks to human rights. Specifically, a rebuttable presumption of bad faith attaches to regulations that fall short of the recommendations and guidelines that meet with general acceptance in the relevant

38 James Crawford, *Brownlie's Principles of Public International Law* (9th ed., Oxford University Press 2019) lxxxiv.
39 Ibid., 363, 434.
40 Vienna Convention on the Law of Treaties.
41 Crawford, *Brownlie's Principles of Public International Law*, 434.
42 Hans Kelsen, *Reine Rechtslehre: mit einem Anhang: das Problem der Gerechtigkeit* (2nd ed., Deuticke 1960) 129–30.
43 Especially the extensive EU legislation in various areas of environmental concern based on Title XX of the Treaty on the Functioning of the European Union.
44 See, for instance, the extensive overview provided by Sands and others, *Principles of International Environmental Law*, xxxvii and Dupuy and Viñuales, *International Environmental Law*, xviii–xvi.

scientific community on a given environmental harm.[45] A rich source for generally accepted standards can be found in the recommendations of international organisations, such as those of the WHO, which are consulted and afforded considerable weight by international courts and tribunals. The WHO is, for instance, highlighted in John Knox's report as a key source for international environmental, health, and safety standards.[46] Reports and evidence from such international organisations also informed the advisory opinion of the International Court of Justice (ICJ) on the legality of nuclear weapons,[47] and experts from international organisations are frequently consulted in trade disputes.[48]

The recommendations and guidelines are developed by independent experts, subjected to peer review, and follow an established scientific methodology that conducts a meta-analysis of studies. They are typically not at the forefront of science precisely because care is taken to corroborate results over longer time periods. This renders their recommendations and guidelines reflective, to the extent possible, of the view generally accepted in the relevant scientific community on a given environmental harm. However, these standards are advisory in nature and not directly legally binding. Therefore, given that the environmental minimum generally respects the outcomes of the democratic process and the necessary trade-offs performed by regulators, the prominence of expert recommendations and guidelines in evaluating good faith requires further justification.

3.2.1.3.1 LEGAL STATUS OF ADVISORY OPINIONS. The proposition that ignoring recommendations of advisory bodies can render administrative and regulatory action illegal is not alien to international and domestic laws. In international human rights law, both the ECtHR and the European

45 Adapted from King, *Judging Social Rights*, 246.

46 Knox, *Final Report of the Special Rapporteur*, para 33.

47 *Effects of Nuclear War on Health and Health Services* (World Health Organization, 1987) cited in *Legality of the Threat or Use of Nuclear Weapons* (Advisory Opinion) [1996] ICJ Rep 226; see further *Case Concerning Pulp Mills on the River Uruguay (Argentina* v. *Uruguay)* (Merits) [2010] ICJ Rep 167; *Nuclear Tests case (Australia* v. *France) Request for the Indication of Interim Measures of Protection* (Order of 22 June 1973) [1973] ICJ Rep 99.

48 *Thailand – Restrictions on Importation of and Internal Taxes on Cigarettes* GATT Panel Report DS10/R, adopted 7 November 1990, BISD 37S/20; *India – Quantitative Restrictions on Imports of Agricultural, Textile and Industrial Products* Complaint by the United States (WT/DS90), Report of the Appellate Body, DSR 1999: IV, 1763; *European Communities – Measures Affecting the Approval and Marketing of Biotech Products* Complaint by United States, Canada, Argentina (WT/DS291, WT/DS292, WT/DS293), Report of the Panel DSR 2006: III, 847 [7.31]; *European Communities – Measures Affecting Asbestos and Asbestos-Containing Products* Complaint by Canada (WT/DS135), Report of the Panel DSR 2001: VIII, 3305 [3.178].

Committee on Social Rights have previously considered and applied advisory standards to rule on compliance with human rights.[49] In the ECHR case of *Fägerskiöld* v. *Sweden*, the applicants complained of noise (between 42 and 45 dB) and light emission from wind turbines constructed on property neighbouring their holiday home.[50] Their application to the ECtHR was rejected, based in significant part on the fact that the WHO guidelines on noise pollution were not exceeded.

This approach to evaluating regulatory conduct is also reflected in framework principle 11 of John Knox's final report. The principle outlines that states must '(. . .) establish, maintain and enforce effective legal and institutional framework for the enjoyment of a safe, clean, healthy and sustainable environment.' While the report recognises that limited resources may prevent states from fully realising this level of environmental protection, it nonetheless contends that discretion on resource allocation is not boundless, with respect to both civil and political, as well as economic, social, and cultural rights.[51] Crucially, the report outlines a strong presumption against retrogressive measures, stating that regulations must '(. . .) not strike an unjustifiable or unreasonable balance between environmental protection and other social goods, in light of its effects on the full enjoyment of human rights.'[52]

At the domestic level, the law of judicial review in England and Wales accepts that ignoring relevant advisory opinions can render administrative action unlawful.[53] Particularly in the exercise of discretionary power, public

[49] *International Federation for Human Rights (FIDH)* v. *Greece* 72/2011 European Committee of Social Rights, 23 January 2013 [42]–[44]; *Fägerskiöld* v. *Sweden* App no 37664/04 (ECtHR, 26 February 2008); *Brincat and others* v. *Malta* App no 60908/11; 62110/11; 62129/11; 62312/11; 62338/11 (ECtHR, 24 July 2014) [37].

[50] *Fägerskiöld* v. *Sweden* App no 37664/04 (ECtHR, 26 February 2008) 8 and 16.

[51] Knox, *Final Report of the Special Rapporteur*, paras 32–33.

[52] Ibid., para 33.

[53] *R (Bradley)* v. *Secretary of State for Work and Pensions* [2008] EWCA Civ 36 [dismissal of the report of the parliamentary ombudsman]; *R (Gallagher)* v. *Basildon District Council* [2010] EWHC 2824 (Admin) [36]–[42] [failure by local council to follow ombudsman recommendation on compensation]; *R (Evans)* v. *Attorney General* [2015] UKSC 21 [66]–[67] [failure to consider upper tribunal ruling in the context of a section 53 Freedom of Information Act 2000 decision]; *R* v. *Secretary of State for the Home Department* [1998] Imm AR 84, Court of Appeal [limiting the ability of the Secretary of State to reject the adjudicator's finding of fact in exercising discretionary power to grant leave to enter]; *R (W)* v. *Chief Constable of Warwickshire Police* [2012] EWHC 406. (Admin) [50] [on the extent to which the decision of an employment tribunal must be considered by the police in the decision to issue an enhanced criminal record certificate]; *R (Harris)* v. *Secretary of State for Justice* [2014] EWHC 3752 (Admin) [on the need to consider parole board decisions in moving a prisoner to open conditions].

bodies must take account of all relevant considerations,[54] and avoid irrelevant considerations,[55] even though both can be difficult to establish in court proceedings.[56] Relevant considerations include the views of advisory bodies, for instance those from parliamentary and local government ombudspersons, which must be afforded consideration in the decision-making process.[57]

In *Bradley* v. *Secretary of State for Work and Pensions* the Court of Appeal held that while the Secretary of State could reject the findings of the parliamentary ombudsman in favour of his own view, he could do so only if '(...) such a view is, itself, not irrational having regard to the legislative intention which underlies the decision [the Parliamentary Commissioner Act 1967].'[58] The Court of Appeal required cogent reasons for a departure from advisory opinions.[59] Similarly, in *Basildon District Council*, the Administrative Court found that while cogent reasons were not required in the context of recommendations from the local government ombudsperson, the local government had nonetheless acted irrationally. Notably, it had failed to take relevant considerations into account and instead relied on irrelevant considerations in rejecting the recommendations.[60] More generally, these rulings suggest that the conclusions and findings of advisory bodies must be taken seriously and not disregarded on irrational grounds, while equally acknowledging a degree of administrative flexibility.

A significantly more restrictive position is taken in Germany under the so-called Doctrine of Errors in the Exercise of Discretion (Ermessensfehlerlehre), which determines how administrative bodies must exercise discretion granted under statutory provisions. An important feature is that the doctrine permits far-reaching judicial review of discretionary decisions based on procedural and substantive grounds. Robert Alexy's article on

54 *Baldwin & Francis Ltd* v. *Patents Appeal Tribunal* [1959] AC 663; *R* v. *Paddington Valuation Officer* [1966] 1 QB 380; *R* v. *Shadow Education Committee of Greenwich BC* (1989) 88 LGR 589 [failure to consider preferences of parents]; *Bristol DC* v. *Clark* [1975] 1 WLR 1443 [on the issue of determining relevant considerations absent a statutory instruction].

55 *R (Sainsbury's Supermarkets Ltd)* v. *Wolverhampton City Council* [2010] UKSC 20; [2011] 1 AC 437; *R* v. *Liverpool University* [1991] 1 QB 124 [on the refusal to permit a meeting on the grounds that apartheid era South African diplomats speaking would provoke violence outside the university premises]; *R* v. *Birmingham Licensing Planning Committee Ex parte Kennedy* [1972] 2 QB 140 [application stalled for failure to meet irrelevant condition].

56 *R* v. *Boundary Commission for England* [1983] 1 QB 600, CA [challenge of the Labour party against a decision of the Boundary Commission].

57 *R (Bradley)* v. *Secretary of State for Work and Pensions* [2008] EWCA Civ 36 [91] and [95]; Harry Woolf and others, *De Smith's Judicial Review* (7th ed., Sweet & Maxwell 2013) [1-087].

58 *R (Bradley)* v. *Secretary of State for Work and Pensions* [2008] EWCA Civ 36 [51].

59 Ibid. [72].

60 *R (Gallagher)* v. *Basildon District Council* [2010] EWHC 2824 (Admin) [36]–[42].

the subject from the late 1980s still accurately reflects the essence of the dominant view.[61] There are three major categories of errors in the exercise of discretion that are recognised, even though the terminology employed by scholars varies:

- **Overstepping the boundaries of discretion** (Ermessensüberschreitung): the administrative body has made a discretionary decision that is at odds with the statutory requirements on which its decision and discretionary power are based. For instance, the body might issue a €40 penalty notice for a parking violation, where legislation permits a penalty between €10 and €30. This primarily constitutes an error of outcome, rather than the administrative procedure that led to the decision.
- **Failure to fully exercise discretion** (Ermessensunterschreitung): the administrative body reaches a decision unaware that it has a duty or the power to exercise discretion. This error attaches to the administrative procedure, often the legal characterisation of facts or the interpretation of statutory requirements, as opposed to the determination of the factual circumstances or the outcome of the discretionary decision as such. For instance, the administrative body might wrongly categorise a bicycle as a motor vehicle and therefore overlook a statutory provision conferring discretion to waive a parking penalty payment.
- **Misuse of discretion** (Ermessensfehlgebrauch): the administrative body fails to consider and afford due weight to information relevant to its decision or conversely, it takes irrelevant information into account. This is a procedural error primarily focused on the determination of relevant facts which nonetheless often involves errors in law.[62] For instance, the administrative body may legally exercise its discretion to waive a penalty payment, but the decision is unlawful if it is based in part on the irrelevant celebrity status of the person driving the car.

For present purposes, the final category of a misuse of discretion is most relevant because we are evaluating regulations and their conformity with scientific evidence as an indicator of bad faith. German administrative courts review regulatory decisions on the basis of the underlying information, including any data, values, figures, and the methodology employed to gather them: in the context of future harm, this notably includes whether realistic assumptions were made and an assessment of whether plausible reasons are offered

61 Robert Alexy, 'Ermessensfehler' (1986) 41 *JuristenZeitung* 701.
62 Ibid., 702.

for the results of projections.[63] Even where all relevant information is considered and duly weighed, the administrative body must ensure that its overall decision meets proportionality requirements in light of competing rights and interests.[64]

Environmental regulations that take the form of primary legislation are a more complicated matter in both the United Kingdom and Germany. On the orthodox account, the UK Parliament is sovereign in its law-making powers.[65] Parliament can in principle pass any law it wishes and take into account any considerations it deems relevant, and the resulting legislation cannot be invalidated through courts. Courts generally interpret legislation in conformity with the rights contained in the Human Rights Act (HRA) and have the power to issue declarations of incompatibility. Even though these declarations leave the legislation nominally intact, government and parliament have a solid track record of swiftly amending statutes in order to remedy the incompatibility.[66] Similarly, the doctrine on administrative errors in Germany does of course not apply to the legislative process. However, the strong form of judicial review practised by the Federal Constitutional Court achieves an at times comparable level of scrutiny for legislation. The Constitutional Court not only routinely invalidates primary federal and state legislation on the basis of fundamental rights violations, but also often prescribes the content and deadlines for the passage of fresh legislation in their place.[67]

Drawing on this general legal rationale of scrutinising regulations in international and domestic laws, the bad faith test of the environmental minimum would appear consistent with a serious commitment to human rights and the rule of law: a failure to consider relevant information, mere disagreement with expert evidence, or a disproportional weight afforded to competing interests should not suffice to disregard generally accepted scientific standards. In line with John Knox's report, the environmental minimum takes retrogression on

63 This is settled case law, Federal Administrative Court, BVerwG 3 B 7811, 11 July 2012, NVwZ 2012, 1175 [11]; Federal Administrative Court, 3 C 2584, 25 July 1985, BVerwGE 72, 38.

64 In an environmental context, see, for instance, Federal Administrative Court, BVerwG 7 C 3607, 27 September 2007, BVerwGE 129, 296 [21]; generally see Jarass and Pieroth, *Grundgesetz für die Bundesrepublik Deutschland*, Artikel 20, paras 112–127.

65 *R (Miller)* v. *The Prime Minister; Cherry and others* v. *Advocate General for Scotland* [2019] UKSC 41 [41].

66 Jeff King, 'Parliament's Role Following Declarations of Incompatibility under the Human Rights Act' in Murray Hunt, Hayley Hooper, and Paul Yowell (eds.), *Parliaments and Human Rights: Redressing the Democratic Deficit* (Bloomsbury Publishing 2015).

67 See, for instance, Federal Constitutional Court, 1 BvL 11/14, 10 April 2018, BverfGE 148, 147 [unconstitutional valuation of property tax, deadline to legislate within specified constitutional parameters].

environmental protection as a core indicator of bad faith: legalising a harmful and previously illegal level of environmental harm can scarcely be a proportional response. An established presumption of bad faith could hence not be rebutted in the following circumstances, which are intended to outline general evaluative principles for courts rather than constitute an exhaustive definition:

- **Failure to consider relevant information**: regulators rely on irrelevant information, fail to consider relevant information in drafting a regulation, or ignore significant changes to generally accepted recommendations and guidelines in the scientific community.
- **Circumvention**: a regulation adopts tokenistic measures in order to avoid addressing a pertinent environmental harm. For instance, regulators might nominally recognise air pollution as a threat, but in the interests of benefitting major polluting industries provide threshold values above current pollution levels for harmful substances.
- **Retrogression**: a regulation legalises harmful and previously unlawful environmental harm, for instance in response to political pressure from special interest groups.
- **Disproportionate limitation**: a regulation disproportionately favours competing interests, such as business opportunities, over environmental protection. A proportionality assessment considers the degree and seriousness of the environmental harm and the weight of competing interests in evaluating the deviation from generally accepted standards.

Providing effective regulations, responding in good faith to environmental harm, updating regulations in line with new scientific insights, and consistently enforcing agreed environmental protection standards is ultimately what environmental human rights, the rule of law, and sound governance require under the environmental minimum framework.

Admittedly, the environmental minimum thus demands a degree of nuanced reasoning from courts reviewing environmental regulations. While there may, for instance, be some doubt in the scientific community as to whether a more restrictive threshold *x* or a more lenient threshold value *y* is appropriate for cyanide in drinking water, such doubt does not extend to values *z* which are generally considered harmful. Regulators have significant leeway to choose from a range of values below *z*, by balancing the demands of environmental protection with other interests. The environmental minimum only attaches a presumption of bad faith to a regulation requiring a threshold equal to or above *z*. Critics might respond that this mechanism constitutes an undue limitation of regulatory flexibility, a challenge we now turn to address.

3.2.1.3.2 REGULATORY FLEXIBILITY. The environmental minimum is unequivocal that a regulatory decision cannot be reached in good faith if: (1) it fails to consider relevant information, (2) is primarily designed to circumvent and undermine generally accepted standards, (3) is retrogressive, or (4) disproportionately favours competing interests. Beyond these requirements the rebuttable presumption does not prejudice the regulatory assessment and balancing of competing interests and risks.

Scientific evidence, including the generally accepted standards of the relevant community – however important – can therefore only be the starting point of a regulatory decision, not authoritatively state its ultimate outcome.[68] Democratic decision making, public perceptions of risk, and underlying values matter, as does the structure and framework provided by the law, including human rights law.[69] Correctly understood, regulation under the environmental minimum therefore does not present itself as a stark and straightforward choice between favouring scientific evidence or the outcome of democratic processes.[70] Rather, the framework reflects the complex interplay between majoritarian preferences and scientific evidence within a democratic system constituted through law and shaped by human rights. The environmental minimum recognises and incorporates this insight into the normative framework through the mechanism of regulatory discretion; within certain boundaries drawn by human rights and specified through the generally accepted scientific standards and practices.

Even if regulators might in practice find it difficult to rebut the presumption of bad faith, this can be justified as a reasonable limitation flowing from general commitments to human rights. From the outset, human rights were designed as limitations on state sovereignty and may thus come into conflict with democratic (majoritarian) preferences. This was well understood by supporters of the post–World War II human rights movement, and indeed, it was the stated raison d'être of human rights. As early as 1950, Hersch Lauterpacht remarked that on the matter of human rights protection, 'no achievements of substance can be brought without actual sacrifices in sovereignty'[71] and a state must 'surrender its freedom of action and its legislative supremacy with regard to a matter relating to the most essential aspect of its law and constitution.'[72] Hence, human rights obligate states to afford a certain

[68] Sheila Jasanoff, 'A World of Experts: Science and Global Environmental Constitutionalism' (2013) 40 *Boston College Environmental Affairs Law Review* 439, 451–2.

[69] Fisher, *Risk Regulation and Administrative Constitutionalism*, 14.

[70] See the persuasive criticism in ibid., 16–18.

[71] Hersch Lauterpacht, *International Law and Human Rights* (Stevens 1950) 452–3.

[72] Ibid., 311.

level of priority to fundamental values of dignity and autonomy over other interests, such as business opportunities, majority preferences, or political expediency. The environmental minimum merely operationalises and extends these commitments to environmental harm. Therefore, any legal order that takes commitments to human rights seriously must recognise the myriad of threats that serious environmental harm poses to the enjoyment of human rights. Undermining in bad faith what the relevant scientific community generally accepts as the best standards and practices would therefore be one such case where a commitment to human rights requires other interests, even those expressing majoritarian preferences through legislation, to be set aside.

Critics may take the general point on board that human rights are designed to check majority preferences, but nonetheless take issue with the fact that it is courts setting and enforcing these boundaries. This argument is best examined through the lens of one of the most influential criticisms that focuses specifically on the institutional capacity of courts to respond to complex policy questions, which we address in Section 3.2.4 of this chapter.

3.2.1.4 Summary

As this section has explained, the result of ascribing bad faith to a regulation is that it no longer counts as a legal standard for the purposes of the environmental minimum and hence does not inform the minimum standards of review. Certainly, regulators are permitted to take competing interests into account when drafting regulations, including business opportunities, the costs and benefits associated with imposing a certain regulation, and the limitations of other rights that regulations would entail.[73] Within the boundaries drawn by generally accepted recommendations and guidelines, regulators enjoy significant discretion. There is certainly no requirement to provide an optimal level of environmental protection under the environmental minimum. The rebuttable presumption of bad faith applies only if a regulation does not meet the standards set by generally accepted recommendations and guidelines, while also falling foul of the evaluative principles of bad faith.

The ECHR case of *Zander* v. *Sweden* provides a good illustration of how the environmental minimum would operate under these circumstances. The case involved the contamination of a drinking water well used by the applicants, likely through government-licenced dump sites.[74] The legislatively prescribed threshold value for cyanide (0.01 mg/litre) had been exceeded

[73] On this point, see *Michigan* v. *Environmental Protection Agency* 576 U.S. 743 (2015).
[74] *Zander* v. *Sweden* App no 14282/88 (ECtHR, 25 November 1993).

and the applicants were accordingly provided with safe drinking water by the local authority. However, amendments were soon passed by the National Food Agency to raise the permitted level of cyanide to 0.1 mg/litre, a tenfold increase over the previous threshold. Following this decision, the municipality stopped providing drinking water to the applicants and renewed the licences for the allegedly responsible dump sites.

In applying the environmental minimum, it would appear obvious that the threat of cyanide poisoning constitutes a specific risk under a right to health. The question then becomes whether the environmental harm meets the applicable minimum standard. At first glance, there are regulations for the permitted concentration of cyanide in water and these are being consistently enforced. However, there is a possibility of bad faith should they fall short of generally accepted guidelines. Current guidelines from the WHO suggest that a threshold value of 0.07 mg/litre is considered safe for both acute and long-term exposure to cyanide.[75] The amended regulatory threshold of 0.1 mg/litre would significantly exceed this safe concentration. The question then arises of whether the presumption of bad faith can be rebutted. For our purposes, it appears most promising to investigate whether the regulation was intended to circumvent a generally accepted guideline and whether it constitutes retrogression of environmental protection standards.

A circumvention is less plausible based on these facts because the regulation, while out of step with generally accepted standards, is not tokenistic. The regulation recognises the threat of cyanide poisoning and establishes some legally enforced threshold values. In the case of a breach of the new threshold, the applicants moreover would have every reason to expect that the provision of drinking water would resume. The charge of retrogression is more difficult to dismiss. The regulation was enacted in bad faith because it falls short of pertinent scientific guidelines and the presumption of bad faith is not rebutted as the regulation legalises a previously unlawful level of environmental harm.

3.2.2 *Established Standards*

Established standards refer to those guidelines and recommendations on environmental harm generally accepted in a relevant scientific community. They determine the minimum standard in the absence of legal standards, including where a legal standard was adopted in bad faith. Established standards are found, for instance, in WHO guidelines on air quality, water,

[75] *Cyanide in Drinking-Water – Background Document for Development of WHO Guidelines for Drinking-Water Quality* (World Health Organization, 2003) 8.

food safety, soil, flora, and fauna, various artificial toxins, radiation, and noise.[76] On some environmental harms there are no such generally accepted guidelines, and determining whether this is the case should be treated as a question of fact in proceedings, thus potentially requiring expert testimony, an issue we shall explore in more depth in Section 3.3 of this chapter.

Ultimately, the application of established standards in the absence of legal standards is justified for the same reasons that support the presumption of bad faith: relevant scientific evidence ought to be duly considered, mere disagreement with its findings ought not to suffice to dismiss them, and irrelevant information should be disregarded. In practice, developed regulatory regimes are less likely to leave significant environmental harm unregulated or produce bad faith regulations. The benefits of a tight regulatory regime are evidenced, for instance, by the low rate of human rights violations under the ECHR for environmental harm in the EU-28 (the 27 EU member states, plus the United Kingdom): it stands at only 25.45 per cent of applications, even though the EU-28 account for more than three quarters of all decisions (165 of 213).[77] This contrasts favourably with the 47.92 per cent rate of human rights violations found against non-EU member states of the Council of Europe. The chief contribution of established standards is to provide a backup for unforeseen regulatory gaps that arise when generally accepted guidelines crystallise in a relevant scientific community before regulation has been enacted.

Critics of this approach might suggest that established standards effectively force states to prioritise environmental interests. A possible response is that under the environmental minimum regulators retain the legal power to diverge from these guidelines. While the environmental minimum insists that avoiding environmental harm ought to enjoy priority over other considerations, this priority is not absolute. The existing mechanisms of human rights law demand a proportionality assessment of the proposed environmental protection measures with competing demands. The environmental minimum does not require, nor does it endorse, any specific outcome of a regulatory process, save for the requirements of good faith. Admittedly, this skews the balance towards environmental protection, but as laid out in Chapter 2 this is

76 See, for instance, the plethora of guidelines available from the WHO: *Review of Evidence on Health Aspects of Air Pollution – REVIHAAP Project* (World Health Organization, 2013); *Guidelines for Drinking-Water Quality* (World Health Organization, 2011); Birgitta Berglund, Thomas Lindvall, and Dietrich H. Schwela, *Guidelines for Community Noise* (World Health Organization, 1999); *Evaluation of Certain Food Additives* (World Health Organization – Food and Agriculture Organization of the United Nations, 2014).

77 See the Regional Breakdown, Results Table in the dataset.

justified by the commitment of states to human rights. The case for upholding an environmental regulation that limits rights, for instance, to property or occupational freedom, is strongest where in the absence of the measure the environmental harm breaches established standards. Conversely, the case is comparatively weaker where the requirements of established standards can be met through means that limit competing rights to a lesser extent.

Consider the previously discussed case of *Zander* v. *Sweden*. Suppose that instead of the original applicant, the operators of the garbage dump site had challenged a denial of an operating licence. Assume further that the licence was specifically withheld on the grounds that further operation would exacerbate water contamination, that the site was responsible for the rise of cyanide levels, and that the legal threshold value had remained unchanged at 0.01 mg/litre. The applicants would argue that the environmental regulation providing cyanide thresholds for drinking water was interfering with their rights. A court should not strike down the regulation entirely, as this would put Sweden in violation of its human rights obligations under the environmental minimum to have some good faith regulation for cyanide in drinking water. However, a court might well arrive at the conclusion that, on balance, the threshold value of 0.01 mg/litre is too onerous. The benefits might not be proportional to the limitations placed on the rights of garbage site operators: a higher threshold value might strike a fairer balance between the conflicting rights. The court might therefore find a violation of the rights of the applicants and require a revised threshold value within the 0.01 and 0.07 mg/litre range determined by the generally accepted guidelines of the WHO.

3.2.3 *Emerging Standards*

In the absence of legal and established standards, the applicable minimum standard is determined by emerging scientific evidence. This is based on the views of the same scientific community as established standards. The key difference is that significant debate persists within the relevant community, overall too few studies have been conducted in order to determine relevant guidelines and best practices, or there is significant uncertainty over whether a given degradation or pollution is harmful at all. Emerging standards thus generally permit greater regulatory flexibility, as there is no firm scientific basis for regulatory and enforcement action.

This greater level of flexibility is limited only if there is evidence that inaction may result in severe and irreversible damage. Whether this is the case is determined through the harm assessment formula, which diverts attention away from the certainty and weight of evidence that supports regulatory intervention. Instead, it embraces the uncertainty inherent in emerging standards and places the potential for harm at the centre of consideration. The formula has been adapted from the German Federal Constitutional Court jurisprudence which specifies when interim measures are appropriate before the main proceedings.[78] Instead of focusing on the likelihood of success of the constitutional complaint, this test emphasises and evaluates the potential harm if no interim order is passed versus the harm caused if an interim order is adopted.[79] While this test is not directly transferable, it can help in dealing with contradictory or inconclusive evidence and the resulting uncertainty over the appropriateness of intervention. This is because it is specifically designed to deal with a comparable uncertainty about the outcome of a subsequent full Constitutional Court evaluation of the case. In particular, the Constitutional Court at the preliminary stage does not yet know what further evidence will be presented during the full hearing by the parties and interveners, nor which lines of inquiry it may wish to pursue: the merits of the case and the likelihood of success are at this stage inherently uncertain.

Adapting this approach to environmental degradation and pollution, the harm assessment formula requires courts to weigh the harm that will be conceivably caused by regulatory inaction *against* the harm conceivably caused through regulatory action, regardless of the relative likelihood of either scenario coming to pass. In short, a court asks whether the harm to the applicant's rights if an intervention does **not** occur (and evidence later emerges that an intervention was in fact warranted) outweighs the harm to competing interests if an intervention does occur (and evidence later emerges that an intervention was in fact **not** warranted). Should the balance of harms favour regulatory intervention, then the environmental minimum would require regulatory action until such a time that a higher level of certainty is achieved one way or the other. This approach is generally in keeping with John Knox's report which states that '(...) the lack of full scientific certainty should not be used to justify postponing effective and proportionate measures to prevent environmental harm, especially when there are threats of serious or irreversible damage.'[80]

78 See § 32 of the Bundesverfassungsgerichtsgesetz of 11 August 1993 (BGBl. I S. 1473), last amended through Article 4 of the Law of 20 November 2019 (BGBl. I S. 1724).

79 Adopted from Federal Constitutional Court, 2 BvR 1027/02, 17 July 2002, BVerfGE 105, 365; this is settled case law, *see* Federal Constitutional Court, 2 BvR 2760/93, 2 BvQ 3/94, 2 BvR 707, 741/94, 3 May 1994, BVerfGE 91, 70; Federal Constitutional Court, 1 BvR 2116/94, 7 February 1995, BVerfGE 92, 126.

80 Knox, *Final Report of the Special Rapporteur*, para 33.

An example where regulatory intervention based on an emerging standard was required by the harm assessment formula, was action taken across South America against mosquitos transmitting the Zika virus. This virus was long suspected to be the cause of microcephaly in new-borns, based on inconclusive and incomplete scientific evidence.[81] This suspicion was only confirmed long after a decision was taken to intervene and actively combat the mosquitos suspected of spreading the virus. A failure to act under these circumstances would have had catastrophic consequences, causing irreparable harm to a generation of new-borns, while the harm to other competing interests through the regulatory intervention was negligible (even had it later emerged that the virus or the mosquito were not to blame). Therefore, measures such as the spraying of insecticides and attempts to eliminate breeding grounds for mosquitos in affected areas were appropriate interventions to reduce the potential harm, even though scientific certainty was not established at the time.

In this sense, the emerging standards and the harm assessment formula have some normative overlaps with the precautionary principle under EU and international environmental law. However, as we shall see, the precautionary principle is a vague concept that requires a specific operational context to be useful. Otherwise, it makes contradictory claims that the environmental minimum notably avoids through its normative foundation in environmental human rights.

3.2.3.1 Precautionary Principle

In international environmental law the precautionary principle suggests that activities which may be hazardous to the environment ought to be avoided, especially where scientific evidence is unclear.[82] It mandates regulatory intervention and rejects the view that scientific uncertainty requires deference to the status quo. The principle was most clearly reflected in Principle 15 of the Rio Declaration on Environment and Development:

> In order to protect the environment, the precautionary approach shall be widely applied by States according to their capabilities. Where there are

[81] 'CDC Concludes Zika Causes Microcephaly and Other Birth Defects' *Center for Disease Control and Prevention* (13 April 2016) www.cdc.gov/media/releases/2016/s0413-zika-microcephaly.html, accessed 1 April 2021.

[82] Meinhard Schröder, 'Precautionary Approach/Principle' in Rüdiger Wolfrum (ed.), *Max Planck Encyclopedia of Public International Law* (Oxford University Press 2014) para 2.

> threats of serious or irreversible damage, lack of full scientific certainty shall not be used as a reason for postponing cost-effective measures to prevent environmental harm.[83]

The origins of the precautionary principle are likely found in domestic law, for instance in the broader German concept of the *Vorsorgeprinzip*,[84] and it has since been integrated expressly or implicitly into a number of international agreements and conventions, such as the Montreal Protocol (1987),[85] and the Cartagena Protocol on Biosafety (2000) to the Convention on Biological Diversity.[86] The precautionary principle thus forms an integral part of contemporary international environmental law,[87] and is enshrined within EU law as a general principle.[88] Its status as customary International Law is still a matter of significant disagreement among scholars, as are the specific implications of the principle.[89] A key issue fuelling the debate is the indeterminacy of the precautionary principle and the resulting diverse normative claims.[90] This has been seized upon by critics, who allege that it is impractical and impedes economic and technological progress by encouraging overly

83 *Rio Declaration on Environment and Development*; see also Antonio Agusto and Cancado Trindade, 'Principle 15: Precaution' in Jorge E. Viñuales (ed.), *The Rio Declaration on Environment and Development* (Oxford University Press 2015) 403.

84 Jonathan B. Wiener, 'Precaution' in Daniel Bodansky, Jutta Brunnée, and Ellen Hey (eds.), *The Oxford Handbook of International Environmental Law* (Oxford University Press 2007) 598; Schröder, 'Precautionary Approach/Principle', para 6; Shawkat Alam and others (eds.), *Routledge Handbook of International Environmental Law* (Routledge 2013) 47.

85 Montreal Protocol on Substances that Deplete the Ozone Layer (adopted 16 September 1987, in force 1 January 1989) 1522 UNTS 3 [mentioned in the Preamble].

86 Cartagena Protocol on Biosafety to the Convention on Biological Diversity (adopted 29 January 2000, in force 11 September 2003) 2226 UNTS 208.

87 Schröder, 'Precautionary Approach/Principle', para 8.

88 See Article 191 (2) Consolidated version of the Treaty on the Functioning of the European Union [2008] OJ C115/01 (TFEU).

89 Ole W. Pedersen, 'From Abundance to Indeterminacy: The Precautionary Principle and Its Two Camps of Custom' (2014) 3 *Transnational Environmental Law* 323; in support of the customary status, see, for instance, Alam and others, *Routledge Handbook of International Environmental Law*, 50; Owen McIntyre and Thomas Mosedale, 'The Precautionary Principle as a Norm of Customary International Law' (1997) 9 *Journal of Environmental Law* 221; David Freestone and Ellen Hey, 'Origins and Development of the Precautionary Principle' in David Freestone and Ellen Hey (eds.), *The Precautionary Principle and International Law: The Challenge of Implementation* (Kluwer Law International 1996) 3; for sceptical accounts, *see* Jonathan B. Wiener and Michael D. Rogers, 'Comparing Precaution in the United States and Europe' (2002) 5 *Journal of Risk Research* 317, 343; Birnie and others, *International Law and the Environment*, chapter 3; Sands and others, *Principles of International Environmental Law*, 228.

90 Crawford, *Brownlie's Principles of Public International Law*, 341.

restrictive environmental regulations.[91] One of the more influential critiques of the precautionary principle comes from Cass Sunstein.[92]

3.2.3.2 Sunstein's Critique

Initially, Sunstein helpfully distinguishes between weak and strong understandings of the precautionary principle. In its weak incarnation, the principle suggests that a lack of evidence of harm ought not to justify a refusal to regulate.[93] In its strong forms, the principle mandates that where there is a significant environmental risk coupled with scientific uncertainty, potentially harmful activities should be prevented until evidence shows that no damage will occur.[94] This yields a spectrum of understanding of the precautionary principle, from mere defence of regulatory intervention to a shift in the burden of proof onto those who wish to engage in risky activities.[95] Sunstein argues that both ends of the spectrum are flawed and provide little practical guidance to regulators.

In its weak form, he argues that the principle spells out mere truisms. Nobody seriously suggests that only complete and total certainty justifies regulatory efforts to tackle environmental harm. If that were the case, scarcely any human activity could legitimately be the subject of regulation. At the same time, this insight alone is insufficient to fully justify intervention. A weak form precautionary principle only holds that it is permissible to enact regulation, but cannot answer the crucial question of whether, and how, regulation should be drafted, because the normative claim appears exhausted in stating that uncertainty is no reason to reject regulation.[96] The practical implication is that there may well be good reasons not to regulate, or to regulate differently, or compelling reasons to proceed with regulation regardless of conflicting evidence: a weak precautionary principle would be neither here nor there in such a debate. To Sunstein, this weak form precautionary principle holds no further insights for the purposes of managing risks through law, the question that preoccupies his book.

[91] Jonathan H. Adler, 'More Sorry Than Safe: Assessing the Precautionary Principle and the Proposed International Biosafety Protocol' (2000) 35 *Texas International Law Journal* 173; Wiener, 'Precaution', 600.

[92] Sunstein, *Laws of Fear*, chapter 1.

[93] Ibid., 18.

[94] Ibid., 19.

[95] Wiener, 'Precaution', 602.

[96] Ibid., 610.

He is therefore more interested in strong forms of the precautionary principle, those that shift the burden of proof onto those engaging in risky activities. Sunstein claims that this strong form paralyses regulators through indeterminacy and incoherent normative claims: both action and inaction have the capacity to produce substantial harm, and therefore applying the precautionary principle leaves regulators paralysed. Should they apply the principle to the risks of action (and thus intervene to prevent the potentially harmful private conduct) or to the risks produced by inaction (and hence decline to regulate potentially harmful private conduct)? Paradoxically, the precautionary principle can advise both simultaneously, depending on where one locates the relevant risk requiring a precautionary approach. This leads Sunstein to reject the principle wholesale as incapable of addressing the management of risks. However, his concerns, at least in part, arise from a misplaced expectation that the precautionary principle should make uniform normative claims in the abstract. In most practical contexts, the risk to which the precautionary principle applies is sufficiently clear and determinable, thus avoiding the dilemma and the essence of Sunstein's critique.

3.2.3.3 Context-Sensitive Nature

As Sunstein surveys different incarnations of strong forms of the precautionary principle, he appears to be guided by an assumption that these accounts ought to form a coherent picture. However, it is not apparent that the definition and role of the precautionary principle in one treaty should be relevant for its operation in other treaties. The principle can be flexible and context sensitive, thus fulfilling different functions in different contexts.[97] It is not clear that drafters ever operated on the understanding that the precautionary principle makes an entirely consistent normative claim across the body of international and domestic laws. Indeed, the context-specific nature has been repeatedly acknowledged – and criticised – in legal scholarship, especially among those that are sceptical of the customary status of the precautionary principle.[98]

Certainly, some common features are necessary in order to permit us to speak of a 'principle' at all, but these are undeniably present. At its core, the

[97] Ole W. Pedersen, 'From Abundance to Indeterminacy', 325, 339.

[98] Wiener and Rogers, 'Comparing Precaution in the United States and Europe', 343; Birnie and others, *International Law and the Environment*, chapter 3; Sands and others, *Principles of International Environmental Law*, 228; Elizabeth Fisher, 'Precaution, Precaution Everywhere: Developing a Common Understanding of the Precautionary Principle in the European Community' (2002) 9 *Maastricht Journal of European and Comparative Law* 7, 8; Bell and others, *Environmental Law*, 72.

precautionary principle deals with competing risks and seeks to provide practical guidelines for regulatory intervention. That these risks require further specification before one can derive normative claims in specific cases need not be a hindrance. Most treaties and statutes that employ the precautionary principle make the necessary specifications on relevant risks and the underpinning values they seek to protect. In the Montreal Protocol and the Cartagena Protocol for instance, the precautionary principle seeks to caution against all actions that can prove detrimental to the ozone layer and the safe handling, transport, and use of living modified organisms, respectively. Thus, applied to a specific risk, the precautionary principle can achieve a shift of the burden of proof that is desirable on normative grounds that stand outside the principle itself, and avoid advocating seemingly contradictory courses of action.

For instance, in the context of the Cartagena Protocol on Biosafety the precautionary principle gains clearer normative contours by specifying certain values and then elevating them above other potentially conflicting considerations and sources of risk. The objective of the Protocol is stated as follows:

> (...) to contribute to ensuring an adequate level of protection in the field of the safe transfer, handling and use of living modified organisms resulting from modern biotechnology that may have adverse effects on the conservation and sustainable use of biological diversity, taking also into account risks to human health, and specifically focusing on transboundary movements.[99]

Hence the precautionary principle, as it operates in the Protocol, seeks to address a specific set of risks arising from living modified organisms, and their potential impact on the values of biological diversity and human health. Because other issues such as food production or business opportunities are not recognised under the Protocol, and do not benefit from this incarnation of the precautionary principle, regulators would not encounter the dilemma identified by Sunstein: the precautionary principle specifies the relevant risks to biological diversity and human health, and does not recognise potentially conflicting demands of food safety or business opportunities. The latter may well be relevant to the overall regulatory approach, or under other incarnations of the precautionary principle, but crucially not within the scope of the same treaty mechanism.

This does not of course eliminate all inconsistent claims of strong forms of the precautionary principle, for instance those that arise between recognised

[99] Article 1 Cartagena Protocol.

risks covered by the same or multiple overlapping treaties and legislation.[100] Balancing the demands of the Cartagena Protocol with those of other treaties and legislation can of course produce conflicting demands, but this is not a feature inherent to or owed to the incoherence of the precautionary principle as Sunstein suggests. Rather, it is the unavoidable outcome of incrementally developing the law on any policy question from multiple valid angles and viewpoints – the essence of democratic contestation and compromise. Wholesale rejection of the precautionary principle therefore appears misplaced. The focus on a few specified risks allows a clearer and more direct application of the principle, and the balancing need only trouble us where legislation and treaties overlap. Sunstein appears to recognise this as well, because he integrates many elements from the precautionary principle into his 'laws of fear.' These include most notably a cost–benefit analysis influenced by his brand of libertarian paternalism.[101] Where specifically the difference lies to the precautionary principle, and whether the suggested analysis merely describes and replicates ordinary democratic policy contestation remains nebulous.

There is no need for present purposes to engage with Sunstein's suggestions in more detail, which may well suffer from many of the problems that render the precautionary principle impractical in his view.[102] Suffice it to say here that some of his concerns with the precautionary principle are points well made:

- understood as a weak principle, the precautionary principle risks becoming devoid of practical value in guiding environmental regulation;
- a strong precautionary principle can be an impediment to economic and technological development to the extent that it encourages overly conservative and onerous environmental regulations; and

[100] Daniel Bodansky, 'Deconstructing the Precautionary Principle' in David D. Caron and H. N. Scheiber (eds.), *Bringing New Law to Ocean Waters* (Brill 2004) 381; Christopher D. Stone, 'Is There a Precautionary Principle?' (2001) 31 *Environmental Law Reporter News and Analysis* 10790; Sands and others, *Principles of International Environmental Law*, 222; Beyerlin and Marauhn, *International Environmental Law*, 55; Wiener, 'Precaution', 610.

[101] He further developed libertarian paternalism in his work on nudging, *see* Cass R. Sunstein, *Why Nudge? The Politics of Libertarian Paternalism* (Yale University Press 2014), which likewise attracted criticism, *see* C. McCrudden and Jeff King, 'The Dark Side of Nudging: The Ethics, Political Economy, and Law of Libertarian Paternalism' [2015] Queen's University Belfast Law Research Paper No 16.

[102] Gregory N. Mandel and James Thuo Gathii, 'Cost-Benefit Analysis versus the Precautionary Principle: Beyond Cass Sunstein's Laws of Fear' [2006] *University of Illinois Law Review* 1037, 1041–2.

- the precautionary principle can lead to contradictory claims, either when it remains unclear on the risks it applies to, its underlying normative determinations (see, for instance, the general principle under EU law), or when it cannot mediate competing regulatory options that both engage the precautionary principle.

3.2.3.4 Contrast to the Environmental Minimum

Many of these concerns can be addressed under the environmental minimum framework. For one, the environmental minimum rejects vague expressions of principle and substitutes them with specific benchmarks for reviewing regulatory conduct. It accepts that scientific evidence is not always conclusive and the harm assessment formula presents one way that harm can be weighed to reach a decision on regulatory intervention: not every anticipated environmental benefit will justify potentially intrusive interventions. More importantly, where the precautionary principle makes contradictory claims, the environmental minimum provides a framework for the resolution of conflicts.

Suppose that regulators are confronted with a pesticide that has been found by some studies to be carcinogenic, while other studies have determined it to be safe pursuant to different methodologies.[103] Regulators might consider options ranging from a blanket ban to unconditional use of the pesticide, and turn to the precautionary principle for guidance. The precautionary principle might on the one hand advise a ban, as humans would otherwise be exposed to a potentially significant health risk. On the other hand, the pesticide may be essential to the viability of domestic food production, and crop yields, along with food security, would suffer from a ban, a consequence that a general precautionary principle might likewise caution against. In such scenarios of uncertainty, the environmental minimum posits the harm assessment formula. Rather than weighing the inherently uncertain and conflicting risk assessments against each other without a commensurable metric, regulators would be encouraged to embrace the scientific uncertainty. They would focus on the potential harm and evaluate whether banning (and later learning that no danger to humans existed) or permitting the pesticide (and later learning that a danger to humans existed) will produce the greater harm. This advocates a broad engagement with all relevant issues surrounding the regulatory intervention and would ultimately promote a measured approach

[103] This example is loosely based on the current regulatory controversy in the EU surrounding the herbicide glyphosate.

that seeks the greatest harm reduction possible while retaining some of the potential benefits.

3.2.4 *Objections to Adjudication*

The question of whether courts are an appropriate forum for resolving all of the environmental questions we have so far considered is a contested one which draws on well-established legal debates on the scope and appropriateness of individual rights enforcement through courts.[104] Based on this broader debate, there are a number of objections that one might raise in opposition to the adjudication of environmental human rights through courts. For the purposes of the environmental minimum, the most salient critique alleges that courts cannot, or at least ought not, adjudicate complex environmental policy questions. Empirically speaking at least, courts are frequently and undeniably called to adjudicate questions related to environmental policy, both at an international and a domestic level.

At the international level, this has been most evident with the ECtHR which has decided hundreds of environmentally themed cases, often with significant policy implications.[105] For instance, in the case of *Jugheli* v. *Georgia*, the applicants were exposed to air pollution and noise from coal fire power plants in the immediate vicinity of their homes.[106] The Court found a violation of Article 8 ECHR due to the 'virtual absence' of a regulatory framework for the operation of the plants.[107] Much the same can be said for domestic courts. In *Mazibuko* v. *City of Johannesburg*,[108] the South African Constitutional Court was called upon to clarify the minimum access required under the right to water but declined to give a specific figure. Likewise, the German Constitutional Court has repeatedly engaged with environmental cases and emphasised legislative discretion in setting environmental policy.[109] To the extent that critics claim that courts are unable to engage with

[104] Hanna Wilberg and Mark Elliott (eds.), *The Scope and Intensity of Substantive Review: Traversing Taggart's Rainbow* (Hart 2015); Lawrence Friedman, 'Testing the Limits: Judicial Enforcement of Positive State Constitutional Rights' (2015) 53 *Duquesne Law Review* 437; Woolf and others, *De Smith's Judicial Review*; C. F. Forsyth, *Judicial Review and the Constitution* (Hart Publishing 2000); Allan, *Law, Liberty, and Justice*; John Hart Ely, *Democracy and Distrust: A Theory of Judicial Review* (Harvard University Press 1980); Dworkin, *A Matter of Principle*.

[105] See the dataset and its summary of relevant facts.

[106] *Jugheli and others* v. *Georgia* App no 38342/05 (ECtHR, 13 July 2017).

[107] Ibid. [75] and [78].

[108] *Mazibuko and Others* v. *City of Johannesburg and Others* [2009] ZACC 28.

[109] See the discussion in Chapter 9.

environmental policy, this is simply not correct in an empirical sense. The deeper objection is that courts ought not to engage with such questions, based on concerns over expertise and democratic legitimacy[110] as well as deference to executive bodies under principles of subsidiarity and the margin of appreciation.[111]

A contentious ruling with respect to climate change emerged from the Netherlands. The Hague District Court had compelled the Dutch state to reduce the greenhouse gas emissions by 25 per cent by 2020, based on the emissions levels of 1990.[112] The Urgenda Foundation had successfully argued that the Dutch state was under an obligation to protect its citizens from climate change on the basis of domestic tort law. The mandated reduction of 17 per cent by 2020 under EU law compared with 1990 levels could not

> (. . .) be deemed as a sufficient and acceptable alternative to the scientifically proven and acknowledged higher reduction path of 25–40% in 2020. (. . .) This leads the court to the conclusion regarding this issue of the dispute that the State, given the limitation of its discretionary power discussed here, in case of a reduction below 25–40% fails to fulfil its duty of care and therefore acts unlawfully.[113]

The higher emission reduction target was justified through the established scientific evidence on climate change and commitments of the Netherlands under international and EU law to reduce its emissions. The decision was hailed as a significant victory for environmentalists.[114] Notably, the District Court rejected the suggestion that the applicants could invoke their rights under the ECHR, considering that they had not established victim status under Article 34 ECHR.

110 On expertise see Jasanoff, 'A World of Experts'; Cranor, *Toxic Torts*; Caroline E. Foster, *Science and the Precautionary Principle in International Courts and Tribunals: Expert Evidence, Burden of Proof and Finality* (Cambridge University Press 2011); on democratic legitimacy see the influential account of Jeremy Waldron, 'The Core of the Case Against Judicial Review' (2006) 115 *Yale Law Journal* 1346.

111 George Letsas, 'Two Concepts of the Margin of Appreciation' (2006) 4 *Oxford Journal of Legal Studies* 705; Eyal Benvenisti, 'The Margin of Appreciation, Subsidiarity, and Global Challenges to Democracy' [2016] Global Trust Working Paper.

112 *Urgenda Foundation* v. *The Netherlands* C/09/456689/HA ZA 13-1396, Den Haag Regional Court, 24 June 2015.

113 Ibid. [4.85]–[4.86].

114 'Netherlands Ordered to Cut Greenhouse Gas Emissions' *BBC News* (24 June 2016) www.bbc.co.uk/news/world-europe-33253772, accessed 1 April 2021; Arthur Nelsen, 'Dutch Government Ordered to Cut Carbon Emissions in Landmark Ruling' *The Guardian* (The Hague, 24 June 2015) www.theguardian.com/environment/2015/jun/24/dutch-government-ordered-cut-carbon-emissions-landmark-ruling, accessed 1 April 2021.

The decision was appealed by the Dutch government and cross-appealed by Urgenda to the Hague Court of Appeal. The court upheld the District Court ruling, and additionally found that the applicants could rely on Articles 2 and 8 ECHR, which specify the domestic duty of care owed by the state.[115] This is a crucial finding, because it offers a persuasive legal basis for imposing a higher emission reduction target and explains how fully meeting Dutch obligations under EU law can nonetheless be insufficient under Dutch law. The Court of Appeal found the emission target set by EU law did not satisfy the independent and separate obligations of the Netherlands under the ECHR. The more exacting requirements can comfortably coexist with EU law because EU law ultimately does not prevent the Netherlands from exceeding mandatory emission reduction targets: it sets a minimum reduction target, not an upper limit. That being the case, the court was free to look to other legal obligations and commitments of the Netherlands to determine whether these required a higher reduction target.

Despite the notably cautious approach taken by both Dutch courts, concerns were raised over judicial activism that usurps the democratic process, in violation of the separation of powers.[116] These criticisms follow a more general line of sceptical literature that objects to courts adjudicating complex policy questions at all.[117]

Lon Fuller presents the strongest and most sophisticated version of this argument in his seminal article on polycentricity.[118] Although Fuller does not reference environmental protection specifically, his concerns are readily transferable. Where, for instance, threshold values for air pollutants are

[115] *The Netherlands* v. *Urgenda Foundation* 200178245/01, Court of Appeal, 9 October 2018.

[116] Josephine van Zeben, 'Establishing a Governmental Duty of Care for Climate Change Mitigation: Will Urgenda Turn the Tide?' (2015) 4 *Transnational Environmental Law* 339, 355–6; Marjan Peeters, 'Urgenda Foundation and 886 Individuals v. The State of the Netherlands: The Dilemma of More Ambitious Greenhouse Gas Reduction Action by EU Member States' (2016) 25 *Review of European, Comparative & International Environmental Law* 123, 129; K. J. de Graaf and J. H. Jans, 'The Urgenda Decision: Netherlands Liable for Role in Causing Dangerous Global Climate Change' (2015) 27 *Journal of Environmental Law* 517, 523–4.

[117] Harrison, 'Reflections on the Role of International Courts and Tribunals in the Settlement of Environmental Disputes', 503; Birnie and others, *International Law and the Environment*, 302; Boyle, 'Human Rights and the Environment', 225; Richard Macrory, 'Environmental Citizenship and the Law: Repairing the European Road' (1996) 8 *Journal of Environmental Law* 219, 232; William Howarth, 'Environmental Human Rights and Parliamentary Democracy' (2002) 14 *Journal of Environmental Law* 377, 388; Chris Miller, 'Environmental Rights in a Welfare State? A Comment on DeMerieux' (2003) 23 *Oxford Journal of Legal Studies* 111, 125.

[118] Fuller and Winston, 'The Forms and Limits of Adjudication'; see also King, 'The Pervasiveness of Polycentricity', 102.

exceeded in a city, regulators may resolve the matter through a vast array of measures ranging from limiting vehicular traffic to long-term shifts towards cleaner energy production. It involves setting a policy goal, allocating limited resources for enforcement, and determining where they can achieve the most impact, as well as the balancing of competing interests. In essence, Fuller argues that such determinations and especially choosing between several seemingly appropriate policy options is a task ill-suited to courts.

Fuller's ideas continue to hold sway, especially in English public law. In their leading textbook, Mark Elliott and Robert Thomas continue to draw on the concept in support of judicial deference in certain cases.[119] The concept has further been accepted as a limit for justiciability, especially in matters of resource allocation, by a host of legal scholars and courts.[120] While some authors support a greater role for courts on environmental issues, this support is often limited to securing procedural access for individuals in the environmental regulation and decision-making process.[121] On this view, enforcing procedural access is less controversial and politically neutral, thereby avoiding some of the legitimacy arguments brought against courts. However, as Chapter 2 already explained, procedural safeguards are not inherently less controversial or more politically neutral than substantive environmental protections. For present purposes, it suffices to note that both the sceptics of policy adjudication and the proponents of procedural protections share concerns over the institutional capacity of courts to engage with substantive policy questions.

In order to engage with the strongest version of this objection, this section begins by (1) framing and analysing Fuller's account, before providing (2) a critique which suggests that Fuller has not offered a convincing argument against adjudication. The section concludes (3) by demonstrating that the case specifically for judicial deference on environmental regulation is untenable, even if one accepts the essence of Fuller's argument.

[119] Elliott and Thomas, *Public Law*, 609–10.

[120] T. R. S. Allan, *Constitutional Justice: A Liberal Theory of the Rule of Law* (Oxford University Press 2003) 187–8; David Feldman, *Civil Liberties and Human Rights in England and Wales* (2nd ed., Oxford University Press 2002) 229–30; Woolf and others, *De Smith's Judicial Review* para [1-042] and [5-140]; *R* v. *Cambridge DHA Ex parte B (No 1)* [1995] 1 WLR 898 [906]; *Council of Civil Service Unions* v. *Minister for the Civil Service* AC 374 [411]; *R* v. *Secretary of State for the Home Department* [1995] All ER 870; *R (Rogers)* v. *Swindon NHS Primary Care Trust* [2006] EWCA Civ 392 [breast cancer treatment with unlicenced drug]; *R (Comer House Research)* v. *Director of the Serious Fraud Office* [2008] UKHL 60 [prosecutorial decisions].

[121] Giorgetta, 'The Right to a Healthy Environment, Human Rights and Sustainable Development', 184; Handl, 'Human Rights and Protection of the Environment', 327.

3.2.4.1 Polycentricity

It is important to note initially that when Fuller references complexity in the context of polycentricity, he does not mean that the facts are difficult to establish. That may be incidental to polycentric cases, but it is not a necessary element of polycentricity.[122] Rather, Fuller posits the interconnection of multiple issues and the vast number of affected individuals as the defining features of polycentricity. One may visualise this interconnection through the metaphor of a spider's web.[123] The instant that an action is taken to address an individual issue – a tug on any part of the web – it reverberates through the entire system, affecting both potential future actions, as well as the parameters that led to the original decision. No issue can be dealt with in isolation, and every action affects the way that all connected interests can be addressed: some decisions may seem appropriate to reach a certain objective but produce adverse effects and unintended consequences.

A good example of a polycentric issue are price negotiations. Suppose a major car manufacturer is seeking to agree the 'best' price for its new Model X car with the dealership network. Regardless of the price ultimately set by the parties, the decision affects the supply and demand for Model X: a lower price presumably leading to higher demand and vice versa. This in turn influences a range of factors that formed the basis of the original negotiations, for instance salaries and wages, overheads, the expected profit margin, production costs, insurance, advertising budgets, and so on.[124] From these then arise further knock-on effects. The sales performance of Model X may affect subsequent salary negotiations with labour unions, which in turn determine discretionary spending of workers in their capacity as consumers, by extension the broader economic development in a state and the world economy, which returns to affect global demand for Model X cars. Depending on the starting point and perspective one adopts, one may place any of these considerations at the heart of the decision-making process, and reach sound, but contradictory conclusions on the 'best' price for Model X cars. According to Fuller, such polycentric issues do not lend themselves well to adjudication, for reasons that are closely related to core characteristics of court proceedings that he describes as follows.

First, cases before courts typically involve only two parties. Second, their respective interests in the outcome of the case are diametrically opposed.

[122] King, 'The Pervasiveness of Polycentricity', 102.
[123] Fuller and Winston, 'The Forms and Limits of Adjudication', 395.
[124] Ibid., 394–5.

Third, a court can ultimately only satisfy one interest, by ruling in favour of one or the other party. Finally, third parties and their interests are typically unable to influence the outcome of a case. In essence, Fuller's objections all have the same root concern: the bipolar, all or nothing, nature of adjudication is contrasted with the multipolar nature of polycentric issues. This, so argues Fuller, ultimately leads to a situation where only a very small group (the parties to a proceeding) has any influence on the outcome of a case, yet the results may affect innumerable individuals.[125] These institutional characteristics led Fuller to identify three problems that arise when courts are asked to adjudicate polycentric issues:

- judges tend to produce unintended results through their rulings by satisfying one interest over another, or through subsequent modification of rulings on appeal;
- courts make assumptions as to the facts, or judges moot possible decisions in informal settings in order to arrive at a decision because they are usually unable to hear from other affected parties;
- courts attempt to recast the issues before them to better suit familiar patterns of judicial decision making.[126]

Fuller thus concludes that polycentricity must set limits on the adjudication of policy questions, especially with respect to resource allocation.[127] When polycentric issues are encountered, he argues, they are best left to the markets to resolve, which employ negotiations, adjustments, and deals to reach decisions.[128] Apart from the economics of a free market, determined by contractual obligations and managerial decision making, Fuller also understood this to include democratic legislatures.[129]

3.2.4.2 Critique

There are a number of objections one might raise against Fuller's argument and this section addresses the most pertinent ones in turn.

First, polycentric issues are abundant in the (legal) world. Taken at face value, Fuller challenges the ability not solely of adjudication, but of any mode of decision making to resolve polycentric issues. On some level, all human

125 Ibid., 364–5.
126 Ibid., 401.
127 Ibid., 400.
128 Ibid.
129 Ibid., 354, see King, 'The Pervasiveness of Polycentricity', 102.

action appears causally interconnected. These relationships are often interdependent and should arguably all be considered before determining any future actions. Even though some actions may appear trivial in their immediate consequences, collectively or individually, they can have profound effects that are missed or dismissed under established decision-making heuristics. The fact that humans discount certain actions as inconsequential speaks more to limitations of human minds than to a property of the natural, or indeed the legal world.

Yet, despite this abundant complexity, decision makers have developed ways to manage polycentricity and ultimately reach reasonable decisions. These include majority voting, managerial decisions, contracting, and adjudication. Even though polycentricity affects and challenges any decision-making process, it is invoked by Fuller only as an objection to adjudication. On a consistent view, decision makers of all stripes frequently provoke unintended consequences by involving non-represented parties, making assumptions as to the facts, and recasting nuanced and complex issues into familiar decision-making patterns. Fuller has adeptly identified a pervasive feature of the legal and political world in polycentricity, but he has not convincingly established that, at the very least, adjudication is affected more strongly than any other decision-making process. In light of this, it appears inconsistent and arbitrary to single out adjudication as problematic.

Second, polycentric issues have been adjudicated with reasonable success for centuries, yet the concerns are only invoked in certain policy areas. Why should polycentricity trouble us solely with regard to human rights and the environment and not equally with respect to taxes and private law?[130] On a consistent reading, Fuller's concerns apply to any policy area: the majority of any population, and certainly the politically, economically, and socially marginalised, are constantly affected by decisions taken without their meaningful input, and they may well be no more represented in the legislature than in the courtroom, especially historically speaking.

Third, courts have arguably proven themselves competent in dealing with polycentric issues and providing minorities with a forum for their concerns. This is evident in human rights cases, but also holds for the legal system more generally. Common law in particular is founded upon an adversarial legal tradition in which hardly any decision is limited exclusively to the parties before a court.[131] It is especially far-reaching in controlling cases, which may

[130] King, 'The Pervasiveness of Polycentricity', 111–12.

[131] King, *Judging Social Rights*, 192; Owen M. Fiss, 'The Supreme Court, 1978 Term' [1979] *Harvard Law Review* 1, 43; Abram Chayes, 'The Role of the Judge in Public Law Litigation' (1976) 89 *Harvard Law Review* 1281, 1289–90.

hold after decades through *stare decisis*: countless individuals who could not possibly have influenced the outcome of the original decision are affected, whether directly in their own proceedings or through political and legal adjustments made in response to such rulings. This temporal dimension only further adds to the plethora of polycentric questions adjudicated on a daily basis. However, despite Fuller's concerns, the track record of courts appears encouraging. The legal system has shown itself to be remarkably capable of addressing countless polycentric issues over time without succumbing to the identified pathologies.

Fourth, the evidence that deficiencies and limitations, which can undoubtedly be found in modern courtroom proceedings, are the result of the polycentric nature of the issues adjudicated is thin, and there is even less evidence to suggest that these would have been mitigated through other decision-making procedures. Fuller was arguably mindful of this objection, as he concedes that, with regard to polycentricity and adjudication, it is primarily a question of '(. . .) knowing when the polycentric elements have become so significant and predominant that the proper limits of adjudication have been reached.'[132] However, even this more measured reading of Fuller does not alleviate this concern completely.

This is because Fuller generally has difficulty accounting for 'good' judgments, where in spite of a morally consistent and desirable outcome, his theory would have us reject them as ill-conceived decisions from an institutionally overwhelmed judiciary. The conventional example is the landmark case of *Brown* v. *Board of Education*.[133] Ostensibly, the case concerned only the enrolment of the expressly mentioned children into a segregated elementary school, a request which stood in contradiction to the state's segregation laws. The implications of the judgment, however, undoubtedly went beyond this limited context. Any state law that required segregation of public schools was henceforth unconstitutional. This case challenges Fuller's theory in two ways: the outcome is morally sound, hence adjudication resolved the polycentric issue in a reasonable manner, and crucially, alternative modes of decision making had demonstrably failed the applicants. They could not abolish segregation through contracting because they had no meaningful economic, political, or social bargaining power, the state and the school board had made managerial decisions to implement segregation policies against

[132] Fuller and Winston, 'The Forms and Limits of Adjudication', 397.

[133] *Brown* v. *Board of Education of Topeka* 347 US 483 (1954) [requiring the desegregation of public schools in the United States].

their wishes, and the marginalised groups had no meaningful influence over the state legislature to effect changes to overtly racist laws.

This exposes the deficiencies of the market environment and the legislatures to which Fuller would have us defer on polycentric questions.[134] A market setting dominated by managerial decisions and contracts tends to operate poorly where grave imbalances in bargaining power persist, which is especially problematic with regard to the politically, economically, and socially marginalised. Democratic legislatures for their part often gravitate towards majority preferences, which is problematic as minority interests are often ignored in the absence of express restrictions and safeguards. Moreover, minorities often simply do not hold sufficient political power to effect legislative action, even if they are formally included in the democratic decision-making process. Voting restrictions and vested interests stood against any resolution of an intolerable status quo in the era of *Brown*, and Fuller's account does not offer a convincing rebuttal to frequent examples of market failure and regulatory inaction. For the purposes of environmental regulation, air pollution demonstrates how these concerns over market and regulatory failure are front and centre in the debate on environmental harm and human rights.

3.2.4.3 Environmental Harm

As many as 95 per cent of all urban areas currently exceed recommended threshold values for daily and annual exposure to one or more regulated air pollutants including nitrogen dioxide (NO_2), particulate matter (PM), ozone (O_3), carbon monoxide (CO) and sulphur dioxide (SO_2), lead (Pb), and benzene (C_6H_6). However, due to the limited availability of suitable measuring stations, some pollutants are not measured at all, and their impact on human health is poorly understood, for instance with regard to ultrafine particles.[135] Among the harmful health effects of air pollution are an increase in instances of lung cancer, cardiovascular mortality, chronic illnesses such as diabetes, links to atherosclerosis, birth defects and childhood respiratory diseases, and detriments to neurodevelopment and cognitive functions, including links to Alzheimer's disease.[136] These burdens are disproportionately borne

[134] Fuller and Winston, 'The Forms and Limits of Adjudication', 398.

[135] Ioar Rivas and others, 'Source Apportionment of Particle Number Size Distribution in Urban Background and Traffic Stations in Four European Cities' (2020) 135 *Environment International* 105345.

[136] Ole Raaschou-Nielsen and others, 'Air Pollution and Lung Cancer Incidence in 17 European Cohorts: Prospective Analyses from the European Study of Cohorts for Air Pollution Effects

by the young, the poor, and the elderly.[137] Overall, the damage to human health is significant, pervasive, and not adequately addressed, either through regulation or enforcement in the surveyed regulatory environments.

The WHO has repeatedly and consistently warned states about the dangers of excessive air pollution.[138] The focal point of this public health crisis is inevitably found in most major European cities. In London, the harm has reached dramatic levels. Within the first week of 2017 some boroughs of London breached the statutory threshold values for pollutants for the entire year.[139] However, at the national level, the regulatory and enforcement response in the United Kingdom to this public health crisis has largely been ineffective. Greater London is currently projected to meet EU air pollution targets at the earliest in 2030 under the current action plan, which in any case only requires NO_2 threshold compliance.[140] There are a number of measures which regulators might choose in order to address air pollution more effectively, broadly following the example of other major cities worldwide:

- Cubatao in Brazil saw reductions of the disastrous levels of air pollution through the regulation of heavy industries;[141]
- The cities of Paris, Athens, Mexico City, and Madrid are committed to banning all diesel engine cars by 2025,[142] and Stuttgart has announced it will prevent all but the latest technology diesel cars from entering the inner city;[143]

(ESCAPE)' (2013) 14 *The Lancet Oncology* 813; Maher and others, 'Magnetite Pollution Nanoparticles in the Human Brain'; *WHO Air Quality Guidelines for Particulate Matter, Ozone, Nitrogen Dioxide and Sulfur Dioxide* (World Health Organization, 2005); *Review of Evidence on Health Aspects of Air Pollution – REVIHAAP Project* 6.

137 'WHO Chief: Air Pollution Link to 600,000 Deaths in Children'; *Global Environment Outlook*, 9.

138 *Review of Evidence on Health Aspects of Air Pollution*; *Economic Cost of the Health Impact of Air Pollution in Europe* (World Health Organization: Regional Office for Europe, 2015).

139 Damian Carrington, 'London Breaches Annual Air Pollution Limit for 2017 in Just Five Days' *The Guardian* (6 January 2017) www.theguardian.com/environment/2017/jan/06/london-breaches-toxic-air-pollution-limit-for-2017-in-just-five-days#comment-90624590, accessed 1 April 2021.

140 *Updated Projections for Nitrogen Dioxide (NO_2) Compliance* (Department for Environment Food & Rural Affairs, 2014).

141 'How Brazil's "Valley of Death" Drastically Cut Pollution' *BBC News* (9 March 2017) www.bbc.co.uk/news/science-environment-39186018, accessed 1 April 2021.

142 *Daring Cities Make Bold Air Quality Commitment to Remove All Diesel Vehicles by 2025 – Mayors of Paris, Mexico City, Madrid and Athens Sign Air Quality Declaration at C40 Mayors Summit, 30 November to 2 December 2016* (C40 Mayors Summit, Mexico City, 2016).

143 'Fahrverbot für Dieselautos in Stuttgart' *Zeit* (21 February 2017) www.zeit.de/mobilitaet/2017-02/feinstaub-stuttgart-dieselfahrzeuge-schadstoffbelastung-fahrverbot, accessed 1 April 2021.

- The city of Paris alternately bans cars with odd- or even-numbered licence plates from entering the city in times of high pollution,[144] and is planning to ban cars from certain areas of the city entirely;[145]
- Stockholm, as part of a specialised EU-funded lighthouse project, is implementing a comprehensive biofuel refit scheme for vans and lorries, and plans to recycle waste heat from data centres to reduce the need for polluting energy production.[146]

Choosing from these and a range of further possible measures plainly involves a series of competing and interconnected interests that affect large numbers of people: a deeply polycentric issue. Just as no single cause can be blamed for air pollution, no single measure will secure improvements and compliance on its own. Rather, it is the interconnected actions of several industries that contribute to varying degrees, especially those involved in energy production – chiefly coal power plants – domestic consumption, transportation, cars and lorries, and heavy industry more generally.[147] However, the fact remains that markets, legislatures, and regulators have, over the course of decades, demonstrably failed to protect citizens from this well-established threat to their human rights.

Air quality standards in the EU are set through Directive 2008/50/EC,[148] which is based on WHO guidelines. It consolidated earlier obligations of member states under air quality directives and contains targets that had been binding in some form since 2005. Chiefly, the Directive requires action plans where a member state exceeds threshold values in order to ensure compliance. However, many member states, including the former member United Kingdom, are projected to breach Directive 2008/50/EC with respect to Greater London and other parts of the country for decades to come.[149] The failures to uphold the Directive have given rise to limited domestic litigation, with the UK Supreme Court ordering the Secretary of State to prepare new air

[144] 'Paris Imposes Car Restrictions to Fight Pollution' *BBC News* (23 March 2015) www.bbc.co.uk/news/world-europe-32016730, accessed 1 April 2021.

[145] David Chazan, 'How Paris Is Stepping Up Its Drive against the Car' *BBC News* (2 May 2016) www.bbc.co.uk/news/magazine-36169815, accessed 1 April 2021.

[146] Jane Wakefield, 'Tomorrow's Cities: Stockholm Turns Green' *BBC News* (8 March 2017) www.bbc.co.uk/news/technology-38404884, accessed 1 April 2021.

[147] Federico Karagulian and others, 'Contributions to Cities' Ambient Particulate Matter (PM): A Systematic Review of Local Source Contributions at Global Level' (2015) 120 *Atmospheric Environment* 475.

[148] Directive 2008/50/EC of the European Parliament and of the Council of 21 May 2008 on ambient air quality and cleaner air for Europe.

[149] *Updated Projections for Nitrogen Dioxide (NO_2) Compliance.*

quality plans under Article 23(1) of Directive 2008/50/EC.[150] The inadequate revision of the action plan has more recently resulted in a further legal challenge, in which the High Court confirmed the need for a new air quality plan that could realistically bring down emissions to within legal limits.[151] Subsequently the court also found the revised 2017 Air Quality Plan unlawful, requiring the production of supplements to reduce nitrogen dioxide concentrations.[152] However, even if these successive court orders are implemented to the letter, the United Kingdom would nonetheless systemically fail to tackle air pollution and thus contribute to thousands of deaths and exacerbated health problems.

Moreover, the moderate progress achieved under action plans is threatened by the departure of the United Kingdom from the European Union. Air quality regulations remain in place through the Air Quality Standards Regulation 2010, which mirrors Directive 2008/50/EC. While the decision to leave the EU is therefore unlikely to have an immediate detrimental impact on air quality in the United Kingdom, it cannot be said that this current legal framework provides for adequate relief for persistently high and dangerous levels of air pollution in major cities. Notably, this is because dangerous pollutants are not limited to nitrogen dioxide: fine particulate matter has recently been linked to brain damage in adults and children and is nonetheless not effectively regulated.[153]

Action plans under EU law appear as a less than satisfactory regulatory response for persistently harmful levels of air pollution and have in any case been the subject of much deference from the European Commission: infringement proceedings over air pollution were repeatedly deferred and postponed despite evident and persistent breaches.[154] Chapter 10 shows that air pollution regulations have also been the subject of widespread industry resistance, especially from the car industry and its use of so-called defeat

[150] *R (ClientEarth)* v. *Secretary of State for the Environment, Food and Rural Affairs* [2015] UKSC 28.

[151] *R (ClientEarth (No 2))* v. *Secretary of State for the Environment, Food and Rural Affairs* [2016] EWHC 2740 (Admin).

[152] *R (ClientEarth (No 3))* v. *(1) Secretary of State for Environment, Food and Rural Affairs (2) The Secretary of State for Transport and (3) Welsh Ministers* [2018] EWHC 315 (Admin).

[153] Maher and others, 'Magnetite Pollution Nanoparticles in the Human Brain'; 'Babies' Brains Damaged by Pollution, Unicef Says' *BBC News* (6 December 2017) www.bbc.co.uk/news/health-42250558, accessed 1 April 2021.

[154] 'Commission Warns Germany, France, Spain, Italy and the United Kingdom of Continued Air Pollution Breaches' *European Commisison* (Brussels, 15 February 2017) http://europa.eu/rapid/press-release_IP-17-238_en.htm, accessed 1 April 2021.

devices which adjust engine performance during benchmarking to artificially reduce emissions.[155]

Overall, neither the markets, legislatures, nor regulators have thus far been able to reduce air pollution to within the limits outlined in WHO guidelines. This seriously undermines the trust that Fuller would have us place in market-based mechanisms, especially their ability to resolve polycentric issues in a manner that is not fundamentally based on the self-interest of economic actors, to the detriment of public health.

3.2.4.4 Conclusion

Fuller's account of polycentricity does not defeat the notion that courts should adjudicate environmental protection on the basis of human rights. Polycentricity is pervasive in modern legal adjudication and there is no clear evidence that courts are any less suitable than alternative methods of resolution. The pervasiveness of polycentricity raises the question of why its cautionary tale is told only in some areas of the law, but not in others. If we accept Fuller's premise, then polycentricity should concern us in constitutional law, human rights law, and environmental law, as much as in tax law and private law. Fuller's account itself does not single out specific areas: it would simply 'render illegitimate almost all adjudication.'[156] Nonetheless, as one of the primary justifications for human rights is the protection of minority interests, it is arguably more suitable to entrust their protection to courts, rather than to elected representatives who are more prone to satisfying majority preferences.[157] Polycentricity may well encourage more judicial self-restraint in certain cases, but it is not appropriate to reject the adjudication of polycentric issues wholesale.[158] Instead, the complexity of the world, and the interconnected nature of any measures, should be managed, and adjudication is a crucial part of the legal and policy-making toolbox.

3.2.5 *Summary*

Based on these minimum standards outlined throughout the chapter, three key implications arise for the scope of environmental protection under the environmental minimum framework.

155 Uwe Tietge and others, *From Laboratory to Road – A 2016 Update of Official and 'Real-World' Fuel Consumption and CO_2 Values for Passenger Cars in Europe* (International Council on Clean Transportation, 2016) 48.

156 Fiss, 'The Supreme Court, 1978 Term'.

157 Dworkin, *Freedom's Law*, 33.

158 For an argument along these lines, see King, *Judging Social Rights*, chapters 7 and 8.

First, regulators naturally retain the ability to draft domestic regulations and international treaties. While the environmental minimum suggests that the generally accepted guidelines and emerging standards ought to guide regulations, there is nothing to prevent states from diverging in good faith. However, regulations that fail to consider relevant information, are designed to circumvent protection, are retrogressive, or manifestly fail to address the underlying environmental harm, do not count as legal standards. Conversely, states are free to adopt stricter environmental regulations that exceed the requirements of the minimum standards.

Second, the requirements of good faith and consistent enforcement of minimum standards place a significant regulatory burden on states. In practice, it might prove difficult to dispel a presumption of bad faith and demonstrate sufficiently consistent enforcement of legal standards. However, it is equally true that tokenistic regulations and mere lip service to environmental protection cannot discharge duties under human rights. The bad faith standard and the default application of established and emerging standards as well as the insistence on their consistent enforcement are ultimately based on commitments that states freely undertook, along with the limitations of sovereignty they necessarily entail. Nonetheless, it would be a mistake to expect the environmental minimum to authoritatively resolve contentious scientific and policy questions, or that it can, on its own, strike an adequate balance between competing interests in environmental regulation. The higher the level of scientific uncertainty, the more leeway the environmental minimum accordingly affords to regulators.

Third, it is important to clarify the implications if an environmental regulation meets the outlined minimum standards. Under these circumstances, the environmental harms it addresses are adequately regulated as far as the environmental minimum is concerned, despite any misgivings about scope and ambition that might linger. This does not render the environmental harm irrelevant, or its complete rectification undesirable, it simply cannot be achieved through the environmental minimum framework. This is both a function of the practical focus of the framework and the inherent respect of the environmental minimum for the outcomes of democratic decision-making processes. While human rights obligations can and, at times, must legitimately set aside majority preferences, this should not be a casually invoked normative mandate. The goal of the environmental minimum is to ensure that basic safeguards are consistently provided and enforced, not to realise a complete and comprehensive vision of environmental justice.

3.3 EVIDENCE

3.3.1 *Burden of Proof and Causality*

As advanced as modern science may at times appear, it can prove difficult to establish that a specific environmental harm has caused a specific detriment to an individual, especially with respect to complex issues surrounding health. While it is for instance comparatively simple to demonstrate that smoking increases instances of lung cancer on a population-wide, epidemiological basis, it is next to impossible to state definitively that an individual developed lung cancer as a result of smoking, much less due to smoking a specific cigarette.[159] This raises complicated legal issues with respect to causality and renders determining who assumes the burden of proof decisive in such cases. Helpfully, these questions have been explored in some detail in the context of asbestos litigation.

Legal complexities generally arise if an individual is exposed to asbestos from multiple sources.[160] The difficulty is that it will be next to impossible for an individual to demonstrate that any single source of exposure is responsible for developing mesothelioma, the cancer of the lining of internal organs caused by asbestos, much less apportion the liability amongst multiple sources.

In England and Wales, the conventional liability test asks whether it was more likely than not that the defendant's breach of a duty of care caused the harm suffered. With respect to asbestos cases, the UK House of Lords (UKHL) noted in *Fairchild* v. *Glenhaven Funeral Services Ltd* that there was no consensus on the mechanism by which an individual developed mesothelioma:

> It is not known what level of exposure to asbestos dust and fibre can be tolerated without significant risk of developing a mesothelioma (. . .). It is accepted that the risk of developing a mesothelioma increases in proportion to the quantity of asbestos dust and fibres inhaled: the greater the quantity of dust and fibre inhaled, the greater the risk. But the condition may be caused by a single fibre, or a few fibres, or many fibres: medical opinion holds none of these possibilities to be more probable than any other, and the condition once caused is not aggravated by further exposure.[161]

159 Stacy M. Carter and Simon Chapman, 'Smoking, Disease, and Obdurate Denial: The Australian Tobacco Industry in the 1980s' (2003) 12 *Tobacco Control* 23, 24.

160 Jane Stapleton, 'Factual Causation and Asbestos Cancers' (2010) 126 *Law Quarterly Review* 351, 354.

161 *Fairchild* v. *Glenhaven Funeral Services Ltd* [2002] UKHL 22 [7].

To resolve the apparent causality challenge, the UKHL accepted proof of a material contribution to the *risk* of a disease as proof of a material contribution to the disease.[162] Thus even relatively minor sources of exposure in the grand scheme of an individual's total exposure to asbestos suffice to establish liability for mesothelioma.[163] The revised causality test was subsequently expanded to cover a second scenario: a single incident of asbestos exposure, combined with other non-tortious and environmental instances of exposure.[164] The claimants in these cases are not required to demonstrate that the tortious exposure had doubled the risk, thereby rendering it more likely than not that the breach of a duty of care resulted in developing mesothelioma. Instead, as with multiple tortious exposures, it sufficed for the claimants to show that the breach resulted in material contribution to the risk of developing mesothelioma.

A similar view is taken by US courts. Despite the absence of firm evidence on the specific development of mesothelioma, for instance whether it may occur through cumulative or individual exposure, courts treat asbestosis as an indivisible injury. This means that any defendant responsible for one of several sources of asbestos exposure is liable for the entire damage suffered by an individual. Accordingly, there is no need for claimants to establish the portion of liability attributable to individual sources of asbestos exposure.[165] This approach essentially removes the need for specific attribution entirely.

By contrast, Germany has no significant history of asbestos litigation to draw upon, as the lawsuits typical for the United States and the United Kingdom from employees against their employers are unlikely to arise. Exposed employees are instead covered under the statutory occupational accident insurance.[166] Once an occupational disease has been recognised by this body, in our case asbestosis and associated lung cancers, the insurance scheme pays a lifelong pension to the employee (or compensation to dependants) based on level of occupational disability and loss of income. Employees and dependants are then legally barred from suing the employer unless the latter acted

162 Stapleton, 'Factual Causation and Asbestos Cancers', 354.

163 *Sienkiewicz, Willmore* v. *Knowsley MBC* [2009] EWCA Civ 1211.

164 *Sienkiewicz* v. *Greif (UK) Ltd* [2011] UKSC 10 [19] and [102].

165 *Borel* v. *Fibreboard Paper Products Corp* 493 F2d 1076 (5th Cir 1973); *Norfolk & Western Ry. Co.* v. *Ayers* 538 US 135, 143 (2003); Jane Stapleton, 'The Two Explosive Proof-of-Causation Doctrines Central to Asbestos Claims' (2008) 74 *Brooklyn Law Review* 1011, 1013.

166 See § 104 of the Seventh Volume of the Social Insurance Code – Statutory Accident Insurance (Siebtes Buch Sozialgesetzbuch SGB VII – Gesetzliche Unfallversicherung) in the version promulgated on 7 August 1996 (BGBl I 1254), last amended by Article 14d of the Law of 24 February 2021 (BGBl I 274).

intentionally to expose the employee, an evidentiary threshold most are unlikely to meet.

The example of asbestos exposure lends itself well to discuss some of the issues encountered in human rights cases that deal with environmental harm. As with asbestos, it will not generally be possible to conclusively demonstrate that an individual environmental harm caused a specific detriment to an applicant, much less how the blame should be apportioned between several sources. In most cases, even cutting-edge science with hypothetically boundless resources would not be able to provide that level of certainty. The best most applicants can hope for is general epidemiological evidence of an increased risk as a result of the environmental harm, which allows certain inferences as to changes in their individual risk profile.[167]

But even if obtaining more specific and individually tailored evidence is not ruled out as a matter of scientific capability, applicants will often not be in a position to furnish such evidence in complex environmental cases. For one, it might well lie beyond their (financial) means to acquire the necessary data. Many applicants simply cannot afford to pay for the specific studies and scientific inquiries necessary, even if the environmental harm is well understood and the necessary technology exists.[168] For instance, consider the well understood and reliably measurable impact of nuclear radiation on the environment and human health.[169] Crucial evidence may be of a historic nature and held by state entities that are unable or unwilling to furnish the applicants with the necessary information to build their case.[170] A prominent environmental example relates to documents pertaining to the UK's nuclear weapons tests in the Pacific, which were said not to exist, or subsequently lost, or destroyed.[171]

[167] Stapleton, 'Factual Causation and Asbestos Cancers', 352.

[168] *Dzemyuk* v. *Ukraine* App no 42488/02 (ECtHR, 4 September 2014) [alleged contamination of groundwater through cemetery].

[169] *Dr S* v. *Germany* App no 715/60 (Commission, 5 August 1960, unpublished) [on the practice of dumping radioactive waste in the North Sea]; *Tauira* v. *France* App no 28204/95 (Commission, 4 December 1995) [on the impact of nuclear weapons test on the food chain and potential exposure of humans].

[170] A lack of cooperation of states with proceedings before courts is not uncommon, see, for instance, on the refusal of Russia to provide documents on forced disappearances in Chechnya, *see* Ole Solvang, 'Russia and the European Court of Human Rights: The Price of Non-Cooperation' (2008) 15 *Human Rights Brief* 14.

[171] *McGinley and Egan* v. *United Kingdom* App nos 21825/93; 23414/94 (ECtHR, 9 June 1998) and *Sinfield and other* v. *United Kingdom* App no 61332/12 (ECtHR, 18 February 2014); *LCB* v. *United Kingdom* App no 23413/94 (ECtHR 9 June 1998).

The environmental minimum reckons with these difficulties. Scientific evidence is most relevant to the framework at two stages: when establishing the specific risk and when determining the applicable minimum standard. The applicants are only responsible for furnishing scientific evidence on the limited question of the specific risk, which significantly alleviates the evidentiary burden. An applicant is only required to establish an environmental harm and put forward a reasonable hypothesis linking it to a specific threat to a recognised human right. The hypothesis need not necessarily reflect the majority view in the relevant scientific community. It is merely a rough filter that excludes unsubstantiated or implausible claims. From there, the burden shifts away from applicants to the respondent state, which must either accept the existence of a specific risk, or provide evidence disproving the environmental harm, or the reasonable hypothesis. In many cases, this will be a difficult burden for a state to satisfy. Given the serious nature of threats, this nonetheless appears consistent with fundamental principles: if a risk is generally accepted to exist, then responsibility should not be limited just because complete certainty over the specific causal source cannot be established. This is the essence of the doctrinal stance courts have taken in asbestos cases and which, by analogy, should apply to environmental harm more broadly. Burdening a state, rather than an individual, with dispelling the specific risk also appears equitable given the superior resources of states and the rationale of human rights as sources of obligations and limits to state conduct. Finally, states can avoid a violation of the environmental minimum: they are only required to provide regulations in good faith and consistently enforce them.

3.3.2 *Evaluation of Scientific Evidence*

The crucial stage for the application of scientific evidence is determining the applicable minimum standards. With respect to legal standards adopted in good faith, this does not present any exceptional difficulty. Courts would merely have to examine relevant regulations and determine whether they were consistently enforced. The complexity increases when bad faith is at issue or no legal standards exist. A court would then determine whether the relevant scientific community has generally agreed on a standard, or whether the matter is contentious enough that it is only subject to an emerging standard. This requires a careful evaluation of scientific evidence, typically aided by expert testimony. The rules and approaches will be context sensitive to the protection regime in question and can therefore only be discussed in general terms in the abstract. The specific approach and potential changes required with respect to the ECHR are discussed in Part II of this book.

Courts and legal scholarship have grappled with scientific evidence in the context of the admissibility and appointment of experts. While most environmental cases that deal with expert testimony do not adjudicate human rights and often operate in the specific context of trade disputes, they nonetheless offer valuable insights.[172] The case *Australia – Apples from New Zealand* revolved around the questions whether the scientific evidence put forward by Australia pertaining to the risk of various pests was sufficient to support a quarantine on New Zealand apples. The WTO Appellate Body upheld the panel report which had found fundamental flaws in the Australian risk assessment: its methodology had exaggerated the risk and hence did not support the measures imposed by authorities. Crucially, there were other less trade-restrictive measures available to Australia, notably '(. . .) restricting New Zealand apples to mature, symptomless fruit, with straight-forward sample inspections at the border.'[173] This conclusion was reached in large parts on the basis of expert evidence admitted following a two-prong test:

> (i) a determination that the scientific basis of the risk assessment comes from a respected and qualified source and can accordingly be considered 'legitimate science' according to the standards of the relevant scientific community; and (ii) a determination that the reasoning of the risk assessor is objective and coherent and that, therefore, its conclusions find sufficient support in the underlying scientific basis.[174]

The key challenge exemplified by this approach is conducting a fair assessment of the quality and reliability of at times conflicting scientific evidence and expert testimony.[175] However, the problem does not necessarily present itself with the same intensity for all sources of minimum standards under the environmental minimum.

3.3.2.1 Legal Standards

Where good faith legal standards exist, the role of the court is comparatively straightforward and familiar to jurisdictions that practise judicial review.[176]

[172] Harrison, 'Reflections on the Role of International Courts and Tribunals in the Settlement of Environmental Disputes', 512; *Australia – Apples from New Zealand* WTO Appellate Body Report WT/DS367/AB/R, 12 November 2010 [on the soundness of scientific evidence]; *US-Continued Suspension of Obligations* WTO Appellate Body Report WT/DS320/AB/R, 19 September 2008 [436]; see also *EC-Hormones* WTO Appellate Body Report WT/DS26/AB/R, 5 January 1998 [148] [due process for the appointment of independent experts].

[173] Cited as per *Australia – Apples from New Zealand* WTO Appellate Body Report WT/DS367/AB/R, 12 November 2010 [331].

[174] Ibid. [220].

[175] Foster, *Science and the Precautionary Principle in International Courts and Tribunals*, 80.

[176] Regardless of whether they subscribe to weak or strong form judicial review, on this distinction, see Waldron, 'The Core of the Case Against Judicial Review', 1354.

Legal standards adopted in good faith serve as a closed-off standard of review for regulatory action that established and emerging standards do not second guess. The need for expert testimony is less pronounced because the court primarily requires evidence on enforcement practices. Should good faith be drawn into doubt, then the court must additionally look to evidence on the guidelines generally accepted in the relevant scientific community. In many cases, this might be a simple matter of determining relevant guidelines, for instance from the WHO. Once the generally accepted standard has been established, a court would then ask regulators to explain and justify any departure: should the reasons fall into the previously outlined suspect categories, the legal standard will likely have been adopted in bad faith. The environmental minimum thus limits the scope of necessary scientific evidence to specific questions that courts are institutionally well equipped to answer.

3.3.2.2 Established and Emerging Standards

In the absence of legal standards, courts must establish the applicable minimum standards themselves. This task is more complicated, as courts must determine whether there are generally accepted standards in the relevant scientific community, or if significant disagreement persists and therefore only emerging standards apply. In extraordinary circumstances, the harm assessment formula might further require regulatory intervention despite the general flexibility of regulators on emerging standards. The harm assessment formula itself involves complex predictions on conflicting courses of regulatory action. To some extent, the environmental minimum alleviates these concerns by shifting the focus to the question of harm, namely the harm caused by regulatory intervention versus the harm caused by inaction. Nonetheless, both established and emerging standards present significant difficulties for courts and necessitate detailed expert testimony. The approaches developed by domestic courts and tribunals can illuminate how a human rights protection regime might approach contentious evidence or the appointment of experts.

3.3.2.2.1 INQUISITORIAL SYSTEMS. In the inquisitorial tradition that prevails in civil law systems, for instance in Germany, courts typically have greater discretion in establishing and determining relevancy, as well as weighing evidence autonomously from party submissions. It is the default position of German civil and public procedure law that the court determines whether it is necessary to seek expert testimony, and it has broad discretion to appoint independent experts.[177] The input of parties is typically limited to expressing their views on suitable experts for the court's consideration. Notably, a court is not limited to the expert evidence presented by the parties and can explore all relevant avenues of inquiry until it has established the scope of scientific certainty and enduring controversy to its satisfaction. Such an approach gives broad discretion to judges and mitigates the 'clash of experts' on contentious evidential questions: research suggests that experts often hold an inadvertent bias towards the party that nominated them.[178] This 'clash' is more common in the course of adversarial proceedings, for instance under the procedural law of the United States and England and Wales. However, even those jurisdictions have enacted reforms that introduce broader discretion and a stronger role for judges.

3.3.2.2.2 ADVERSARIAL SYSTEMS. In the United States, the case of *Frye v. United States* provided a test that long prevailed as the standard for scientific evidence (the test still applies in many state courts), which significantly limited the ability of courts to curtail the admissibility of expert testimony.[179] The key element of the test was whether the basis for expert testimony, for instance, studies, tests, and techniques, are generally accepted in the relevant field.[180] The approach took a largely agnostic view on the content and conclusions of the testimony and thus arguably encouraged a 'clash of experts.' The *Frye* test was significantly modified, and largely superseded, by the landmark case of *Daubert*, at least with respect to federal jurisdictions.[181] The case involved the litigation of serious birth defects allegedly caused through a medication prescribed for the management of morning sickness during pregnancy. In that decision, the US Supreme Court held that the test developed in *Frye* had been superseded by an intervening revision of the Federal Rules of Evidence, specifically Rule 702. Pursuant to this rule, courts must consider whether

(a) the expert's scientific, technical, or other specialized knowledge will help the trier of fact to understand the evidence or to determine a fact in issue;
(b) the testimony is based on sufficient facts or data;

177 Hans-Jürgen Ahrens, '§ 402' in Bernhard Wieczorek and Rolf A. Schütze (eds.), *Zivilprozessordnung und Nebengesetze: Großkommentar*, vol 6 (4th ed., De Gruyter 2015) 516–17.
178 E. Allan Lind and Tom R. Tyler, *The Social Psychology of Procedural Justice* (Plenum Press 1988) 115; Foster, *Science and the Precautionary Principle in International Courts and Tribunals*, 98.
179 *Frye* v. *United States* 54 App DC 46, 293 F 1013 (1923), 1013.
180 Ibid., 1014.
181 *Daubert* v. *Merrell Dow Pharmaceuticals* 509 US 579 (1993) [588]–[589].

(c) the testimony is the product of reliable principles and methods; and
(d) the expert has reliably applied the principles and methods to the facts of the case.[182]

While this addressed some of the concerns with the permissive *Frye* test, it also likely made it more difficult for plaintiffs to succeed. Judges and the parties now have more grounds to challenge the relevance and credibility of expert testimony, and legal scholarship suggests that this has led to significantly lower success rates, especially in cases challenging environmental harm.[183]

With respect to England and Wales, certain limitations on the admissibility of expert evidence were introduced following the final report of Lord Woolf.[184] Most importantly for our purposes, Part 35 of the Civil Procedure Rules requires the parties to seek permission before calling any experts or submitting their reports in evidence. Crucially, the parties must justify the relevance of the expert evidence to the case,[185] and the court can specify the issues that the expert evidence will address.[186] The court may require parties to settle on a joint expert, or select one on their behalf from a list of experts,[187] as well as direct one party to disclose information to the other party.[188]

3.3.2.3 Conclusion

Human rights courts do not usually operate in an adversarial legal tradition. Therefore, admissibility concerns with respect to expert evidence are less pronounced. Indeed, a court might admit evidence from a wide variety of sources, and scrutinise more closely when it comes to affording weight to the evidence.[189] Building on the prevailing legal position in civil law systems and the emerging trends in adversarial common law traditions, we might well go significantly further.[190] Depending on the governing procedural rules, a court could itself appoint experts to determine the relevant scientific evidence for the specific risk and minimum standards, and thus alleviate many of the concerns that experts nominated by the parties might bring about. Provided that impartial expert assistance is obtainable, courts could certainly evaluate

182 US Federal Rules of Evidence, 702.
183 Cranor, *Toxic Torts*, 5.
184 Woolf, *Access to Justice Final Report* (Department for Constitutional Affairs, 1996).
185 Civil Procedure Rules Part 35.4 (2) (a).
186 Ibid., 35.4 (3).
187 Ibid., 35.7.
188 Ibid., 35.9.
189 King, *Judging Social Rights*, 246.
190 Foster, *Science and the Precautionary Principle in International Courts and Tribunals*, xiv.

the studies, tests, and techniques that underlie any scientific evidence with respect to three basic questions: (1) whether the assumptions and methodology are in line with those generally employed in the relevant field; (2) whether experts with broadly relevant credentials and experience conducted them; and (3) the scope and limitations of the results and their conclusions with respect to the case.

3.4 REMEDIES

The remedies afforded by a court depend on the powers available under the human rights protection regime and the position of the court in a broader constitutional hierarchy. In this sense, it is not possible to answer the question of appropriate remedies conclusively in the abstract. Nonetheless, certain guidelines can be derived from existing legal principles and the practice of domestic and international courts. Broadly speaking, where a violation of a human right occurs under the environmental minimum, appropriate remedies ought to follow relative to the applicable minimum standard.

3.4.1 *Legal Standards*

With respect to legal standards, the matter of remedies is fairly straightforward: nothing short of an order that requires effective enforcement of the domestic or International Law obligations will suffice. Such strong remedies for breaches of legal standards are justified on the basis of fundamental legal principles, namely the rule of law in a domestic setting, as well as the principle of *pacta sunt servanda* in International Law. For most courts, appropriate remedies chiefly involve directions to executive and legislative bodies, or even the state as such, notably including:

- **Directions to executive bodies**: domestic courts can typically issue directions to executive bodies and administrative agencies to take specified measures that remedy violations of rights, often by bindingly interpreting legislative provisions, including on a provisional basis. This is generally a power available to courts in legal systems that permit judicial review of executive action, whether in a comparatively weak (England and Wales)[191] or strong form (Germany);[192]

[191] For instance, under section 8 of the Human Rights Act 1998.

[192] On the distinction between weak and strong forms of judicial review, see Waldron, 'The Core of the Case Against Judicial Review', 1354–5.

- **Directions to legislative bodies**: this describes the power of courts to disapply, strike down, and require legislative amendments within a given timeframe, often with a specified content, to satisfy obligations under human rights. This is particularly common in legal systems with constitutional rights and strong form judicial review, for instance in Germany.[193] In systems with weak form judicial review, and a strong constitutional position of the legislature, courts might be limited to issuing declarations of incompatibility, as is the case in the United Kingdom under the Human Rights Act 1998;[194]
- **Directions to states**: directions to states are common in international human rights protection regimes, and require a state to take all measures necessary to remedy a violation of rights and hence fulfil its obligations under International Law, including provisional measures and a duty to address structural problems. Broader powers for courts and tribunals are common in the more recent regional human rights protection regimes, for instance under Article 27 of the Protocol to the African Charter on Human and Peoples' Rights on the Establishment of the African Court on Human and Peoples' Rights,[195] and the similarly worded Article 63 ACHR.[196] Although not expressly provided, such powers have also been read into Article 46 ECHR, especially through the practice of so-called pilot judgments.[197]

It is important to note that at all stages, the remedies need not require a state to completely eliminate the environmental harm: it is sufficient under the environmental minimum to remove the concerned individuals or groups from harm's way, or to provide them with appropriate resources to mitigate the environmental harm. The latter may either reduce the threat to a point where it no longer constitutes a specific risk, or where it meets the applicable minimum standard.

[193] Jarass and Pieroth, *Grundgesetz für die Bundesrepublik Deutschland*, Artikel 20, 532 [46]–[49].
[194] See section 4 of the Human Rights Act 1998.
[195] Protocol to the African Charter on Human and Peoples' Rights on the Establishment of the African Court on Human and Peoples' Rights (adopted 10 June 1998, in force 1 January 2004) OAU Doc. OAU/LEG/MIN/AFCHPR/PROT.1 rev.2 (1997).
[196] American Convention on Human Rights.
[197] The practice of directing the state to resolve a structural or systemic human rights violation through reforms of the underlining legislative framework or administrative practice on the basis of a representative 'pilot' judgment, while similar pending and future cases are adjourned pending the adoption of remedial measures, see Rule 61 of the Rules of the European Court of Human Rights.

3.4.2 *Established Standards*

The key difference between remedies for legal and established standards is that principles such as the rule of law cannot be brought to bear to justify the enforcement of established standards. Regulations that require enforcement simply do not exist. Instead, the normative force of established standards lies in the level of scientific certainty they represent and their intimate connection to environmental human rights. Permitting unregulated environmental harm to humans amounts to wilful ignorance that is incompatible with any serious commitment to human rights. Depending on the powers available to courts under the engaged human rights protection regime, remedies can involve the same directions to legislative and executive bodies, and states as discussed in the context of legal standards. In addition, they can include:

- **Declaratory findings**: the most common remedy available under most international human rights protection regimes is a declaratory finding of a violation. This is the only remedy available under regimes that rely on monitoring bodies, such as the International Covenant on Civil and Political Rights (ICCPR),[198] International Covenant on Economic, Social and Cultural Rights (ICESCR),[199] and the European Social Charter;[200]
- **Compensatory damages**: these can be awarded under some regional human rights protection frameworks following findings of a violation, for instance under Article 41 ECHR and Article 63 ACHR.

The remedies available are less important than their underlying objective: to incentivise states to regulate the environmental harm in question by raising both the public profile and the cost (monetary and political) of ignoring environmental harm. Crucially, through the process of active regulation, states retain the ability to balance environmental protection with competing considerations. Regulation in good faith can avoid the comparatively blunt application of established standards that otherwise occurs.

3.4.3 *Emerging Standards*

By contrast, emerging standards leave states with broad discretion to determine the scope and desirability of regulatory intervention. The scientific evidence is by definition scarce or contradictory, making it difficult to derive any firm

198 International Covenant on Civil and Political Rights.
199 International Covenant on Economic, Social and Cultural Rights.
200 European Social Charter (adopted 18 October 1961, in force 26 February 1965) 529 UNTS 89.

conclusions. An exception applies only pursuant to the conditions of the harm assessment formula. It is designed to address circumstances where potentially dire consequences would ensue if an intervention does not occur. Under such exceptional circumstances, the environmental minimum would require some appropriate intervention that protects individuals from a specific risk, until such a time that some superior level of certainty permits a more nuanced regulatory intervention. Interventions may thus be justified despite persisting uncertainties over:

- the extent of the threat posed by the environmental harm, provided that sufficient evidence exists to support a specific risk to a recognised human right;
- the specific root causes, their interaction, and the degree and distribution of responsibility among several actors that might be burdened by regulatory measures, provided that sufficient evidence exists to enact some regulations;
- the appropriate extent and degree of regulatory intervention, provided that the proposed intervention is not so invasive as to be disproportionate to the reduction in harm;
- whether the proposed intervention will achieve an optimal reduction in harm, provided that it is not unreasonable on the basis of available evidence that some significant reduction can be achieved.

However, it is important to recall that emerging standards likewise do not invariably require that a state prevent the environmental harm in its entirety. As long as individuals are removed from immediate danger and are provided with the necessary resources to mitigate the specific risk, no issue arises under the environmental minimum. The specific measures have to be decided on a case-by-case basis and in consideration of all relevant circumstances.

3.5 SUMMARY

The environmental minimum presents a normatively attractive framework, due to its compatibility with environmental human rights, the core characteristics of human rights, and its alignment with the normative claims of the rule of law and *pacta sunt servanda*. The framework thus translates general and abstract commitments to human rights and environmental protection into specific and practical regulatory measures.

The principles of the environmental minimum chiefly require a good faith engagement with scientific evidence on the harmfulness of environmental degradation and pollution. This is achieved in two steps: First, it must be determined whether the environmental harm represents a specific risk to human rights. To that end, the environmental minimum is only triggered if an individual can establish a threat to a previously recognised right under the invoked protection regime. This is necessary for the framework to remain practically viable.

Second, an environmental harm that constitutes a specific risk must fall short of the applicable minimum standard of review, which is determined either based on legal norms, established or emerging scientific evidence. Legal standards comprise domestic and International Law obligations adopted in good faith, and generally demand consistent enforcement. Good faith is determined by reference to the guidelines generally accepted in the relevant scientific community. Should the regulations diverge from these guidelines, bad faith would typically attach if the outlined indicative elements are present. Established standards apply in the absence of legal standards and are comprised of the guidelines and best practices generally accepted in the relevant scientific community. Failing to meet established standards generally equates to a violation of the underlying right identified at the specific risk stage. In the absence of established standards, emerging standards apply. These exist where significant scientific uncertainty and controversy prevail, rendering firm conclusions on regulatory action impossible. In such cases, regulators enjoy significant flexibility in determining whether an intervention is required and what constitutes an appropriate response. In exceptional circumstances, the harm assessment formula can nonetheless require regulatory intervention in the absence of clear evidence.

This approach raises complicated issues for the burden of proof and evaluating scientific evidence, as well as potential remedies for violations. Applicants might not be in a position to provide conclusive evidence of harmfulness and therefore burdening the states with dispelling an established specific risk appears equitable. With respect to the evaluation of scientific evidence, some lessons can be drawn from inquisitorial and adversarial legal traditions. Remedies for their part are generally contingent on the context of the specific human rights protection regime, and they should be tailored to address violations of the environmental minimum dependent on the applicable minimum standard. Legal standards generally require compliance with domestic regulations or obligations under International Law. Established and emerging standards essentially rely on whatever enforcement mechanism is available: these can range from declarations, to specific orders for legislative amendments, especially under certain domestic constitutional rights.

In Part II of the book we now turn towards the practical case in favour of the environmental minimum, notably the level of compatibility of the framework with the environmental human rights doctrine of the ECHR.

PART II

The Environmental Minimum under the ECHR

Part II of the book focuses on the practical case in favour of the environmental minimum. It addresses two key gaps in the relationship between environmental protection and human rights law as conventionally presented:

- a lack of acknowledgement of the substantial and promising engagement with environmental harm under the ECHR; and
- the absence of a framework that can translate a plethora of general and abstract commitments to environmental protection and human rights into specific and practical regulations.

Environmental protection has found its way into several constitutions,[1] as well as into the case law of domestic and international courts and tribunals. This is most evident under the ECHR, which has dealt with a comparatively high volume of cases.

Ostensibly, both the former European Commission of Human Rights (EComHR) and the European Court of Human Rights (ECtHR) long held that there is no right to environmental protection under the Convention.[2] Likewise, legal commentators widely state this position as a starting point in discussions of environmental case law.[3] Indeed, there is no provision in the ECHR that safeguards or so much as mentions the natural environment. The protections that have been developed under other provisions are often

[1] See the overview provided by May and Daly, *Global Environmental Constitutionalism*.

[2] The EComHR was abolished pursuant to Protocol No. 11 to the Convention for the Protection of Human Rights and Fundamental Freedoms, restructuring the control machinery established thereby (Protocol No. 11 to the ECHR) (adopted 5 November 1994, entered into force 1 November 1998) ETS 155.

[3] Pätzold, 'Artikel 8 EMRK', 263 [38]; Harris and others, *Harris, O'Boyle & Warbrick*, 341; Daniel Garcia San José, *Environmental Protection and the European Convention on Human Rights* (Council of Europe Publishing 2005) 7; Beyerlin, 'Umweltschutz und Menschenrechte', 528.

presented as piecemeal addendums to the conventional catalogue of civil and political rights. However, this historically well-founded scepticism has been largely superseded through the development of a nuanced doctrine on environmental harm. The absence of widespread recognition of this shift in many scholarly accounts arises primarily from entrenched narratives about significant 'landmark' environmental cases and the superficial examination of a comparatively small sample of decisions.[4]

Even authors claiming to provide a comprehensive overview fall well short of recognising the over 300 cases identified in the dataset.[5] As a result, even the more critical and in-depth examinations lack an appreciation of the full doctrinal context.[6] This is because the assessment of what ought to count as an important case is based on an incomplete sample and is therefore liable to misrepresenting doctrinal significance.[7]

When instead conducting a comprehensive evaluation of all available cases, one is struck by the fact that the vast majority of 'landmark' cases do not appear out of nowhere: they have a rich developmental history and there is a broader context of cases that is crucial to a sound doctrinal appraisement. Acknowledging the predecessors and evaluating the effects on subsequent decisions is not just a matter of accurate record-keeping but promises a clearer and more accurate picture of the law. This is where a methodology based on both doctrinal and data analysis of decisions can close a significant knowledge gap. This approach situates cases in their proper context and thus provides fresh insights that stimulate critical engagement with conventional interpretations. With this methodological background established, the following chapters investigate the extent to which the environmental minimum fits with the prevailing doctrinal approach of the ECHR to environmental harm.

4 Cursory engagement is common and notable especially in otherwise comprehensive treatises, see, for instance, Schabas, *The European Convention on Human Rights*, 387–8; Alston and others, *International Human Rights*, 1536; De Schutter, *International Human Rights Law*, 372–3, 457; Smith, *Textbook on International Human Rights*, 395, remarkably does not mention the environmental case law, nor do Batekas and Oette, *International Human Rights Law and Practice*, 516–7; Amos, *Human Rights Law*, 425, 508–9, 515.

5 See the dataset gathered for this book.

6 See, for instance, Gearty, 'Do Human Rights Help or Hinder Environmental Protection?', 18 dismissing the potential for a 'green' reading of the Convention on the basis of four environmentally themed cases; similar to Churchill, 'Environmental Rights in Existing Human Rights Treaties', 89.

7 For an in-depth case study of this phenomenon under the ECHR, see Stefan Theil, 'Excavating Landmarks – Empirical Contributions to Doctrinal Analysis' (2020) 32 *Journal of Environmental Law* 221.

Overall, there is a significant degree of compatibility between the environmental minimum and the doctrinal position taken by the ECtHR. With some notable exceptions, the framework fits with the case law on specific risks, minimum standards, and the high-priority principle. Certainly, many will remain sceptical. It is not widely accepted that human rights and environmental protection are sufficiently compatible to allow for a comprehensive and practically viable framework. As we have seen, scepticism often centres on the vagueness of environmental human rights as well as the conflicts between rights that they might bring about. Environmentalists, for their part, may lament that the application process before the ECtHR continues to rely on individuals and excludes harms from protection that cannot be directly linked to humans. Contrary to this scepticism, the rich case law of the ECHR can be incrementally developed towards greater environmental protection in line with existing commitments of states and without radical departures from established legal categories and principles. The following chapters are modelled on the principles of the environmental minimum and highlight notable decisions that both support and draw the compatibility with the framework into question.

Chapter 4 commences with the elements of the specific risk principle and introduces the doctrinal data analysis used to determine whether the ECHR generally recognises all environmental harm as a threat to human rights. Chapter 5 then examines whether a reasonable hypothesis of harm is generally sufficient to establish victim status under the Convention. Chapter 6 probes the doctrinal position on the minimum standards put forward by the environmental minimum, specifically whether the ECtHR recognises the crucial importance of enforcing legal standards and following established and emerging guidelines to determine human rights violations. Chapter 7 then turns to the high-priority principle, which interrogates the ECHR approach to balancing Convention rights with competing interests and notably discusses economic justifications and the margin of appreciation doctrine.

4

Specific Risk Principle

This chapter introduces the doctrinal data analysis to explore the compatibility of the ECHR with the specific risk principle of the environmental minimum. The specific risk principle is the trigger of the framework, determining whether the minimum standards of review apply to a given environmental harm. The principle requires that an environmental harm is successfully linked to a threat to a specific human right recognised under the invoked protection regime.

The primary question for this chapter is whether the ECtHR accepts that an environmental harm can give rise to an infringement of a recognised Convention right. Compatibility can be determined through the extent the prevailing doctrine subscribes to two key claims: (1) no environmental harm is excluded a priori from protection under the Convention and (2) under pertinent case law, environmental harm has been successfully linked to infringements of Convention rights.

Despite the generally open-minded approach of the ECHR, it must be acknowledged that the Convention only demands limited environmental protection. If the environmental minimum were applied, a much greater degree of environmental protection would be afforded than is typically accepted both in academic scholarship and the case law of the Court. However, based on an appraisal of the case law trajectory and an exhaustive analysis of all cases, this chapter concludes that the environmental minimum is already compatible with the prevailing doctrinal positions to a significant extent, even before considering the possibility of amendments or additional protocols.

4.1 ENVIRONMENTAL CASES

4.1.1 *Data Analysis*

At this stage, it is not relevant whether the Court accepts that environmental harms can give rise to violations of Convention rights, a question the chapter addresses below.[1] Rather the question here is a narrow, empirical one: whether the Court is presented and substantively engages with a wide range of environmental cases. Following a review of the case law, the answer is clear. In principle, the Court does not exclude any environmental degradation or pollution from substantive consideration. Nonetheless, environmental cases only make up a fraction of total cases, and the number of environmental cases per thousand applications filed with the Court has declined considerably over the past two decades despite pervasive environmental harm, as visualised in Figure 4.1. Protocol 11 to the ECHR, which extended the right of individual applications to all states, led to a sharp rise in the number of applications filed as well as the time taken from initial application to final conclusion. That may explain some of the relative decline in the number of environmental cases, see Figures 4.2 and 4.3, respectively.

The sharp rise in time from initial application to final decision accounts for the limited data currently available for the years after 2011 in Figure 4.1. Currently, two-thirds of applications require at least four years before a final decision, and a majority will be resolved in eight years. It can therefore be projected that a majority of cases originating from between 2012 and 2018 are not be resolved before 2020 and 2026, respectively. This requires a degree of caution when interpreting the data, especially with respect to all data after 2014. It is unlikely a majority of cases from that year have been resolved at the time of writing and hence have likely not yet been reflected in the dataset. Therefore, whether the ECHR is indeed experiencing a strong ebb of environmental cases is difficult to determine, as Figure 4.1 currently suggests. It is equally possible that the figures average out and will remain significantly above the 0.2 case per thousand mark seen in previous years, with a potential increase in 2014 due to the overall dip in the number of applications filed, shown in Figure 4.2. In any case, in terms of absolute numbers, there are many environmental cases, and they generally deal with a broad array of environmental harms, without any discernible pattern of exclusion. The cases in category one (protection function), where applicants invoke the ECHR to seek protection from environmental harm, may be categorised as follows:

- *Noise pollution*: this sub-category includes any man-made noise pollution, defined as 'unwanted or undesirable sound, particularly if it is loud and disturbs people's communication, sleep, study, or recreation',[2] as well as vibrations, whether directly or indirectly caused by human

[1] See Section 4.2.
[2] 'Noise', in Park (ed.), *A Dictionary of Environment and Conservation*.

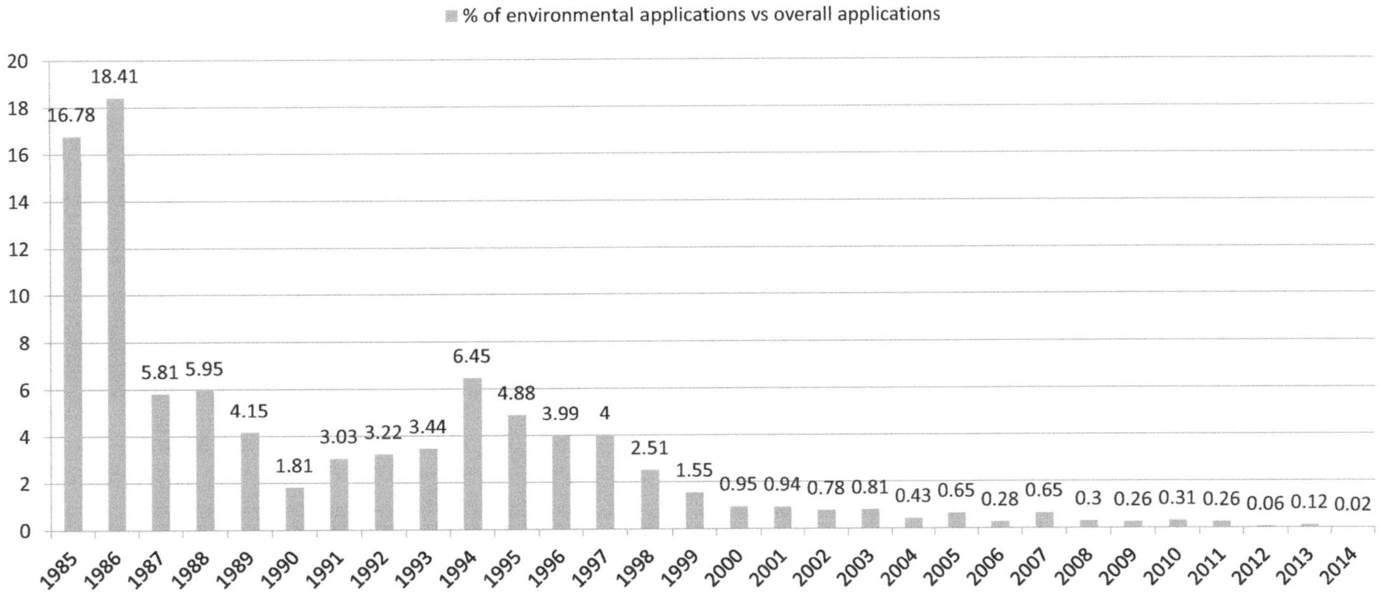

FIGURE 4.1 Number of environmental applications filed per thousand applications 1985–2014

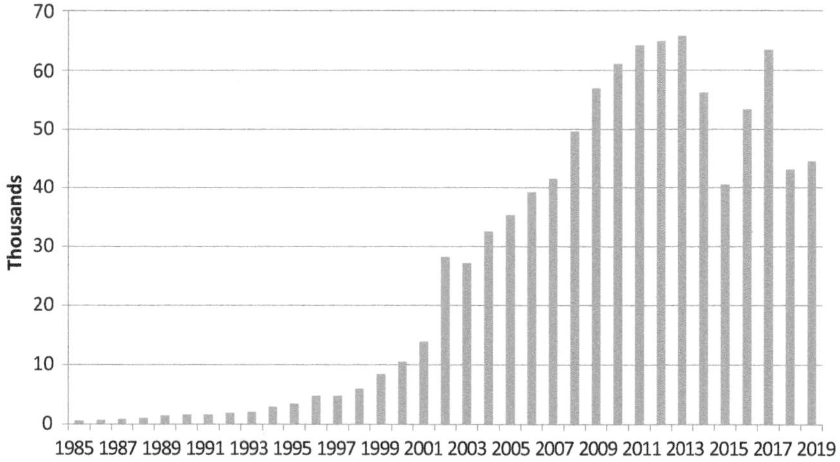

FIGURE 4.2 Total number of applications filed per year 1985–2019

actions. Noise and vibrations have been consolidated in the same category due to their similar physical characteristics. On a very basic level, sound is the interpretation given by the human ear and brain to specific vibrations in the air.[3] Typical cases involve complaints of airport noise in residential areas which affect daytime activities and sleep.[4] There are further examples involving noise emissions from a ferry service and nightclubs.[5]

- *Air pollution*: this sub-category includes any man-made pollution of the air and atmosphere, particularly through industrial activities. One prominent case in this category is *Fadeyeva*, which we consider in more detail later.[6]
- *Procedural*: this sub-category comprises all cases where applicants sought access to information on environmental projects and access to courts, as well as all cases where applicants invoked procedural violations in the course of public hearings and court proceedings. It is common to all of these cases that no substantive environmental claims were made, with the

3 Berglund and others, *Guidelines for Community Noise*.

4 *Arrondelle* v. *United Kingdom* App no 7889/77 (Commission, 13 May 1982); *Rayner* v. *United Kingdom* App no 9310/81 (Commission, 16 July 1986); *Baggs* v. *United Kingdom* App no 9310/81 (Commission, 16 October 1985); *Hatton and others* v. *United Kingdom* App no 36022/97 (Grand Chamber, 8 July 2003).

5 GA v. *Sweden* App no 12671/87 (Commission, 13 March 1989) [ferry service]; *Moreno Gomez* v. *Spain* App no 4143/02 (ECtHR, 16 November 2004); *Oluic* v. *Croatia* App no 61260/08 (ECtHR, 20 May 2010); *Udovicic* v. *Croatia* App no 27310/09 (ECtHR, 24 April 2014).

6 *Fadeyeva* v. *Russia* App 55723/00 (ECtHR, 9 June 2005).

applicants instead relying on procedural rights under the Convention arising from Articles 6 and 13 ECHR.[7]

- *Soil pollution*: this sub-category comprises the pollution and degradation of soil, as well as a failure to prevent unsustainable depletion of nutrients.[8] This includes the exploitation of natural resources through mining and extraction,[9] the construction and maintenance of earth works and dams, as well as the depositing of toxic substances. Additionally, it comprises the immediate consequences of these actions, such as landslides and flooding.[10]
- *Water pollution*: this sub-category primarily encompasses man-made pollution of water, defined as 'the pollution of fresh water (for example through eutrophication, acidification, and groundwater pollution), which decreases its purity and often makes it unsuitable for use as a water resource if not dangerous to human health.'[11] Typical cases involve the contamination or threat to ground water through heavy metals, chemicals, and biological agents.[12]
- *Radiation*: radiation is used as a general sub-category for radioactive pollution,[13] electromagnetic radiation,[14] and light disturbances.[15] Typical cases falling within this category concern themselves with nuclear radiation, particularly the risks involved with the military use and the civilian production of nuclear power, as well as the disposal of

7 *Kleyn and others* v. *Netherlands* App nos 39343/98; 39651/98; 43147/98; 46664/99 (ECtHR, 6 May 2003) [mistakes in environmental planning procedure]; *Zimmermann and Steiner* v. *Switzerland* App no 8737/79 (ECtHR, 13 July 1983) [length of proceedings in challenging pollution from Zurich airport].

8 Defined as 'the thin layer of disintegrated rock particles, organic matter, water, and air that covers most of the land surface to an average depth of about 20 centimetres', see 'Soil', in Park (ed.), *A Dictionary of Environment and Conservation*.

9 *Kjellberg* v. *Sweden* App no 13724/88 Commission (Commission, 5 March 1991).

10 See, for instance, the cases of *Budayeva and others* v. *Russia* App nos 15339/02; 21166/02; 20058/02; 11673/02; 15343/02 (ECtHR, 20 March 2008) and *G and E* v. *Norway* App no 9278/81; 9415/81 (Commission, 3 October 1983), respectively.

11 'Water', in Park (ed.), *A Dictionary of Environment and Conservation*.

12 *Dzemyuk* v. *Ukraine* App no 42488/02 (ECtHR, 4 September 2014) [*E.coli* bacteria]; *Atanasov* v. *Bulgaria* App no 12853/03 (ECtHR, 2 December 2010) [heavy metal contamination of groundwater].

13 Defined as 'contamination by a radioactive material', see 'Radioactive Pollution', in Park (ed.), *A Dictionary of Environment and Conservation*.

14 Defined as 'the energy that is transmitted through space in the form of electromagnetic waves, which include light, radio waves, X-rays, and gamma rays', see 'Electromagnetic Radiation', in Park (ed.), *A Dictionary of Environment and Conservation*.

15 *Fägerskiöld* v. *Sweden* App no 37664/04 (ECtHR, 26 February 2008).

spent fuel rods and the subsequent handling of nuclear waste.[16] Electromagnetic radiation is typically a concern for applicants living in the vicinity of cell phone towers and antennas.[17]

- *Flora and fauna*: this sub-category encompasses complaints over harms to plants and wildlife. Typical cases revolve around hunting rights.[18] Applicants also attempt to enjoin deforestation for the purposes of housing construction,[19] protect marshland for the purposes of preserving natural beauty,[20] preserve wildlife diversity in swampland,[21] and the entirety of a regional ecosystem.[22]
- *Amenity*: typically, applicants in these cases allege that changes in the natural environment, usually through construction, have caused a loss of visual amenity.[23]
- *Other*: this sub-category covers all cases that do not readily fit into any of the preceding categories and where the environmental nature of the complaint is at times debatable. Applicants, for instance, claimed general diminishments to their environment without specifying this in more detail,[24] a failure of authorities to step in during a neighbourhood dispute involving the dumping of manure on the applicant's doorstep,[25] danger from a live electrical wire left hanging into a river after a storm,[26] and falling debris from a yearly firework display.[27]

16 *Balmer-Schafroth and others* v. *Switzerland* App no 22110/93 (ECtHR, 26 August 1997); *L, M, R* v. *Switzerland* App no 30003/96 (ECtHR, 1 July 1996).

17 *Gaida* v. *Germany* App no 32015/02 (ECtHR, 3 July 2007); *Tudor* v. *Romania* App no 42820/09 (ECtHR, 3 June 2014); *Luginbühl* v. *Switzerland* App no 42756/02 (ECtHR, 17 January 2006).

18 *Chassagnou and others* v. *France* App nos 25088/94; 28331/95; 28443/95 (Grand Chamber, 29 April 1999); *Schneider* v. *Luxembourg* App no 2113/04 (ECtHR, 10 July 2007); *Herrmann* v. *Germany* App no 9300/07 (Grand Chamber, 26 June 2012).

19 *Ogloblina* v. *Russia* App no 28852/05 (ECtHR, 26 November 2013).

20 *X and Y* v. *Germany* App no 7407/76 (Commission, 13 May 1976).

21 *Kyrtatos* v. *Greece* App no 41666/98 (ECtHR, 22 May 2003).

22 *Jonsson and others* v. *Iceland* App no 41242/98 (ECtHR, 21 October 1998); see also similar attempts at protecting a nature reserve, *Ünver* v. *Turkey* App no 36209/97 (ECtHR, 26 September 2000).

23 *M* v. *Italy* App no 14563/89 (Commission, 7 October 1991) [building code exemptions for USSR Consulate in Milan]; *Ortenberg* v. *Austria* App no 12884/87 (ECtHR, 25 November 1994) [construction of a terrace adjacent to the applicant's home]; *L'Association des Amis de St-Raphael et Fréjus and others* v. *France* App no 45053/98 (ECtHR, 29 February 2000) [opposition to the construction of tourist apartments].

24 *Olsson* v. *Sweden* App no 22977/93 (Commission, 7 September 1995); *Berglund and others* v. *Sweden* App no 24191/94 (Commission, 19 October 1995); *Gounaridis, Iliopoulos, Pappostolou* v. *Greece* App no 41207/98 (ECtHR, 21 October 1998).

25 *Surugiu* v. *Romania* App no 48995/99 (ECtHR, 20 April 2004).

26 *Fedina* v. *Ukraine* App no 17185/02 (ECtHR, 2 September 2010).

27 *Maempel and Others* v. *Malta* App no 24202/10 (ECtHR, 22 November 2011).

The empirical data speaks to the general admissibility of environmental cases. This is important for the purposes of the environmental minimum, as a central claim is that threats to human rights must be treated equally: human dignity and autonomy require protection regardless of whether the threat comports with traditionally accepted threat scenarios. The next section will examine whether this initially liberal approach carries through to a successful connection with individual provisions of the Convention.

4.1.2 *Convention Articles*

A central claim that underlies the environmental minimum is that, on some level, environmental harm poses a threat to all human rights. Therefore, a green interpretation is key to the specific risk principle, as it demonstrates that the ECHR not only engages with environmental cases but accepts that it may violate Convention rights. At this stage, it is not immediately relevant whether the protection supports all the elements of the environmental minimum. At an empirical level, a closer look at the different categories of cases reveals that the ECtHR consistently finds violations. Of the 371 environmentally themed cases more than half (213) dealt with complaints under category one. The median violation rate of all cases in category one stands at 33 per cent, and Figure 4.3 provides an overview of the number of violations by category of environmental harm.

The doctrinal movement towards a more open approach to environmental harm is borne out by the historic overview of the number of cases and violations found per given year from 1960 to 2014 (Figure 4.4). The discrepancy in the figure between applications filed and decisions issued arises

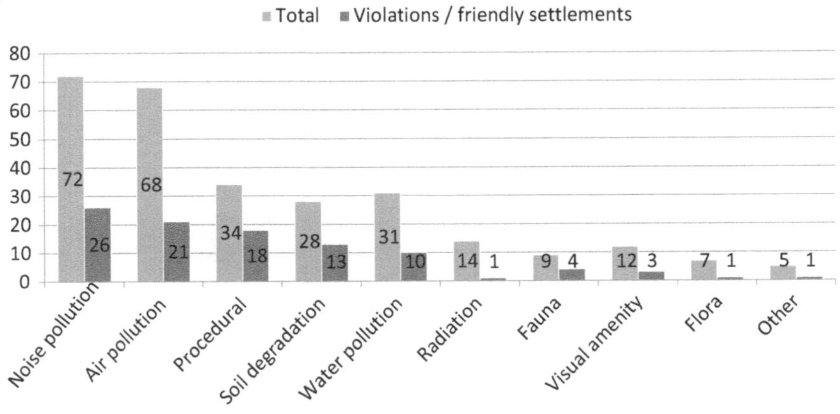

FIGURE 4.3 Environmental issues of cases in category 1

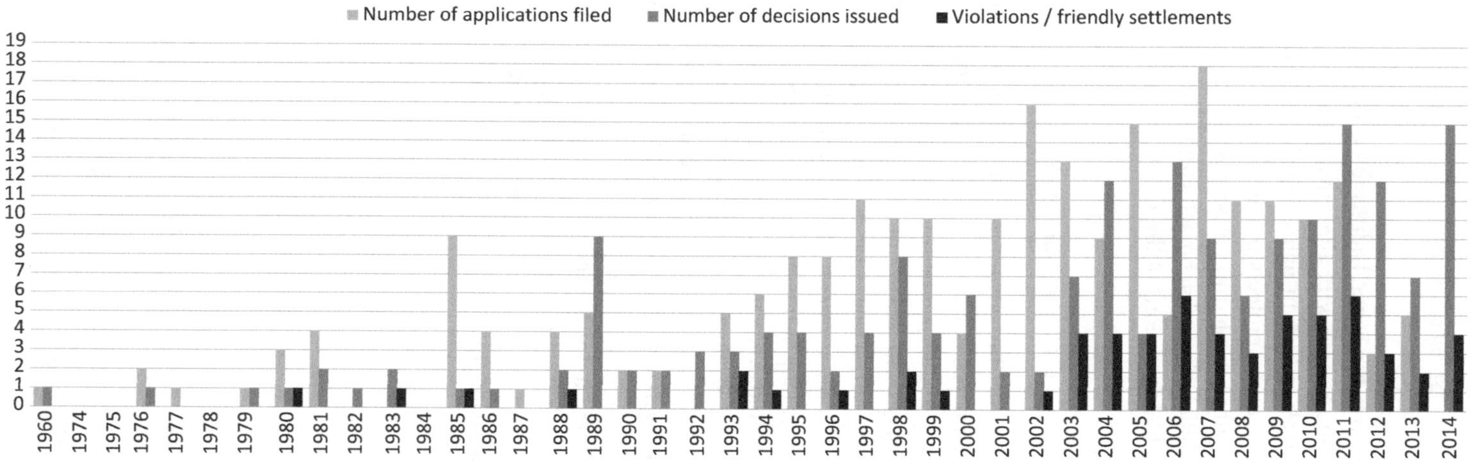

FIGURE 4.4 Number of environmental cases and violations per year in category 1 1960–2014

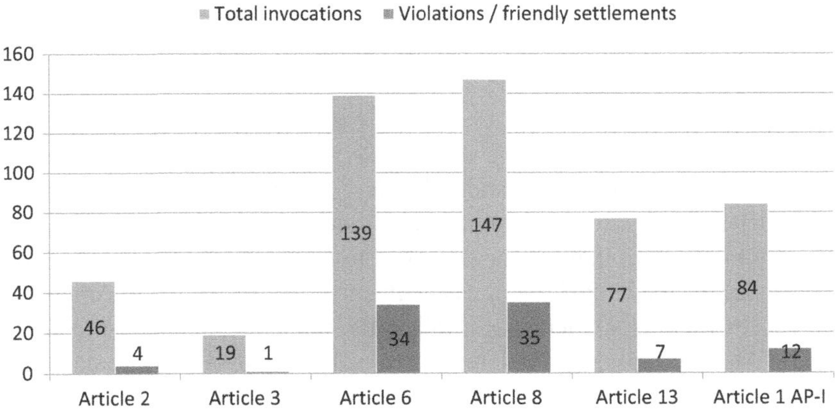

FIGURE 4.5 Number of successful invocations by Article in category 1

because the ECtHR routinely joins applications together into one decision: therefore, one decision may resolve multiple applications that were originally filed independently.

Representation of the data ends in 2014 for reliability purposes related to the time taken to reach final decisions and the associated problems, outlined in Chapter 1 and the explanatory notes in the Appendix.

When environmental harm is considered under the ECHR, the Court typically engages with Articles 2 (Right to Life), 8 ECHR (Private Life), and 1 of Additional Protocol I (AP-I) (Property). Articles 6 (Fair Trial) and 13 ECHR (Effective Remedy) are often invoked but will not be considered separately as they deal with procedural safeguards that are not of primary relevance to the environmental minimum. Figure 4.5 provides an overview of the number of times an Article was invoked and the number of corresponding violations.

However, these figures alone cannot answer the question on what basis a particular decision was reached. The findings may be based on an innovative 'green' reading, or on a piecemeal interpretation of Convention rights that does not recognise the environmental harm as significant in its own right. Likewise, the Court might dismiss an application on a consistent 'green' doctrine or alternatively because it is unwilling to expand the protection of the Convention to all environmental harms. Data analysis cannot replace in-depth doctrinal interrogation: both approaches work best in tandem with delineated but overlapping spheres of inquiry.

4.2 DOCTRINAL ANALYSIS

At the outset, it appears astonishing that Article 2 ECHR has not risen to greater prominence in environmental cases. This phenomenon can partially be accounted for by the relationship between Articles 2 and 8 ECHR in the environmental case law. The Court appears to have established Article 2 ECHR as *lex specialis* to Article 8 ECHR in environmental cases. Should the applicant be deceased or exposed to a serious, immediate threat to life, the Court will typically engage with Article 2 ECHR. Failing this threshold, applications are considered under Article 8 ECHR only, which is initially broader in scope but where infringements are easier to justify. Article 1 AP-I (Right to Property) stands somewhat apart from the other Articles. It arguably plays only a very limited role in category one, as whenever either Article 2 or 8 ECHR is engaged, there is generally little Article 1 AP-I adds to the discussion. A curious exception is found in several cases involving the mandatory transfer of hunting rights by property owners, which we discuss in Section 4.2.2.3.3.

The section begins with a comprehensive analysis of early environmental cases to provide a solid grounding for the evaluation of the modern doctrinal position. The historical development of the Convention towards recognising environmental harm is best illustrated through five cases identified through the comprehensive data analysis. Although these cases often receive little attention in contemporary scholarship, they laid important groundwork, notably for the decision in *López Ostra*, a ruling that is commonly seen as pivotal to environmental protection under the ECHR.

In part *López Ostra* gained much attention due to its factual circumstances. The applicant lived in a city with a high density of factories specialising in the treatment of leather. A waste treatment plant had begun operating in 1988 in only 12 metres distance from the applicant's home, without a licence and contrary to domestic law. Several incidents at the plant led to significant contamination, constituting severe public health problems through gas fumes and 'pestilential smells.'[28] Local residents, including the applicant, were evacuated and the plant was ordered to partially shut down. The remaining operation continued to pose significant health risks, as evidenced by expert opinions in the years from 1991 through 1993.[29] The applicant complained that the inaction of local authorities violated her rights under Articles 3 and 8 ECHR. The government argued that the environmental pollution endured by the applicant had been exaggerated.[30] The ECtHR rejected this view, particularly as the emissions exceeded the domestic legal limits and posed a substantial threat to human health. The Court noted that the state had failed to strike

28 *López Ostra* v. *Spain* App no 16798/90 (ECtHR, 9 December 1994) [8].
29 Ibid. [9].
30 Ibid. [40].

a fair balance between the economic well-being of the town and the respect for the applicant's rights, resulting in a violation of Article 8 ECHR.[31]

There is widespread agreement in legal scholarship that the decision in *López Ostra* was highly significant. Most general environmental and international law scholars argue that it broke new ground for environmental protection under the Convention,[32] and most specialised scholarship gives the case significant attention.[33] This level of attention admittedly has some merit: *López Ostra* is an important piece of the modern doctrinal approach and was influential as the first case to find a violation of Article 8 ECHR. Since the decision was handed down, *López Ostra* has been cited in 60 subsequent environmental cases. This makes the ruling stand out as the average number of subsequent citations for environmental decisions is only approximately 4.46.[34] However, it is worth looking beyond the mere frequency of references for deeper evidence of doctrinal significance.

As the section demonstrates, the significance of *López Ostra* has generally been overstated: both in terms of its historic role in bringing about changes to environmental doctrine and its enduring relevance.[35] *López Ostra* reflects three principles that were first established in earlier environmental cases:

- environmental harm as such is relevant under the Convention;
- the state is responsible for the actions of private industry; and
- Article 8 ECHR is generally applicable to environmental cases.

31 Ibid. [55]–[58].

32 Dupuy and Viñuales, *International Environmental Law*, 307; Sands and others, *Principles of International Environmental Law*, 783; Beyerlin and Marauhn, *International Environmental Law*, 400.

33 Karen Morrow, 'The ECHR, Environment-Based Human Rights Claims and the Search for Standards' in Stephen J. Turner and others (eds.), *Environmental Rights: The Development of Standards* (Cambridge University Press 2019) 43; Dinah Shelton, 'Human Rights and the Environment: Substantive Rights' in Malgosia Fitzmaurice, David M. Ong, and Panos Merkouris (eds.), *Research Handbook on International Environmental Law* (Edward Elgar 2010), 275–6; Boyle, 'Human Rights and the Environment', 204; Chris Hilson, 'The Margin of Appreciation, Domestic Irregularity and Domestic Court Rulings in ECHR Environmental Jurisprudence: Global Legal Pluralism in Action' (2013) 2 *Global Constitutionalism* 262, 270; Giorgetta, 'The Right to a Healthy Environment', 179; Garcia San José, *Environmental Protection and the European Convention on Human Rights*, 25, 59 [assumes it was the first case and informed all subsequent ones]; Richard Desgagné, 'Integrating Environmental Values into the European Convention on Human Rights' (1995) 89 *American Journal of International Law* 263, 282.

34 See the dataset gathered for this book.

35 The section is based in part on the in-depth evaluation of *López Ostra* conducted in Theil, 'Excavating Landmarks, 221.

The doctrinal position that an applicant need not demonstrate specific health detriments is also often attributed to *Lopéz Ostra* but was likewise foreshadowed, if not expressly recognised, by the decisions in the cases of *Rayner*, *Baggs*, and *Arrondelle*.[36]

4.2.1 *Early Environmental Cases*

4.2.1.1 The First Environmental Case

The first, albeit seldom recognised, environmentally themed case under the Convention was *Dr S* in 1960.[37] It lends early credence to the argument that environmental issues are irrelevant under the Convention, a view that would generally hold over the following decades.

In its decision, the Commission rejected any form of environmental protection under the ECHR regardless of the particular circumstances. The applicant complained of threats arising from nuclear technology under Articles 2 and 5 ECHR, namely the dumping of nuclear waste in the North Sea,[38] the stockpiling of nuclear material and the stationing of ballistic missiles in Germany. The Commission summarily dismissed the application, holding that '(. . .) neither the examination of the allegations of the applicant, nor the direct examination of the case-file as it stands, disclose the appearance of a violation (. . .).'[39]

From the brief decision, it is not immediately clear on which grounds the Commission is dismissing the application. One possibility is to interpret the decision as a comment on the applicant's victim status under the Convention. Had that been the case however, the Commission would have at least accepted that there was a case to answer under the Convention. The question of victim status can logically only arise if it is acknowledged that, for instance, improper handling of nuclear waste engages the Convention in the first place. Instead, it would appear that the Commission held that the facts as stated

36 *Rayner* v. *United Kingdom* App no 9310/81 (Commission, 16 July 1986), 12; *Baggs* v. *United Kingdom* App no 9310/81 (Commission, 16 October 1985), 20; *Arrondelle* v. *United Kingdom* App no 7889/77 (Commission, 13 May 1982).

37 *Dr S* v. *Germany* App no 715/60 (Commission, 5 August 1960, unpublished) on file with author.

38 A practice outlawed through the Convention on the Prevention of Marine Pollution by Dumping of Wastes and Other Matter (adopted 29 December 1972, in force 30 August 1975) 1046 UNTS 120 and the 1996 Protocol to the Convention on the Prevention of Marine Pollution by Dumping of Wastes and Other Matter (adopted 7 November 1996, in force 24 March 2006) 36 ILM 1 (1997).

39 *Dr S* v. *Germany* App no 715/60 (Commission, 5 August 1960, unpublished).

cannot give rise to a Convention violation under any circumstances, presumably because it seeks environmental protection not available under the Convention.

4.2.1.2 A Statement of Principle

The case of *X and Y* was one of the first opportunities for a reconsideration of this doctrinal position. The case concerned objections to destructive military activity on marshlands neighbouring the applicants' property.[40] The Commission considered Articles 2, 3, and 5 ECHR and ultimately held that '(. . .) no right to nature conservation is, *as such* included among the rights and freedoms guaranteed by the Convention and in particular by Arts. 2, 3 or 5 as invoked by the applicant.'[41]

The decision by the Commission is again quite brief. It remains unclear whether the Commission declined to engage with Article 8 ECHR, or if the applicants failed to invoke the provision. Whatever the reason, the decision by the Commission amounted to the first unequivocal statement of a legal principle to the effect that the Convention does not protect the environment. While this principle appears uncompromising at first glance, it notably does not close the door on environmental harm that affects the enjoyment of Convention rights. A more thoughtful decision may have taken the opportunity to grapple with this question in more detail.

In any case, *X and Y* did not have an enduring impact on doctrine beyond the very narrow idea that 'pure' nature conservation cases are not admissible under the Convention. The Court continues to invoke this principle, but it remains a largely rhetorical starting point rather than doctrinally significant. There are two reasons for this: for one, the vast majority of environmental cases before the ECtHR contain express allegations of harm to individual applicants and second, there are inherent difficulties in separating harm to nature from harm to humans on a principled basis.[42] Deep ecologists have convincingly argued since the early 1970s that there is, in fact, no meaningful distinction: both humans and their environment form part of the same global ecosystem and there is at most a separation by degrees and intensity of harm

[40] *X and Y* v. *Germany* App no 7407/76 (Commission, 13 May 1976).
[41] Ibid. (emphasis added).
[42] The number of cases where the principle made a discernible difference are exceedingly rare, see *Schiopu and Verzescu* v. *Romania* App no 26040/06 (ECtHR, 26 November 2013) [failure of authorities to remove oil tanks on land returned after expropriation no violation]; *Ogloblina* v. *Russia* App no 28852/05 (ECtHR, 26 November 2013) [dismissed application challenging deforestation].

but not a distinction in principle.[43] Thus, even while rejecting environmental protection for its own sake under the Convention, the Court routinely engages with the substance of applications that have a tentative connection to individual harm and has no trouble in finding violations of the Convention.[44]

4.2.1.3 Airport Noise Emission Cases

Several decisions soon qualified the somewhat sceptical view taken by the Commission in *X and Y*, especially a string of cases related to airport noise emissions in the early 1980s. One key reason for this gradual shift is changes to the interpretational doctrine in the late 1970s. The Court adopted a so-called evolutive approach to interpretation, which was first applied in the cases of *Tyrer* and *Marckx*.[45] This approach broadly holds that the Convention rights are subject to evolution and change over time. Dubbed the 'living instrument' approach, it has led the ECtHR to adapt the Convention rights to keep pace with modern developments and changes in the prevailing societal attitudes.[46] In its own words, the Court thus intends to provide for an interpretation of the Convention that '(...) upholds individual rights as practical and effective, rather than theoretical and illusory protections.'[47] This allowed the ECtHR to consider environmental threats to human rights within the purview of the traditional civil and political rights of the Convention.

In the case of *Arrondelle* the runway of Gatwick Airport was originally located approximately one mile from the applicant's home.[48] Both the runway and the M23 motorway were extended in the years preceding the application and encroached to within a few hundred feet of the home. An independent inspector determined that this led to '(...) intolerable stress by reason of the intensity, duration and frequency of the noise primarily from low flying aircraft passing almost overhead.'[49] Not surprisingly, attempts to sell the

43 Naess, 'The Shallow and the Deep, Long-Range Ecology Movement'.

44 *Kyrtatos* v. *Greece* App no 41666/98 (ECtHR, 22 May 2003) [successful application against destruction of swampland]; *Schneider* v. *Luxembourg* App no 2113/04 (ECtHR, 10 July 2007) [successful application based on ethical opposition to hunting]; *Andersson and others* v. *Sweden* App no 29878/09 (ECtHR, 25 September 2014) [successful challenge against infrastructure project in nature reserve].

45 *Tyrer* v. *UK* App no 5856/72 (ECtHR, 25 April 1978); *Marckx* v. *Belgium* App no 6833/74 (ECtHR, 13 June 1979).

46 Letsas, 'The ECHR as a Living Instrument'; Theil, 'Is the "Living Instrument" Approach'.

47 *Mamatkulov and Askaraov* v. *Turkey* App no 46827/99; 46951/99 (Grand Chamber, 4 February 2005) [121].

48 *Arrondelle* v. *United Kingdom* App no 7889/77 (Commission, 13 May 1982).

49 Cited as per ibid.

property proved futile, and the applicant invoked Articles 6, 8, 13, 14 and Article 1 AP-I before the Commission. The case was ultimately settled during the course of these proceedings.[50] The local council agreed to purchase the home at its current market value and the difference to the unblighted value of the property was paid by the UK Government. While the settlement resolved the case in question, the underlining legal issues with airport noise emissions remained salient.

The facts in *Baggs* were largely similar to those in *Arrondelle*, although the noise levels endured by the family were considerably worse and originated from Heathrow Airport.[51] The independent Eyre Report on Heathrow Air Noise summarised the situation as follows:

> The conditions which Mr. and Mrs. Baggs and their family have been forced to endure for years without respite or redress are truly shocking and deplorable. Of necessity, the Baggs spend many hours in the open air. There is no escape when they go indoors. Insulation is ineffective in their well built bungalow. The Baggs will have to serve out their lives in these intolerable conditions which the introduction of less noisy aircraft will never wholly eradicate. Nobody wishes to buy their hell and I am not surprised.[52]

The applicant invoked Articles 8 and 13 ECHR as well as Article 1 AP-I. The UK government argued that it was not responsible for the noise emissions caused through the conduct of third-party aircraft operators. The Commission instead deemed the direct cause of the noise emissions immaterial and held the state responsible by way of its planning and regulatory authority. The proceedings in *Baggs* were likewise settled, but not before being joined with two other applications considered by the Commission: *Powell* and *Rayner*. The latter two applications proceeded to the Court, but as the claims under Article 8 ECHR had been previously dismissed by the Commission this aspect was not examined further.[53] Nonetheless, the Commission decisions in both *Powell* and *Rayner* acknowledged that environmental issues can give rise to infringements under Article 8 ECHR, but ultimately deemed them justified under paragraph 2.[54] Collectively, the airport noise cases were important precursors to the decision in *Lopéz Ostra* in three respects.

50 Ibid. [4].

51 *Baggs* v. *United Kingdom* App no 9310/81 (Commission, 16 October 1985).

52 Inquiry headed by the then Inspector for the London Airports, Sir Graham Eyre, from 1981 to 1984, which published its report in December 1984, cited as per ibid. 3.

53 *Powell and Rayner* v. *United Kingdom* App no 9310/81 (ECtHR, 21 February 1990) [25].

54 *Rayner* v. *United Kingdom* App no 9310/81 (Commission, 16 July 1986), 12.

First, they represent the first cases where the Commission engages in detail with environmental harm under the Convention. While *Baggs* and *Arrondelle* were settled before a definite ruling was issued, and the claims of *Powell* and *Rayner* were ultimately not upheld, the cases nonetheless ensured that environmental concerns became firmly established as relevant and potential infringements of Convention rights. The dismissive attitude in *Dr S* and *X and Y* was overcome and states began to take notice. They could no longer rely on an outright refusal to hear environmental cases and at least some friendly settlements appear to be motivated by a desire to avoid an adverse precedent.

Second, in all airport cases, states were consistently held responsible for environmental harm arising from the conduct of third parties. It was largely immaterial that the aircraft were operated by private airlines: states have oversight and regulatory authority. Before the aircraft noise cases, this was far from a settled view. Indeed in *Rayner*, the Commission cites *Marckx* as the sole authority for the proposition that states are directly responsible for private actors under the Convention.[55] However, *Marckx* does not deal with environmental harm and is better understood as supporting the more general notion of positive obligations arising from Article 8 ECHR: that is to say, the idea that the state has certain regulatory obligations with respect to the conduct of private actors. The case of *Marckx* was not concerned, nor did it assign direct responsibility for individual actions of private actors by virtue of these general regulatory obligations.

Third, the airport cases held Article 8 ECHR admissible and accepted, to varying degrees, that infringements could arise from environmental harm. Even if no violations were ultimately found, these unprecedented decisions opened the door to the modern environmental protection under the Convention. Notably absent from all of the rulings was any legal requirement for specific health detriments to the applicants as a result of the aircraft noise. Collectively, these cases of course do not diminish the significance of *López Ostra* as the first case expressly finding a violation through environmental harm. However, in the end, it was likely the strategic decision to settle that prevented earlier findings of violations in the airport noise emission cases.

4.2.1.4 Beyond Airport Noise

The case of *G and E v. Norway* expanded and confirmed the doctrinal approach to environmental harm beyond airport noise cases. In the decision,

[55] Ibid., 8, citing *Marckx v. Belgium* App no 6833/74 (ECtHR, 13 June 1979) [31].

the Commission accepted that even a lack of nature conservation with limited harm to individuals may constitute a violation of the Convention.[56] The applicants were two indigenous reindeer farmers who objected to the construction of a hydro-electric dam which threatened to flood land traditionally used by the Sami people.[57] The Commission accepted that this may constitute an infringement of Article 8 ECHR when read in conjunction with minority rights.[58] However, it held that

> (...) without ascertaining the exact extent and nature of the interference with the applicants' rights under Article 8, para. 1, after careful consideration of the necessity of the project by the national organs, the interference could reasonably be considered justified under Article 8, para. 2, as being in accordance with the law, and necessary in a democratic society in the interests of the economic well-being of the country.[59]

Drawing strong inferences from such cautiously worded decisions is difficult. However, it appears that the Commission accepted that the environmental harm amounted to an infringement of the Convention, which in turn could give rise to violations given appropriate circumstances. *G and E* therefore arguably constitutes a further early example of Article 8 ECHR proceeding to the justification stage. In any case, the ruling confirms two of the principles established in the airport noise cases: environmental harm is relevant to the Convention and constitutes a potential source of violations, particularly of Article 8 ECHR. This apparently holds even if there is no tangible health risk to the applicant, and even where the emphasis of the environmental harm is on nature conservation rather than discrete individual detriments.

The scepticism towards environmental harm expressed in earlier cases appears short-lived. The ECtHR engages with the substance of the application, even where the link is tentative and nature conservation takes a leading role in the application, overshadowing alleged harm to individuals.

4.2.1.5 Conclusion

As we have seen, the airport noise emission cases laid important groundwork for the decision in *López Ostra*. They set the precedent for three important

56 *G and E* v. *Norway* App no 9278/81; 9415/81 (Commission, 3 October 1983).
57 The members of the 'indigenous people of the extreme north of Scandinavia, traditionally associated with the herding of reindeer', see *Oxford Dictionary* (Oxford University Press 2018).
58 *G and E* v. *Norway* App no 9278/81; 9415/81 (Commission, 3 October 1983), 3.
59 Ibid., 7.

principles that later formed the foundation for the decision and the modern doctrinal approach to environmental harm more generally: (1) environmental harm may constitute an infringement of the Convention; (2) states can be held responsible for the environmental harm caused by third parties under their regulatory control; and (3) Article 8 ECHR is particularly relevant to evaluating environmental harm under the Convention. The specific innovations attributed to *López Ostra* appear more ambivalent given this doctrinal context, but especially so when considering the modern environmental interpretation of the Convention.

4.2.2 *Modern Interpretation*

4.2.2.1 Article 2

The ECtHR tends to handle environmental claims under Article 2 ECHR restrictively. Even though Article 2 ECHR is generally interpreted as encompassing a positive obligation on the state to take necessary actions to protect human life, the Court has been reluctant to engage with the provision, particularly in the context of environmental harm.[60] Partially, this may be because infringements of Article 2 ECHR can only be justified in defence of any person from unlawful violence (paragraph 2 lit. a), in the context of lawful arrest or to prevent escape of a lawfully detained individual (paragraph 2 lit. b) and in action lawfully taken to quell a riot or insurrection (paragraph 2 lit. c). Environmentally themed cases are unlikely to engage any of these exceptions and consequently, the dataset records no cases where an infringement of Article 2 ECHR was ultimately justified.[61] Article 2 thus risks becoming a comparatively blunt instrument that eschews the middle ground often sought by the Court. When the provision was invoked in category one, the ECtHR found a violation in only 8.7 per cent of environmental decisions. This limited number of cases has one common denominator: the Court requires the death or at least a mortal health risk to the applicant. All cases falling short of this threshold are typically considered under Article 8 ECHR, which operates as *lex generalis* to Article 2 in environmental cases.

[60] This is settled case law, see *LCB* v. *United Kingdom* App no 23413/94 (ECtHR, 9 June 1998) [36]; *Akkoc* v. *Turkey* App nos 22947/93; 22948/93 (ECtHR, 10 October 2000) [77]; *Osman* v. *United Kingdom* App no 23452/94 (Grand Chamber, 28 October 1998) [115]; *Dodov* v. *Bulgaria* App no 59548/00 (ECtHR, 17 January 2008).

[61] See the dataset gathered for this book.

The case of *Öneryildiz* is a rare example where the Court did not forego Article 2 in an environmental context.[62] The applicant lived in a dwelling constructed without official authorisation in the vicinity of a garbage dump site. The area around the site developed into a slum over the years.[63] Expert reports noted repeatedly that the site gave rise to major health risks for the inhabitants. In spite of these findings, garbage depositing and residency continued until a methane explosion rocked the site in 1993.[64] Refuse erupted, covering approximately 10 dwellings and leaving 39 people dead. The surviving applicant complained to the Court over the death of nine close relatives, invoking Article 2 ECHR, as well as the loss of his house and property under Article 1 AP-I.[65] The Court held that the authorities had failed to take even basic preventative measures, such as installing a gas extraction system. Moreover, the criminal proceedings against officials for apparent failures were deemed ineffective to secure full accountability.[66] Consequently, the Court found a violation of Article 2 ECHR and did not rule on Article 8 ECHR as it raised no separate issue.[67]

Thus, a violation of Article 2 ECHR appears only possible in cases of death or a mortal health risk. This view is further supported by cases where the court considered an environmental harm that exposed the applicant to a mortal health risk, but where that risk did not materialise, for instance in the case of *Kolyadenko*.[68] Following major rainfall, the authorities caused a flash flood by releasing large amounts of water from an upstream reservoir without any prior warning to the inhabitants of the village.[69] In this context, the Court confirmed that Article 2 ECHR covers '(...) not only situations where certain action or omission on the part of the State led to a death complained of, but also situations where, although the applicant survived, there clearly existed a risk to his or her life (...).'[70]

This contrasts with the handling of cases where the applicant invoked Article 2 ECHR but was not threatened with an immediate mortal health risk. In *Taskin* the applicants were neighbours of a gold mine which employed

62 *Öneryildiz* v. *Turkey* App no 48939/99 (Grand Chamber, 30 November 2004).
63 Ibid. [11] and [13].
64 Ibid. [18].
65 Ibid. [63] and [119].
66 Ibid. [117].
67 Ibid. [118]; similar to the approach taken in *Budayeva and others* v. *Russia* App nos 15339/02; 21166/02; 20058/02; 11673/02; 15343/02 (ECtHR, 20 March 2008) [160].
68 *Kolyadenko and others* v. *Russia* App nos 17423/05; 20534/05; 20678/05; 23263/05; 24283/05; 35673/05 (ECtHR, 28 February 2012).
69 Ibid. [130].
70 Ibid. [151].

cyanide leaching in the mining process.[71] When the Ministry of the Environment issued a permit for the continued operation of the mine, the applicants challenged this decision before domestic courts. They based their complaint on the risk cyanide posed to human health and the groundwater, as well as the destruction of local flora and fauna.[72] On appeal, the Supreme Administrative Court ruled in their favour, holding that the '(...) risk posed to the local ecosystem and to human health and safety by sodium cyanide use (...)' supported the conclusion that the permit was not in the public interest and that safety measures were insufficient.[73] Thus, the operating permit was annulled. In an apparent effort to circumvent the ruling, authorities nonetheless granted the mine leave to continue its operation by issuing administrative decisions not subject to judicial review.[74] This resulted in an effective circumvention of the domestic court order over a four-month period. Having determined a violation of Article 8 ECHR through the continued operation of the gold mine and use of cyanide, the ECtHR declined to examine the complaint under Article 2 ECHR, presumably because there was no mortal health risk.[75] The Court drew an identical conclusion in the related cases of *Öckan*, and *Lemke*, which also involved the use of cyanide in mining, and involved no mortal health risks or deaths.[76]

The apparent *lex specialis* relationship with Article 8 ECHR is further reflected in the more recent decision in *Brincat*.[77] The applicants worked in a state-owned shipyard and were exposed to asbestos for decades during their employment from the 1950s, up until approximately the year 2000. Information concerning the fatal health risks associated with asbestos exposure became publicly available in Malta at the latest in 1979 through a court case. This incident failed to prompt any change in the health and safety standards for workers, who were assured that ventilation and fabric masks provided sufficient protection.[78] Following the death of a colleague, the applicants underwent tests to determine whether their lungs showed signs of a malignant cancer linked to asbestos exposure (mesothelioma).[79] The results showed

71 *Taskin and others* v. *Turkey* App no 46117/99 (ECtHR, 10 November 2004).
72 Ibid. [23].
73 Ibid. [26].
74 Ibid. [62]–[63], [71], and [75].
75 Ibid. [140].
76 *Lemke* v. *Turkey* App no 17381/02 (ECtHR, 5 June 2007); *Öckan and others* v. *Turkey* App no 46771/99 (ECtHR, 10 November 2004).
77 *Brincat and others* v. *Malta* App no 60908/11; 62110/11; 62129/11; 62312/11; 62338/11 (ECtHR, 24 July 2014).
78 Ibid. [11].
79 Ibid. [12].

extensive damage to the lungs and significant health issues in all applicants. The applicants claimed that their exposure to asbestos and the failure of the government to protect their health constituted violations of Articles 2 and 8 ECHR.[80] The Court noted, however, that while the applicants all suffered from respiratory problems, they had not been diagnosed with malignant mesothelioma. Therefore, they could not conclusively show the life-threatening nature of their disease, nor prove a sufficient link to asbestos exposure.[81] The Court therefore dismissed the claim under Article 2 ECHR for all but one of the applicants, who had died as a result of asbestos exposure. With regard to the other applicants, the Court instead ruled that the state had failed to satisfy its obligations under Article 8 ECHR, which we now turn to consider in more detail.

4.2.2.2 Article 8

A holistic account of the case law demonstrates that Article 8 ECHR has been moulded into a de facto right to health, and that a myriad of cases jointly contributed to the modern doctrine.

The hallmarks of this modern approach are: (1) the broad scope of application; (2) the limited connection required to an applicant's private and family life; (3) the low minimum threshold of severity required to engage Article 8 ECHR; (4) the absence of a requirement that applicants demonstrate a direct causal link between an environmental harm and a detriment to their health, including taking possible future harm into account; and finally (5) the comparatively broad range of justifications available to the state.

4.2.2.2.1 SCOPE. The subject matters that fall within the scope of Article 8 ECHR are broad owing to a wide interpretation: the ECtHR has determined that a final and complete definition of 'private life' is not possible.[82] Among the areas touched upon beyond traditional privacy and family life concerns are bodily integrity and self-determination, both physical and psychological (insofar as it does not amount to a violation of Article 3 ECHR),[83] as well as medical treatment, pregnancy, sexuality, gender, and of course environmental

80 Ibid. [73].
81 Ibid. [84].
82 See *Van Kück* v. *Germany* App no 35968/97 (ECtHR, 12 June 2003) [69]; *PG and JH* v. *United Kingdom* App no 44787/98 (ECtHR, 25 September 2001) [56].
83 See, for instance, *Van Kück* v. *Germany* App no 35968/97 (ECtHR, 12 June 2003) [69]; *Juhnke* v. *Turkey* App no 52515/99 (ECtHR, 13 May 2008) [70].

protection.[84] Apart from protecting individuals from infringements through the state, the provision also contains positive obligations. These require the state to take appropriate measures to safeguard Article 8 rights from infringements by third parties.[85] Especially with respect to dangerous activities, states must put in place a regulatory framework that is adapted to specific activities and alleviates the risks they produce.[86]

4.2.2.2.2 CONNECTION. In order to engage Article 8 ECHR, the Court requires some connection to family and private life, although this is rarely decisive to a case.[87] For one, the Court does not appear to attach a great evidentiary burden on establishing this connection. This is because once a detriment, such as a health condition, is established and linked to an environmental harm, it becomes difficult to imagine how this would not in some way impact private and family life. Indeed, a multitude of cases exist where the connection to the family and private life has been tenuous, but the Court found no difficulty moving to the justification stage.[88] In the past, the ECtHR has required that an environmental harm manifests itself in the vicinity of an applicant's home. However, this is a feature of the overwhelming majority of environmental applications in any case, regardless of outcome.

There are also some signs in the case of *Otgon* that the court may be willing to forego that connection requirement entirely when an applicant suffers

[84] This is settled case law, see *Dudgeon* v. *United Kingdom* App no 7525/76 (ECtHR, 22 October 1981) 18–19; *B* v. *France* App no 13343/87 (Plenary, 25 March 1992) [63]; *Burghartz* v. *Switzerland* App no 16213/90 (ECtHR, 22 February 1994) [24]; *Pretty* v. *United Kingdom* App no 2346/02 (ECtHR, 29 April 2002) [61]; *Y F* v. *Turkey* App no 24209/94 (ECtHR, 22 July 2003) [33].

[85] This is settled case law, see *W* v. *United Kingdom* App no 9749/82 (ECtHR, 9 July 1988) [60], more recently, *Schüth* v. *Germany* App no 1620/03 (ECtHR, 23 September 2010) [55]; specifically on environmental harm, see the instructive case of *Arrondelle* v. *United Kingdom* App no 7889/77 (Commission, 13 May 1982); *Baggs* v. *United Kingdom* App no 9310/81 (Commission, 16 October 1985); *Guerra* v. *Italy* App no 14967/89 (Grand Chamber, 19 February 1998) [58] and *Hatton and others* v. *United Kingdom* App no 36022/97 (Grand Chamber, 8 July 2003) [98].

[86] *Bacila* v. *Romania* App no 19234/04 (ECtHR, 30 March 2010) [61]; *Öneryildiz* v. *Turkey* App no 48939/99 (Grand Chamber, 30 November 2004) [90].

[87] This is settled case law, see *Fadeyeva* v. *Russia* App 55723/00 (ECtHR, 9 June 2005) [68]–[69]; *Fägerskiöld* v. *Sweden* App no 37664/04 (ECtHR, 26 February 2008) [1]; *Borysiewicz* v. *Poland* App no 71146/01 (ECtHR, 1 July 2008) [51]; more recently in *Koceniak* v. *Poland* App no 1733/06 (ECtHR, 17 June 2014) [58].

[88] From the abundant case law, see *G and E* v. *Norway* App no 9278/81; 9415/81 (Commission, 3 October 1983) [indigenous farm land]; *Andersson and others* v. *Sweden* App no 29878/09 (ECtHR, 25 September 2014) [use of cyanide in a mining operation].

serious health detriments.[89] The case involved the contamination of drinking water through a ruptured sewage pipe. The Moldovan authorities had failed to adequately regulate the private actor responsible, who neglected to properly maintain and replace the pipes. A domestic court established the private actor's liability and awarded the applicant minimal damages. The ECtHR held that Article 8 ECHR nonetheless applied in this case because the applicants' '(...) physical integrity has been affected by an unhealthy environment' and found a violation because the damages awarded by the domestic court were insufficient.[90] This conclusion was specifically lamented in the sole dissenting opinion of Judge Lemmens, who suggested that a detriment to physical integrity was on its own incapable of engaging Article 8 ECHR without the applicant also demonstrating some impact on her private or family life.[91] The fact that the majority of the Court disagreed perhaps hints at a broader shift in the doctrinal approach.

4.2.2.2.3 MINIMUM THRESHOLD. The Court further requires that both the environmental harm and resulting detriment rise above a certain minimum threshold.[92] The requirement is not further specified in most decisions, with the Court stating only that '[t]he assessment of that minimum is relative and depends on all the circumstances of the case, such as the intensity and duration of the nuisance, and its physical or mental effects.'[93] While this may appear as a principle advocating a case-by-case analysis, on a broader view certain patterns begin to emerge.

The threshold is typically used by the Court to exclude applications it considers trivial, although such examples are rare. In the case of *Fägerskiöld*, the noise (between 42 and 45 dB) and light emission from wind turbines failed to meet the minimum threshold.[94] Nonetheless, the Court will hold

89 *Otgon* v. *The Republic of Moldova* App no 22743/07 (ECtHR, 25 October 2016).

90 Ibid. [15], [17], and [19].

91 Ibid., Dissenting opinion of Judge Lemmens [3].

92 This is settled case law, see *Fadeyeva* v. *Russia* App 55723/00 (ECtHR, 9 June 2005) [68]–[69]; *Fägerskiöld* v. *Sweden* App no 37664/04 (ECtHR, 26 February 2008) [1]; *Borysiewicz* v. *Poland* App no 71146/01 (ECtHR, 1 July 2008) [51]; more recently *Koceniak* v. *Poland* App no 1733/06 (ECtHR, 17 June 2014) [58].

93 *Gaida* v. *Germany* App no 32015/02 (ECtHR, 3 July 2007) [1]; this is settled case law, *see Fadeyeva* v. *Russia* App 55723/00 (ECtHR, 9 June 2005) [68]–[69]; *Fägerskiöld* v. *Sweden* App no 37664/04 (ECtHR, 26 February 2008) [1]; *Borysiewicz* v. *Poland* App no 71146/01 (ECtHR, 1 July 2008) [51]; more recently *Koceniak* v. *Poland* App no 1733/06 (ECtHR, 17 June 2014) [58].

94 *Fägerskiöld* v. *Sweden* App no 37664/04 (ECtHR, 26 February 2008) [1]; *Koceniak* v. *Poland* App no 1733/06 (ECtHR, 17 June 2014); *Borysiewicz* v. *Poland* App no 71146/01 (ECtHR, 1 July 2008) [56].

Article 8 engaged when there is at least some evidence on the harmfulness of an environmental harm (mostly to the health of applicants) and almost invariably when authorities fail to uphold domestic environmental regulations and court rulings.[95] In practice, the threshold is rarely decisive because the overwhelming majority of applicants make credible claims of serious detriments to their health through environmental harm.

4.2.2.2.4 HEALTH DETRIMENT. Although applicants often invoke detrimental health impacts, the ECtHR does not generally require much evidence to support these claims. This was an important contribution of *López Ostra* to the modern doctrine: it expressly recognised that Article 8 does not invariably require the applicants to demonstrate a direct causal link between the environmental harm and a detriment to their health.[96] This may at first appear odd, given that the chapter suggested that Article 8 operates as a right to health under the Convention. However, it is important to distinguish between the function and scope of a right, what it protects in practice, and what applicants must establish in order to successfully claim a violation. Generally, it is difficult, and at times impossible, to demonstrate that a specific health condition was caused by a specific environmental harm.[97] Therefore, the Court's approach to health detriment evidence is best understood as re-distributing the evidentiary burden between the individual and the state. In other words, the Court allows applicants to bring cases based on a reasonable hypothesis of harm which may be satisfied through future harm, particularly where the potential consequences are serious and irreversible.[98] This is an aspect of the case law that we investigate in more detail in Chapter 5. In brief, risks of future threats may give rise to a violation, particularly where the consequences are serious and irredeemable. This argument featured prominently in the case of *Taskin*, where the Court accepted that serious environmental harm may give

95 *Gaida* v. *Germany* App no 32015/02 (ECtHR, 3 July 2007), 11; this is settled case law: *Moreno Gomez* v. *Spain* App no 4143/02 (ECtHR, 16 November 2004) [57]–[58]; *Oluic* v. *Croatia* App no 61260/08 (ECtHR, 20 May 2010) [48]–[49]; *Deés* v. *Hungary* App no 2345/06 (ECtHR, 9 November 2010) [23].

96 This is settled case law, see *López Ostra* v. *Spain* App no 16798/90 (ECtHR, 9 December 1994) [51]; *Taskin and others* v. *Turkey* App no 46117/99 (ECtHR, 10 November 2004); *Fägerskiöld* v. *Sweden* App no 37664/04 (ECtHR, 26 February 2008) [1]; more recently see *Maempel and Others* v. *Malta* App no 24202/10 (ECtHR, 22 November 2011) [36] and *Koceniak* v. *Poland* App no 1733/06 (ECtHR, 17 June 2014) [57].

97 *Asan and others* v. *Turkey* App no 56003/00 (ECtHR, 31 July 2007) [69]; *Avsar* v. *Turkey* App no 25657/94 (ECtHR, 10 July 2001) [282]; *Ireland* v. *United Kingdom* App no 5310/71 (ECtHR, 18 January 1978) [161]; *Cyprus* v. *Turkey* App no 25781/94 (Grand Chamber, 10 May 2001) [112].

98 *Taskin and others* v. *Turkey* App no 46117/99 (ECtHR, 10 November 2004) [113] and [126].

rise to violations when the well-being of individuals is at stake, even if there is no immediate health risk involved.[99]

4.2.2.2.5 JUSTIFICATION. Any infringement of Article 8 ECHR naturally remains open to justification, provided that the interference is

> (...) in accordance with the law and is necessary in a democratic society in the interests of national security, public safety or the economic well-being of the country, for the prevention of disorder or crime, for the protection of health or morals, or for the protection of the rights and freedoms of others.[100]

In practice, the Court does not draw a sharp distinction between state action and positive obligations in the context of Article 8 ECHR. It accepts that the elements of the test are broadly similar,[101] and in practice the distinction has little relevance:

> The applicable principles are broadly similar whether the case is analysed in terms of a positive duty on the State to take appropriate measures to secure the applicants' rights or, conversely, in terms of an interference by a public authority to be justified in accordance with paragraph 2. In both contexts regard must be had to the fair balance that has to be struck between the competing interests of the individual and of the community as a whole; and in both contexts the State enjoys a certain margin of appreciation in determining the steps to be taken to ensure compliance with the Convention.[102]

Therefore, this analysis likewise foregoes this distinction and considers both jointly with respect to the requirements of Article 8 (2) ECHR.

In practice, the broad catalogue allows states to justify much environmental harm: economic considerations play a considerable role. This is perhaps most obvious in the Grand Chamber decision in *Hatton*.[103] The applicants lived near Heathrow Airport, where daytime noise emissions rose to between 57 and 66 dBA Leq. They were particularly concerned with the night flight policy, which led to noise levels of up to 80 dBA Lmax. A majority (12–5) of the Grand

99 Ibid. [113] and [126].

100 Article 8 (2) Convention for the Protection of Human Rights and Fundamental Freedoms (European Convention on Human Rights, as amended) (ECHR).

101 This is settled case law, see *Stjerna* v. *Finland* App no 18131/91 (ECtHR, 25 November 1994) [38]; *A, B and C* v. *Ireland* 25579/05 Grand Chamber, 16 December 2010 [248].

102 *Eckenbrecht and Ruhmer* v. *Germany* 25330/10 (ECtHR, 10 July 2014) [30]; see also *López Ostra* v. *Spain* App no 16798/90 (ECtHR, 9 December 1994) [51]; *Hatton and others* v. *United Kingdom* App no 36022/97 (Grand Chamber, 8 July 2003) [119]; *Orlikowscy* v. *Poland* App no 7153/07 (ECtHR, 4 October 2011) [92].

103 *Hatton and others* v. *United Kingdom* App no 36022/97 (Grand Chamber, 8 July 2003).

Chamber rejected the complaint under Article 8 ECHR, citing the state's margin of appreciation in striking a fair balance between environmental and economic interests. However, the exclusion of aircraft noise emissions from the scope of the Noise Abatement Act 1960 effectively rendered aggrieved individuals powerless in challenging the night flight policy and hence resulted in a violation of Article 13 ECHR.[104] The joint dissenting opinion of Judges Costa, Ress, Türmen, Zupancic, and Steiner[105] pointed out that the majority did not further probe the government's claim that the flight policy at Heathrow was in the economic interest of the country. This general and unsubstantiated proposition should, at the very least, have been examined further and would have required closer scrutiny under the environmental minimum, especially with respect to establishing whether the regulation was adopted in good faith. We revisit the implications for the environmental minimum arising from economic considerations and the margin of appreciation doctrine when discussing the high-priority principle.[106]

4.2.2.3 Article 1 of AP-I

The subsequent inclusion of property rights into the Convention (Article 1 AP-I) was largely the result of an initial failure of states to agree on the content of a 'right to property.'[107] The provision protects from expropriation, both in a formal and de facto sense. It also covers the limits placed on the use of property and the peaceful enjoyment of possessions.[108] With respect to environmental harm, the ECtHR holds that there is no general right to enjoy property in a pleasant environment. This is most relevant with respect to real estate property, and it often overlaps with Article 8 ECHR to the extent that the home of an individual is concerned. Article 1 AP-I is primarily interpreted to protect the financial value and integrity of property, but not necessarily to protect it from environmental harm. There is little distinctly environmental

[104] See Section 1 (7) of Noise Abatement Act 1960: 'Nothing in this section or the said section sixty-six shall apply to noise or vibration caused by aircraft.'

[105] *Hatton and others* v. *United Kingdom* App no 36022/97 (Grand Chamber, 8 July 2003), Joint Dissenting Opinion of Judges Costa, Ress, Türmen, Zupancic and Steiner 41.

[106] See Chapter 7.

[107] *Collected Edition of the 'Travaux préparatoires' of the European Convention on Human Rights – Consultative Assembly, Committee of Experts, Committee of Ministers, Legal Committee, Advisers to the Ministers, 1 October 1951–20 March 1952* (Martinus Nijhoff, volume 8, 1985) 4.

[108] Karen Kaiser, 'Artikel 1 ZP I' in Ulrich Karpenstein and Franz C. Meyer (eds.), *Europäische Menschenrechtskonvention – Kommentar* (2nd ed., C. H. Beck 2015) 415 [27].

about this protection, as far as one is willing to disregard the curious anomaly of hunting rights cases.

4.2.2.3.1 SCOPE. An infringement of Article 1 AP-I requires a significant loss in property value that the applicants must establish.[109] When the first allegations of property violations were raised in environmentally themed cases, the EComHR stated that Article 1 of AP-I does not grant a right to enjoy property in a 'pleasant environment', and requires applicants to demonstrate a detriment to property value.[110] The Court later endorsed this position in several cases, for instance in *Galev*.[111] The applicants complained of noise and other emissions from a dentist surgery situated in their apartment building, but were unsuccessful because they did not provide any evidence of a detriment to the value of their flats.[112] In *Fägerskiöld*, the Court held that the enjoyment of the property was not significantly diminished: noise emissions of 45 dB and light reflections from the blades of a wind turbine alone would not suffice.[113] The Court reached a similar conclusion in a case where an applicant failed to show how an initially illegally installed heating system in his apartment block caused any adverse effects to his specific flat.[114] Though formally illegal, it was in compliance with all relevant safety regulations and no causal link could be established between the heating system and the alleged air pollution suffered by the applicant.[115] When a substantial loss is established, the question turns to compensation, where the state enjoys a considerable margin of appreciation. In the case of *S* v. *France*, compensation of 250,000 French Francs was considered proportionate, as the applicant had failed to show a greater detriment to the property value.[116]

[109] *Ashworth and others* v. *United Kingdom* App no 39561/98 (ECtHR, 20 January 2004) 14; *Fägerskiöld* v. *Sweden* App no 37664/04 (ECtHR, 26 February 2008) 19.

[110] *Rayner* v. *United Kingdom* App no 9310/81 (Commission, 16 July 1986) 14; Desgagné, 'Integrating Environmental Values into the European Convention on Human Rights', 266.

[111] *Galev and others* v. *Bulgaria* App no 18324/04 (ECtHR, 29 September 2009) 7; see also: *Ünver* v. *Turkey* App no 36209/97 (ECtHR, 26 September 2000); *Walkuska* v. *Poland* App no 6817/04 (ECtHR, 29 April 2008); *Kuchta and Kuchtova* v. *Slovakia* App no 33658/05 (ECtHR, 31 August 2010).

[112] *Galev and others* v. *Bulgaria* App no 18324/04 (ECtHR, 29 September 2009) 5–7; confirmed in *Atanasov* v. *Bulgaria* App no 12853/03 (ECtHR, 2 December 2010) [83] [referencing the earlier case of *Rayner* v. *United Kingdom* App no 9310/81 (Commission, 16 July 1986) 14.

[113] *Fägerskiöld* v. *Sweden* App no 37664/04 (ECtHR, 26 February 2008) 19.

[114] *Darkowska and Karkowski* v. *Poland* 31339/04 Court, 15 November 2011.

[115] Ibid. [68].

[116] S v. *France* App no 13728/88 (Commission, 17 May 1990) 262, approximately 38,000 Euros under the official exchange rate adopted by the European Central Bank at the introduction of the EURO.

This does not, however, permit a wholly arbitrary determination of compensation. Should it be deemed insufficient, there is the possibility of an infringement of Article 1 AP-I: states are required to take into account the impact of environmental harms when calculating the compensation payment. For instance, in *Bistrovic*, part of the applicant's property had been expropriated for the construction of a motorway, which was poised to cause significant noise emissions and pollution.[117] The Court found that by failing to indemnify the applicants for the expected decrease in value, the state had violated Article 1 AP-I.[118] Once a loss in property value and hence an infringement has been satisfactorily established, the Court moves on to the justification stage. There, it considers primarily whether the applicant has borne a disproportionate and excessive burden, despite the compensation offered.

4.2.2.3.2 JUSTIFICATION. Environmental harm may give rise to an infringement of Article 1 AP-I, provided that it leads to a substantial loss in property value. However, absent a disproportionate and excessive burden, the infringement is generally justified, provided the applicant is adequately compensated.[119] Despite the considerable environmental hazard and property value loss arising from leaking oil tanks that were left behind on property returned to the applicants (and that threatened to pollute a nearby river), the ECtHR was not convinced that this was sufficient to constitute a disproportionate and excessive burden on the applicants.[120]

The doctrine remains vague on how significant a burden must be for the Court to conclude otherwise. In environmental cases, this is exceptionally difficult to determine because the vast majority of cases that involve Article 1 AP-I also deal with Article 8 ECHR. The provisions overlap significantly and rarely lead to distinct outcomes: of the 20 cases where the ECtHR reached a distinct conclusion between Article 8 ECHR and Article 1 AP-I, only three decisions ultimately found Article 1 AP-I violated and Article 8 ECHR satisfied.[121] It is therefore unlikely that a case arises where the environmental harm lowers property value and constitutes a disproportionate and excessive burden,

[117] *Bistrovic* v. *Croatia* App no 25774/05 (ECtHR, 31 May 2007) [6].
[118] Ibid. [44].
[119] *Sporrong and Lönnroth* v. *Sweden* App nos 7151/75; 7152/75 (Plenary, 23 September 1982); *Brumărescu* v. *Romania* App no 28342/95 (Grand Chamber, 28 October 1999).
[120] *Schiopu and Verzescu* v. *Romania* App no 26040/06 (ECtHR, 26 November 2013).
[121] See the dataset gathered for this book, Results table; the cases were *Öneryildiz* v. *Turkey* App no 48939/99 (Grand Chamber, 30 November 2004); *Chassagnou and others* v. *France* App nos 25088/94; 28331/95; 28443/95 (Grand Chamber, 29 April 1999); *Allen and others* v. *United Kingdom* App no 5591/07 (ECtHR, 6 October 2009).

but does not simultaneously engage the comparatively wide scope of Article 8 ECHR. This severely limits the practical impact and relevance of Article 1 AP-I in environmental cases, and perhaps goes some way to explain why the doctrine is comparatively underdeveloped.

4.2.2.3.3 HUNTING RIGHTS CASES. Exceptionally, the Court departed from these general doctrinal positions on scope and justification in a series of cases concerning the mandatory transfer of hunting rights. Owing to the size of their estate, the applicants in the case of *Chassagnou* were legally required to grant hunting rights to a municipal hunting association: hunters were thus permitted to hunt game on the applicants' property.[122] The applicants were opposed to hunting on ethical grounds and challenged the practice, alleging that the compulsory transfer without compensation violated their rights under Article 1 AP-I.[123] The Court agreed, holding that a forced transfer of hunting rights which ran counter to the owner's ethical beliefs was not justified under Article 1 (2) AP-I.[124] Remarkably, the Court did not at any point require, nor did the applicants demonstrate, any detriment to the value of their property. Indeed, given their ethical opposition to hunting, it seems unlikely the applicants would have economically benefitted at all from the ability to retain the hunting rights to their estate.

The Court confirmed this stance in the similar case of *Schneider*, equally not requiring a detriment to the property value,[125] nor holding it relevant that Luxembourg law expressly provided automatic compensation to the applicant.[126] This line of cases was condensed into a general principle in *Herrmann*. The Court held that, irrespective of compensation available under domestic law or a detriment to the property value, '(…) the obligation to tolerate hunting on their property imposes a disproportionate burden on landowners, who (…), are opposed to hunting for ethical reasons.'[127] However, the Court requires genuine ethical concerns. In *Chabauty* the applicant unsuccessfully invoked Article 1 AP-I in opposition to a transfer of hunting rights, but as it transpired, was himself the owner of a valid hunting licence.[128] The applicant apparently intended to hunt game himself to the

[122] *Chassagnou and others* v. *France* App nos 25088/94; 28331/95; 28443/95 (Grand Chamber, 29 April 1999) [13].
[123] Ibid. [70].
[124] Ibid. [85].
[125] *Schneider* v. *Luxembourg* App no 2113/04 (ECtHR, 10 July 2007) [51].
[126] Ibid. [49].
[127] *Herrmann* v. *Germany* App no 9300/07 (Grand Chamber, 26 June 2012) [93]–[94].
[128] *Chabauty* v. *France* App no 57412/08 (Grand Chamber, 4 October 2012) [57].

exclusion of the hunting association. Curiously, this is likely the only hunting case where establishing an economic detriment was fairly straightforward: the applicant clearly intended to hunt and lost the economic value of this right.

Beyond the hunting rights cases, Article 1 AP-I has not been construed in a way that provides much substantial environmental protection over that already offered by Articles 2 and 8 ECHR.

4.2.2.4 Article 3

Article 3 ECHR protects individuals from torture, as well as inhumane and degrading treatment. The provision requires that a certain minimum threshold of severity is met, dependent on the particular circumstances of individual cases.[129] In environmentally themed cases, Article 3 ECHR is rarely raised and its relevance is limited. The provision was only invoked 19 times in the 213 cases of category one and held violated only once. Applications that invoke Article 3 ECHR typically include some form of detention facility and the living conditions experienced by inmates, which have been largely excluded from the environmental dataset as not meeting the definition of natural environment. An exception is the case of *Branduse*, where the applicant endured an unlicenced garbage dump site in close proximity to the prison, which also posed the threat of a methane gas explosion.[130] The Court found both Articles 3 and 8 ECHR violated in this case, owing to the proximity to the garbage dump site and the overall condition of the applicant's detention, including overcrowding of the cell, the absence of hot water, and insufficient toilet facilities. Beyond this singular case, the Court has not accepted a violation of Article 3 ECHR in environmental cases and the relevance of the provision is thus very limited.

4.2.3 *Limitations*

The general openness of the Convention towards environmental harm is challenged to an extent by a few cases where the Court has resisted an interpretation towards a general protection of the environment. The settled

[129] *Pretty* v. *United Kingdom* App no 2346/02 (ECtHR, 29 April 2002) [52]; *Juhnke* v. *Turkey* App no 52515/99 (ECtHR, 13 May 2008) [69] [gynaecological examination of detainee]; *A, B and C* v. *Ireland* 25579/05 Grand Chamber, 16 December 2010 [164] [limitations placed on legal abortion].

[130] *Branduse* v. *Romania* App no 6586/03 (ECtHR, 7 April 2009) [11].

case law expressly requires environmental harm to be linked to harmful effects on humans:

> (...) although there is no explicit right in the Convention to a clean and quiet environment, where an individual is directly and seriously affected by noise or other pollution an issue may arise under Article 8 of the Convention.[131]

Despite the qualification of this doctrinal position in some respects, it continues to serve as the basis to reject applications that primarily seek protections of the environment for its own sake. The case of *Ogloblina* arguably yields the clearest statement of this position.[132] The applicant had brought several cases against local authorities, seeking protection of local forests from housing developments. Favourable decisions initially reached by domestic courts were quashed on appeal. The applicant complained before the ECtHR that these rulings breached Article 6, as well as Articles 2, 8, and 13 ECHR, as an expression of a right to a clean environment. The Court recalled that environmental harm does not necessarily imply a violation of Article 8 ECHR, as there is no right to preservation of nature as such. Rather, the interference must directly affect the home, family, or private life, and the environmental harm must rise above a certain minimum level.[133] Having found insufficient evidence to support this connection, the Court rejected the application.

This position was confirmed by the subsequent decision in *Kyrtatos*, albeit with a curious argumentative nuance.[134] It is not immediately clear whether the Court intended to qualify its position, or merely made a passing reference owing to the peculiarities of the case. The applicants opposed the development of properties, as the construction threatened to destroy a swamp close to their home. Indigenous plant and wildlife suffered to such an extent that, according to the applicants, '(...) the area had lost all of its scenic beauty and had changed profoundly in character from a natural habitat for wildlife to a tourist development.'[135]

The Court did not find a violation, holding that the applicants had not shown that the destruction of the swamp infringed on their Article 8 rights. However, the Court indicated, rather paradoxically, that while the destruction of a swamp was insufficient to find a violation, the destruction of a forest

[131] *Udovicic* v. *Croatia* App no 27310/09 (ECtHR, 24 April 2014) [137].
[132] *Ogloblina* v. *Russia* App no 28852/05 (ECtHR, 26 November 2013).
[133] Ibid. [26]–[27].
[134] *Kyrtatos* v. *Greece* App no 41666/98 (ECtHR, 22 May 2003).
[135] Ibid. [46].

located closer to the applicants' property could have given rise to a violation.[136] It is difficult to discern from the passing reference what precisely gives rise to this distinction between the swamp and the forest, as neither was apparently in the immediate vicinity of the applicants' home, and the effect on their private life was arguably equally limited. Perhaps the Court was suggesting that a forest would have some recreational value that was absent from a swamp. It is worth pointing out that distinctions of this nature have not been drawn by the Court in subsequent cases.

As previously discussed, the case of *Otgon* has given some indication that the ECtHR now foregoes the need for a connection to Article 8 ECHR in environmental cases relating to health, but the general position continues to require a limited, albeit direct, connection. Overall, the Court has resisted attempts to read broader protections of the environment beyond harm to humans into the ECHR.

4.3 SUMMARY

Early cases on aircraft noise emissions largely foreshadowed important elements of the modern doctrinal approach to environmental cases, most notably the crucial role of Article 8 ECHR. Against this backdrop, the impact of *López Ostra* appears more ambivalent than scholarship and subsequent references in court rulings would suggest. There is, of course, scope for reasonable disagreement on this point. Nonetheless, considering only a limited sample of decisions when conducting doctrinal research can and should be avoided through a methodology that includes basic data analysis. Otherwise, we risk narrowing attention to a small set of cases and obscure other decisions that may be crucial to a complete picture of the law.

In its modern case law, the Court generally takes environmental threats seriously. The nature of the environmental harm is typically not indicative of whether a violation is found: states are generally held responsible for specific risks. The reasoning that a specific risk to a human right ought to be sufficient is largely compatible with the case law of the Court as it stands today. Environmental harm is not only considered by the Court, but also frequently found to violate Convention rights. Articles 2 and 8 ECHR feature most prominently, while the role for Article 1 AP-I and Article 3 ECHR is significantly more limited.

The Court does not require a serious endangerment of health to find a violation,[137] and deems a serious risk of future violations sufficient, particularly

[136] Ibid. [53], a claim criticised in the dissenting opinion of Judge Zagrebelsky.

[137] *López Ostra* v. *Spain* App no 16798/90 (ECtHR, 9 December 1994) [51].

where the consequences were serious and potentially irreversable.[138] This is a crucial observation for the purposes of the environmental minimum, as it vindicates a central requirement. The doctrinal position broadly supports the specific risk principle that the environmental minimum employs to trigger the framework. That being said, there are limitations, most notably where environmental harm cannot be easily and directly tied to an individual applicant. The question therefore arises whether the Court is content when applicants establish a reasonable hypothesis, as opposed to providing definitive proof that links environmental harm to a specific risk to a Convention right. Chapter 5 now turns to exploring this crucial question.

[138] *Taskin and others* v. *Turkey* App no 46117/99 (ECtHR, 10 November 2004) [113].

5

Reasonable Hypothesis

This chapter examines whether the ECtHR recognises that a reasonable hypothesis of harm is sufficient to establish a link between environmental harm and the invoked Convention right. Under the Convention, this is primarily determined through the rules of evidence and victim status under Article 34 ECHR. This chapter recognises the difficulties obtaining relevant evidence and records under the control of the state, but overall, the rules comport with the requirements of the environmental minimum to a significant extent. The ECtHR does not require applicants to demonstrate a direct scientific link between environmental harm and an impact on their rights in their specific circumstances: in many recent decisions the ECtHR is content with evidence of a general population level risk.

This chapter first sets out the general principles developed by the ECtHR with respect to the rules of evidence and victim status (Section 5.1), then points to some encouraging developments in more recent case law under Articles 2 and 8 ECHR (Section 5.2), while acknowledging enduring inconsistencies, especially the difficulties encountered by applicants seeking access to relevant evidence held by states (Section 5.3).

5.1 VICTIM STATUS

The victim status requirement of Article 34 ECHR was introduced in its current form through Protocol 11,[1] which made the jurisdiction of the ECtHR mandatory for all signatory states upon accession. Victim status requires that the applicant is '(. . .) directly affected by the impugned measure' or omission, and claims a violation of Convention rights.[2] An applicant will

[1] Protocol No. 11 to the ECHR.
[2] Patrick Schäfer, 'Artikel 34 EMRK' in Ulrich Karpenstein and Franz C. Meyer (eds.), *Europäische Menschenrechtskonvention – Kommentar* (2nd ed., C. H. Beck 2015) 551 [61].

generally not be considered a victim if the act or omission has no legal force,[3] although intermittent legal force is sufficient.[4] At this stage, it is not necessary for applicants to show any damage or disadvantage, as the question only becomes relevant with respect to compensation under Article 41 ECHR.[5] The victim status requirement is investigated *ex officio* at every stage,[6] and hence applicants may lose their victim status in the course of the proceedings.[7] Substantiating victim status with respect to potential future violations is generally more challenging. The applicant must convincingly demonstrate the likelihood of an infringement[8] and mere speculations will generally not suffice to meet this burden,[9] especially when infringing state measures remain hypothetical.[10] The primary exceptions to this rule fall into one of two categories: criminal law provisions are actionable by applicants to the extent that they are forced to adjust their behaviour or be directly affected,[11] and cases involving deportation.[12]

3 *Sisojeva and others* v. *Latvia* App no 60654/00 (Grand Chamber, 15 January 2007) [92]–[93].

4 *López Ostra* v. *Spain* App no 16798/90 (ECtHR, 9 December 1994) [42]; *Monnat* v. *Switzerland* App no 73604/01 (ECtHR, 21 September 2006) [33].

5 This is settled case law, see *Marckx* v. *Belgium* App no 6833/74 (ECtHR, 13 June 1979) [27]; *Brumărescu* v. *Romania* App no 28342/95 (Grand Chamber, 28 October 1999) [50]; *İlhan* v. *Turkey* App no 22277/93 (Grand Chamber, 27 June 2000) [52]; *Moscow Branch of the Salvation Army* v. *Russia* App no 72881/01 (ECtHR, 5 October 2006) [65]; Harris and others, *Harris, O'Boyle & Warbrick*, p. 85; Schäfer, 'Artikel 34 EMRK', 551 [65].

6 *Scordino* v. *Italy (No. 1)* App no 36813/97 (Grand Chamber, 29 March 2006) [179]; *Burdov* v. *Russia* App no 59498/00 (ECtHR, 7 May 2002) [30]; *Şandru and others* v. *Romania* App no 22465/03 (ECtHR, 8 December 2009) [50].

7 *Scordino* v. *Italy (No. 1)* App no 36813/97 (Grand Chamber, 29 March 2006) [180]; *Gäfgen* v. *Germany* App no 22978/05 (Grand Chamber, 1 June 2010) [115].

8 *Kjeldsen and others* v. *Denmark* App nos 5095/71; 5920/72; 5926/72 Court (ECtHR, 7 December 1976); *Senator Lines GmbH* v. *Austria, Belgium, Denmark, Finland, France, Germany, Greece, Ireland, Italy, Luxembourg, the Netherlands, Portugal, Spain, Sweden and the United Kingdom* App no 56672/00 Grand Chamber (Grand Chamber, 10 March 2004).

9 *Segi and others & Gestoras Pro-Amnistia and others* v. *15 States of the European Union* App nos 6422/02; 9916/02 (ECtHR, 23 May 2002); *Senator Lines GmbH* v. *Austria, Belgium, Denmark, Finland, France, Germany, Greece, Ireland, Italy, Luxembourg, the Netherlands, Portugal, Spain, Sweden and the United Kingdom* App no 56672/00 Grand Chamber (Grand Chamber, 10 March 2004) [11]; *Tauira* v. *France* App no 28204/95 (Commission, 4 December 1995) 131–2; Schäfer, 'Artikel 34 EMRK', 552 [69].

10 *Ahmed Zaoui* v. *Switzerland* App no 41615/98 Court (ECtHR, 18 January 2001) [2]; *Willis* v. *United Kingdom* App no 36042/97 (ECtHR, 11 June 2002) [49].

11 *Klass* v. *Germany* App no 15473/89 (ECtHR, 21 May 1992) [34]; *Marckx* v. *Belgium* App no 6833/74 (ECtHR, 13 June 1979) [27]; *Sejdić and Finci* v. *Bosnia and Herzegovina* App nos 27996/06; 34836/06 (Grand Chamber, 22 December 2009) [28]; Schäfer, 'Artikel 34 EMRK', 552 [70].

12 *Soering* v. *United Kingdom* App no 14038/88 (Plenary, 7 July 1989) [91]; *D* v. *United Kingdom* App no 30240/96 (ECtHR, 2 May 1997) [53]; *Saadi* v. *Italy* App no 37201/06 (Grand Chamber, 28 February 2008) [146]; Schäfer, 'Artikel 34 EMRK', 553 [71]; Harris and others, *Harris, O'Boyle & Warbrick*, pp. 84–5.

The rules of evidence apply to the factual basis for an application, while the legal evaluation follows the continental European principle of *iura novit curia*, literally meaning that the 'court knows the law.' Pursuant to this principle, the parties to the case are not required to plead or evidence the legal rules that apply to the facts of the case. An exception applies to the rules of domestic law, which the ECtHR treats as questions of fact, provided that they do not relate directly to the interpretation of the ECHR.[13] An alleged violation through the state must generally be established by the applicant 'beyond reasonable doubt.'[14] This standard operates similarly to criminal law cases and has been criticised as too exacting for a human rights context.[15] In practice, the requirements are softened through the inquisitorial nature of the Court, with judges, rather than the parties, controlling the investigation of facts and the evaluation of evidence. Notably, the ECtHR considers that 'proof may follow from the coexistence of sufficiently strong, clear and concordant inferences or of similar unrebutted presumptions of fact.'[16] In practice, the Court often relies heavily on the evidence produced by the parties, and the facts established in domestic court proceedings, unless they were the result of inadequate investigations.[17]

The Court operates under a 'free evaluation of all evidence, including such inferences as may flow from the facts and the parties' submissions', without any formal procedural or admissibility constraints.[18] This applies in particular to the 'level of persuasion necessary for reaching a particular conclusion and,

[13] Nicola Wenzel, 'Artikel 38 EMRK' in Ulrich Karpenstein and Franz C. Meyer (eds.), *Europäische Menschenrechtskonvention – Kommentar* (2nd ed., C. H. Beck 2015) 629 [2]; Tobias Thienel, 'The Burden and Standard of Proof in the European Court of Human Rights' (2007) 50 *German Yearbook of International Law* 543, 558.

[14] This is settled case law, see *Asan and others* v. *Turkey* App no 56003/00 (ECtHR, 31 July 2007) [69]; *Avsar* v. *Turkey* App no 25657/94 (ECtHR, 10 July 2001) [282]; *Ireland* v. *United Kingdom* App no 5310/71 (ECtHR, 18 January 1978) [161]; *Cyprus* v. *Turkey* App no 25781/94 (Grand Chamber, 10 May 2001) [112].

[15] Although the ECtHR rejects the notion it intended to borrow the approach from domestic law, see *Nachova and other* v. *Bulgaria* App nos 43577/98; 43579/98 (Grand Chamber, 6 July 2005) [147]; Rüdiger Wolfrum, 'The Taking and Assessment of Evidence by the European Court of Human Rights' in Stephan Breitenmoser (ed.), *Human Rights, Democracy and the Rule of Law: liber amicorum Luzius Wildhaber* (Nomos 2007) 921; Thienel, 'The Burden and Standard of Proof in the European Court of Human Rights', 579; Wenzel, 'Artikel 38 EMRK,' 630 [3].

[16] *Avsar* v. *Turkey* App no 25657/94 (ECtHR, 10 July 2001) [282]; *Stoica* v. *Romania* App no 42722/02 (ECtHR, 4 March 2008) [63]; *Fadeyeva* v. *Russia* App 55723/00 (ECtHR, 9 June 2005) [79].

[17] *Avsar* v. *Turkey* App no 25657/94 (ECtHR, 10 July 2001) [283]–[284]; *Stoica* v. *Romania* App no 42722/02 (ECtHR, 4 March 2008) [64]; *Cobzaru* v. *Romania* App no 48254/99 (ECtHR, 26 July 2007) [65].

[18] See especially Rules A1, A2, and A5 of the Annex to the Rules of the Court; *Nachova and other* v. *Bulgaria* App nos 43577/98; 43579/98 (Grand Chamber, 6 July 2005) [147].

in this connection, the distribution of the burden of proof.'[19] The Court nonetheless often burdens the applicant with providing evidence of a violation through a state,[20] but draws inferences from publicly available sources and medical records if the government does not disprove presumptive health detriments.[21] This becomes problematic if most evidence for a violation is held by the respondent state or third parties, and consequently is not available to applicants.[22] In addressing this problem, the Court has developed conditions under which the burden is shifted to the respondent state.[23] The Court draws inferences from the behaviour of the state, and treats allegations as proven in the absence of cooperation from the state, provided that the applicant has established a prima facie case.[24]

5.2 ENVIRONMENTAL CASES

Following these general principles, the Commission also commonly required applicants to establish an infringement beyond reasonable doubt in early environmental cases. In the case of *Dr S*, the applicant complained of the threats to his well-being stemming from the use of nuclear technology.[25] The EComHR dismissed the application, holding that the Convention did not offer protection from generalised dangers and arguably concluded that the applicant had failed to establish victim status. This position remained largely unchanged throughout the 1990s. In *Tauira*, the state argued that a scientific link between health risks to the applicants and nuclear tests conducted by France had not been established: the testing as such was legal, barring any evidence proving a causal link to an environmental harm. The EComHR

19 *Nachova and other* v. *Bulgaria* App nos 43577/98; 43579/98 (Grand Chamber, 6 July 2005) [147].

20 Thienel, 'The Burden and Standard of Proof in the European Court of Human Rights', 553.

21 *Grimkovskaya* v. *Ukraine* App no 38182/03 (ECtHR, 21 July 2011) [60]–[61]; [evidence required on noise levels and air pollution]; *Fadeyeva* v. *Russia* App 55723/00 (ECtHR, 9 June 2005) [79] and [88] [inferences drawn from medical records and publicly available data].

22 Wenzel, 'Artikel 38 EMRK', 632 [10].

23 *Orhan* v. *Turkey* App no 25656/94 (ECtHR, 18 July 2002) [266].

24 See Rule 44c of Rules of the Court; this is settled case law, see *Shamayev and others* v. *Georgia and Russia* App no 36378/02 (ECtHR, 12 April 2005) [503]; *Tepe* v. *Turkey* App no 27244/95 (ECtHR, 9 May 2003) [135]; *Orhan* v. *Turkey* App no 25656/94 (ECtHR, 18 July 2002) [274]; *Imakayeva* v. *Russia* App no 7615/02 (ECtHR, 9 January 2006) [124]–[125] and *Togcu* v. *Turkey* App no 27601/95 (ECtHR, 31 May 2005) [95] [both regarding forced disappearances]; *Selmouni* v. *France* App no 25803/94 (Grand Chamber, 28 July 1999) [87] and *Akkum and others* v. *Turkey* App no 21894/93 (ECtHR, 24 March 2005) [210] [both regarding injuries sustained in police custody].

25 *Dr S* v. *Germany* App no 715/60 (Commission, 5 August 1960, unpublished) 3.

agreed, concluding that the mere risk associated with the tests was insufficient on its own to find a violation.[26] Likewise, in the case of *Asselboourg*, which concerned dangerous emissions from steel mills, the ECtHR held that victim status would be considered established for future violations only in exceptional circumstances:

> (. . .) [the] exercise of the right of individual petition cannot have the aim of preventing a violation of the Convention. It is only in wholly exceptional circumstances that the risk of a future violation may nevertheless confer the status of 'victim' on an individual applicant (. . .).[27]

An early indication that the Convention could better engage with evidentiary difficulties came a few years later through the decision in *Gounaridis*.[28] The applicants could not provide any evidence of existing environmental harm, as the subject of the case was an anticipated expansion of a motorway in the vicinity of their home.[29] The Commission found that while the mere invocation of environmental harm was insufficient to establish victim status, it would have been satisfied had the applicants demonstrated 'in an arguable and detailed manner the *probability* of direct harm sufficient to rise to the level of a violation of the Convention.'[30] Crucially, the EComHR seemed open to considering the future prospect of environmental harm as a violation of the ECHR, provided that it has been established in sufficiently clear terms.

The more recent case law of the ECtHR appears to have adopted and accepted this relaxing of the burden of proof and victim status requirements for environmental harm. This appears to be, at least in part, inspired by the precautionary principle familiar from general environmental law. The principle is mentioned in a limited number of ECHR decisions, but it has not yet been decisive to finding a violation. This limited role was lamented in the joint dissenting opinion of Judges Zupancic and Gyulumyan in *Tatar*.[31] Their dissent argued that where an applicant cannot demonstrate with certainty that exposure has caused a specific disease, it should suffice that the applicant demonstrate that the exposure has led to a statistically significant

[26] *Tauira* v. *France* App no 28204/95 (Commission, 4 December 1995) 133; see also *Bernard and 47 others individuals as well as the association Greenpeace-Luxembourg* v. *Luxembourg* App no 29197/95 (ECtHR, 29 June 1999) 6–7.

[27] *Asselboourg and 78 others and Greenpeace Association-Luxembourg* v. *Luxembourg* App no 29121/95 (ECtHR, 29 June 1999) 6.

[28] *Gounaridis, Iliopoulos, Pappostolou* v. *Greece* App no 41207/98 (ECtHR, 21 October 1998).

[29] Ibid. [1].

[30] Ibid. [2], emphasis added, my translation.

[31] *Tatar* v. *Romania* App no 67021/01 (ECtHR, 27 January 2009).

increase in the risk of contracting the disease or aggravating an existing ailment.[32] Although the majority opinion in *Tatar* did not follow this reasoning, the modern ECtHR doctrine nonetheless accepts this standard when applicants furnish sufficient evidence of environmental harm. Provided that the environmental harm is a plausible cause, or at least an aggravating factor, the Court accepts their victim status under Article 34 ECHR.

In the case of *Bacila*, the Court held that states were under a general obligation to regulate dangerous activities, especially through licensing and monitoring, as well as prescribing suitable measures to ensure the effective protection of citizens.[33] The applicant had brought the case in opposition to a metal factory that produced dangerous lead and sulphur dioxide emissions. The ruling is notable because it does not rely primarily on whether the risk posed by the industrial activity has materialised. The positive obligation of the state exists regardless, although it should be noted that the respondent government had not disputed the existence of the environmental harm or the alleged health effects.[34] In *Öneryildiz*, the Court expanded on *Bacila* and found within the context of Article 2 ECHR:

> The [positive obligation to take all appropriate steps to safeguard life for the purposes of Article 2] indisputably applies in the particular context of dangerous activities, where, in addition, special emphasis must be placed on regulations geared to the special features of the activity in question, particularly with regard to the level of the potential risk to human lives. They must govern the licensing, setting up, operation, security and supervision of the activity and must make it compulsory for all those concerned to take practical measures to ensure the effective protection of citizens whose lives might be endangered by the inherent risks.[35]

Curiously, the Court also found violations where the applicants were unable to determine, due to a lack of information provided by authorities, whether an environmental harm had occurred.[36] The Court appears to suggest that regulation of risks is part and parcel of the positive obligations arising from Convention rights, regardless of whether these risks have materialised in an individual case.

[32] Ibid. 46.

[33] *Bacila* v. *Romania* App no 19234/04 (ECtHR, 30 March 2010) [61].

[34] Ibid. [63].

[35] *Öneryildiz* v. *Turkey* App no 48939/99 (Grand Chamber, 30 November 2004) [90].

[36] *Tatar* v. *Romania* App no 67021/01 (ECtHR, 27 January 2009) [107].

This principle was confirmed in the subsequent case of *Fadeyeva*.[37] The applicant lived in close proximity to a steel mill, the largest iron smelter in Russia. A buffer zone of a five-kilometre radius around the plant was nominally established to limit the number of people affected by excessive pollution. Nonetheless, several thousand people, including the applicant, were at the time living within this zone. The ECtHR heard that the overall air quality in the city resulted in significantly higher mortality rates among the residents. In particular, the rates of respiratory, blood, and skin diseases among children were significantly higher than the national average.[38] The applicant herself had developed various illnesses, which an independent medical report concluded were at least exacerbated by the widespread environmental harm.[39] Even though the applicant was unable to show a direct causal link between the environmental harm and the detrimental effects on her health, the ECtHR accepted that

> (...) the very strong combination of indirect evidence and presumptions makes it possible to conclude that the applicant's health deteriorated as a result of her prolonged exposure to the industrial emissions from the Severstal steel plant. Even assuming that the pollution did not cause any quantifiable harm to her health, it inevitably made the applicant more vulnerable to various illnesses.[40]

It therefore concluded that the pollution and the detriment to the applicant's health constituted a breach of Article 8 ECHR. In spite of the cautious phrasing, the applicant was ultimately not required to demonstrate that the environmental harm from the steel mill had caused her illnesses. At least for the purposes of Article 8 ECHR, the Court is willing to hold a state responsible for a significant increase of risk alone.

In modern environmental cases, the applicants are not generally required to establish a direct, irrefutable causal link between the environmental harm and the detriment suffered. If both elements have been established independently, and a link between them is at least conceivable, it falls on the respondent state to rebut the assumed link.[41] This doctrine mirrors the environmental

[37] *Fadeyeva* v. *Russia* App 55723/00 (ECtHR, 9 June 2005).
[38] Ibid. [15].
[39] Ibid. [45].
[40] Ibid. [88].
[41] Ibid. [45]; *Ledyayeva and others* v. *Russia* App nos 53157/99; 53247/99; 53695/00; 56850/00 (ECtHR, 26 October 2006) [95] and [100]; *Bacila* v. *Romania* App no 19234/04 (ECtHR, 30 March 2010) [63]; *Grimkovskaya* v. *Ukraine* App no 38182/03 (ECtHR, 21 July 2011) [60]–[61]; *Brincat and others* v. *Malta* App no 60908/11; 62110/11; 62129/11; 62312/11; 62338/11 (ECtHR, 24 July 2014) [83]–[85].

minimum to a significant extent, specifically the reasonable hypothesis that must link the environmental harm with a threat to human rights. That is not to say that environmental harm and a relevant detriment to a Convention right are effortlessly established. There are significant caveats to the compatibility of the environmental minimum with this doctrinal position.

5.3 LIMITATIONS

Establishing a specific risk is comparatively straightforward for applicants in cases where the detriment and the environmental harm are easily evidenced. For instance, in the case of *Vearncombe* the applicants complained of noise emissions, originating from a shooting range.[42] Applicants in such cases can reasonably acquire expert evidence that demonstrates that an environmental harm exists, and that it constitutes a detriment to their well-being.[43] Accordingly, applicants have often prevailed in cases where the evidence can be established with comparatively little effort, or where evidence is publicly available, for instance in the form of studies and official public health reports.[44]

Matters become more complicated when such evidence is not reliable, not publicly available, or requires complex scientific analysis that exceeds the (financial) capabilities of applicants. In *Atanasov*, the applicant alleged and produced evidence that sludge used by a third party in a reclamation scheme

42 *Vearncombe and others* v. *Germany and United Kingdom* App no 12816/87 (Commission, 18 January 1989).

43 The applicants notably failed in demonstrating a detriment in a number of cases, even though the environmental harm was accepted by the Court, see *Orlikowscy* v. *Poland* App no 7153/07 (ECtHR, 4 October 2011) [98]–[99] [failure to show health detriment arising from air pollution]; *Tauira* v. *France* App no 28204/95 (Commission, 4 December 1995) [failure to show detriment from nuclear weapons tests]; *Stockton and others* v. *United Kingdom* App no 36053/97 (Commission, 15 January 1998) [failure to show further detriment after domestic proceedings regarding liquid aluminium contamination of drinking water]; *Gounaridis, Iliopoulos, Pappostolou* v. *Greece* App no 41207/98 (ECtHR, 21 October 1998) [construction of motorway]; *Bernard and 47 others individuals as well as the association Greenpeace-Luxembourg* v. *Luxembourg* App no 29197/95 (ECtHR, 29 June 1999) [failure to demonstrate detriment from steel mill].

44 *Surugiu* v. *Romania* App no 48995/99 (ECtHR, 20 April 2004); *Moreno Gomez* v. *Spain* App no 4143/02 (ECtHR, 16 November 2004); *Fadeyeva* v. *Russia* App 55723/00 (ECtHR, 9 June 2005); *Ledyayeva and others* v. *Russia* App nos 53157/99; 53247/99; 53695/00; 56850/00 (ECtHR, 26 October 2006); *Giacomelli* v. *Italy* 59909/00 (ECtHR, 2 November 2006); *Mileva and others* v. *Bulgaria* App nos 43449/02; 21475/04 (ECtHR, 25 November 2010); *Dubetska and others* v. *Ukraine* App no 30499/03 (ECtHR, 10 February 2011); *Di Sarno and others* v. *Italy* App no 30765/08 (ECtHR, 10 January 2012); *Brincat and others* v. *Malta* App no 60908/11; 62110/11; 62129/11; 62312/11; 62338/11 (ECtHR, 24 July 2014).

was contaminated with heavy metals that threatened soil and groundwater.[45] However, these claims could not be further substantiated, as the lack of supervision through competent authorities failed to ensure that accurate reports on the environmental harm were produced. Instead of holding this failure against the respondent state, the ECtHR invoked the uncertainty on whether the risk of heavy metal poisoning had materialised to dismiss the application.[46] Applicants simply do not always have the necessary resources to obtain credible and complete evidence to fill the gaps left by third party or state failures.

There are also a host of other practical reasons that might prevent obtaining relevant evidence. This is most acute where the evidence is historical, and either lost or confined to secrecy by a state. In such cases, applicants may be forced to prove the existence of documents that states deny ever existed, or be burdened with meeting a standard of proof that eludes even cutting-edge research. Most jurisdictions make it difficult to compel a government to investigate the existence of a potential environmental harm, at least in the absence of any strong evidence that it is linked to harm endured by applicants. This leaves applicants in a double bind: the scientific evidence of a link that is required to proceed to court is in effect the same evidence they wish to compel the government to produce.

A string of cases revolving around UK nuclear weapons tests illustrates this dilemma. The applicants had little difficulty establishing their individual, health-related detriments, but were, for a variety of practical reasons, unable to demonstrate that a relevant environmental harm had existed. The two applicants were part of a contingent of 20,000 military servicemen who were exposed to nuclear radiation between 1952 and 1967, following atmospheric nuclear tests conducted by the United Kingdom in the Pacific Ocean and Maralinga, Australia. Particularly notorious were a number of tests conducted at Christmas Island between November 1957 and September 1958. Both applicants 'were ordered to line up in the open and face away from the explosions with their eyes closed and covered until twenty seconds after the blast', allegedly in an attempt to deliberately expose them to radiation, a charge denied by the respondent state.[47] The question before the ECtHR was ostensibly a very narrow one: does the lack of monitoring of radiation exposure and failure to release medical records decades after the incident except through a specialised procedure satisfy the requirements of Articles 6

45 *Atanasov* v. *Bulgaria* App no 12853/03 (ECtHR, 2 December 2010) [10].
46 Ibid. [78].
47 *McGinley and Egan* v. *United Kingdom* App nos 21825/93; 23414/94 (ECtHR, 9 June 1998) [10].

and 8 ECHR? By a narrow majority of five to four and six to three, respectively, the Court did not find a violation. The majority ruled it particularly relevant that the applicants were unable to produce any evidence that either relevant documents recording their exposure existed, or that the government had withheld them.[48] This is a problematic allocation of the burden of proof, especially given the court's modern environmental case law. Whether further relevant documents pertaining to the applicants' exposure exist, and whether their withholding is due to reasons justified through the interests of national security or motivated by a desire to avoid responsibility, can be determined only through information to which the authorities alone are privy.

The Court took at face value the government assertion that making use of an information disclosure procedure before the domestic Pension Tribunal would yield access to all relevant documents. This assumes, without evidence, that all relevant documents have been declassified, not destroyed in the meantime, and that the Tribunal was in a position to compel the release of all documents from military archives. All of these assumptions made by the Court can scarcely be disproved by the applicants.[49] It is left to the imagination as to how the applicants might have convinced the Court that documents existed and were inappropriately withheld.

As pointed out in the dissenting opinion of Judges De Meyer, Valticos, and Morenilla, there is sound circumstantial evidence that the United Kingdom was at the time interested in gathering data on the effects of radiation exposure, particularly on personnel with various types of protective gear and at various distances to the blast. The implementation of the lining up procedure with respect to both applicants hints that there was perhaps more interest in monitoring than the absence of medical records suggests. Military leaders appeared concerned not to reveal information that might support the notion that the nuclear tests had caused health detriments in servicemen.[50] Specifically, they rejected blood tests for personnel for fear that a subsequent development of leukaemia might make it difficult to refute a causal connection to radiation exposure. There is also further evidence of attempts to downplay any harmful effects, even to close allies such as the Australian government.[51] Whether medical records were created at all, lost, destroyed, or confined to permanent secrecy enters the realm of speculation. However,

48 Ibid. [88].
49 Ibid. [89].
50 Ibid., Joint Dissenting Opinion of Judges De Meyer, Valticos and Morenillia [footnote 8].
51 Ibid., Joint Dissenting Opinion of Judges De Meyer, Valticos and Morenillia [footnote 1].

it is not apparent that the Court should burden the applicants with dispelling this uncertainty, especially when it is not even remotely in their power to do so.

This problematic approach of the ECtHR to the burden of proof was also a feature in a related decision regarding the effects on the descendants of test veterans. In *L.C.B.* v. *United Kingdom*, a lack of evidence on documents relating to health monitoring was equated by the Court to evidence of their absence.[52] The child of a test veteran had developed leukaemia at age four, and there were serious allegations that the radiation exposure of her father, a catering assistant with the Royal Air Force based on Christmas Island between 1957 and 1958, who had also participated in the subsequent clean-up efforts, was causally linked. The Court nonetheless concluded that

> [h]aving examined the expert evidence submitted to it, the Court is not satisfied that it has been established that there is a causal link between the exposure of a father to radiation and leukaemia in a child subsequently conceived.[53]

Consequently, the United Kingdom was not found to have violated Articles 2 and 8 ECHR. The alleged absence of any health monitoring in both *McGinley* and *L.C.B.* also raises a deeper question, quite apart from any alleged suppression of documents. Assuming that records were not kept, does this omission not constitute an infringement in its own right? In other words, wilful ignorance on the part of the state in the face of the well-established health hazards of nuclear explosions could be seen as an entirely separate nexus of relevant infringements that the majority decision in *McGinley* chose not to address.[54] The question was similarly not addressed in *L.C.B.* as the applicant had not raised that question before the Commission, although there is an argument that following the principle of *iura novit curia*, the ECtHR might have examined these issues regardless: they are, after all, questions of the legal qualification of the presented facts, and hence there should be no need for the applicants to plead on them.[55] A more recent application involving test veterans has not brought any significant changes to this restrictive approach. The case involved an evaluation of legal aid before domestic courts under Article 6 ECHR, and an alleged failure to hold a public investigation into the health effects of the nuclear tests under Article 2 ECHR. The former

52 *LCB* v. *United Kingdom* App no 23413/94 (ECtHR 9 June 1998) [37].

53 Ibid. [39].

54 *McGinley and Egan* v. *United Kingdom* App nos 21825/93; 23414/94 (ECtHR, 9 June 1998), Joint Dissenting Opinion of Judges De Meyer, Valticos and Morenillia 34.

55 *LCB* v. *United Kingdom* App no 23413/94 (ECtHR 9 June 1998) [35].

complaint was dismissed as ill-founded by the ECtHR,[56] and the latter was dismissed because the relevant facts of the claim arose before 14 January 1966, the day the United Kingdom made declarations with respect to Articles 25 and 46 of the Convention on the ability of the EComHR and the ECtHR to hear individual complaints.[57]

The saga of the test veteran cases is instructive in several ways. Especially in more complex environmental cases, applicants will often be unable to furnish the evidence that the Court requires: (1) it might be simply beyond their means to acquire existing information, (2) gathering necessary data might be beyond the reach of cutting-edge science, even given unlimited resources, or (3) the evidence is of a historical nature, and held by state entities that are unwilling or unable to furnish the necessary information. Given such circumstances, it appears inequitable and counter-productive to burden the applicants with proving all relevant elements of a violation.

5.4 SUMMARY

The review of the case law demonstrates that in many respects, the ECtHR has lowered the evidentiary burden placed on applicants, at least where there is sufficient basis for linking the environmental harm to a detriment of Convention rights. The Court requires active risk management of states to mitigate environmental harm, regardless of whether it has materialised in a specific case. Both trends are closer to the environmental minimum than one might have thought at the outset when considering only the general case law on victim status.

In other respects, the doctrinal position is more problematic. The threat of the environmental harm in *Bacila* was well established in numerous reports and in any case not contested by the respondent state. Likewise, the environmental harm in *Fadeyeva* was well documented and in *Öneryildiz* the harm had even resulted in the death of several people. Especially *Öneryildiz* can therefore also be interpreted as the logical conclusion of applying the longstanding case law on positive obligations under Article 2 ECHR to environmental harm. That would, however, amount to denying the environmental dimensions of the case and the regulatory and enforcement failures of local authorities that distinguish the case from the overwhelming majority of Article 2 ECHR decisions. *Tatar* further softens the evidentiary requirements by demonstrating that an inability of applicants to secure relevant information

[56] *Sinfield and other* v. *United Kingdom* App no 61332/12 (ECtHR, 18 February 2014) [46].
[57] Ibid. [36].

from authorities can itself give rise to a Convention violation, even though this case and its conclusions are drawn into question by the rulings on test veterans. Then again, divergence from the environmental minimum is not surprising. Any adoption of the framework would introduce and require changes, even if the ECtHR has laid much of the necessary groundwork.

What is further needed is a clear commitment to the modern case law. Should an applicant establish an environmental harm and a detriment to Convention rights, then an unequivocal link beyond all reasonable doubt should not be necessary. Instead, a reasonable hypothesis ought to suffice in those cases, and a complete shift in the burden of proof ought to be considered where states are potentially withholding relevant information. The question that Chapter 6 now turns towards is whether, given a specific risk, the ECtHR applies minimum standards of review akin to the requirements of the environmental minimum.

6

Minimum Standards

This chapter explores the ECHR doctrinal approach to the minimum standards of the environmental minimum. These standards apply once a specific risk has been established and are derived either from legal norms, or in lieu of these, established and emerging scientific evidence. Notably, unclear or contradictory scientific evidence affords states broad discretion in formulating a regulatory response. Should a violation of minimum standards be established, it furnishes individuals with a claim based on the right identified at the specific risk stage.

The question is whether the rights guaranteed under the Convention require the consistent enforcement of legal norms (Section 6.1), and whether the Court takes subsidiary recourse to established and emerging scientific evidence in their absence (Section 6.2). This requirement is generally met, as a clear majority of cases where domestic environmental regulations were ignored resulted in the ECtHR finding a violation of the Convention. However, the chapter also acknowledges some enduring concerns, notably the questionable ECtHR approach to some cases with an absence of legal norms and its approach to a lack of consistent enforcement (Section 6.3).

6.1 LEGAL STANDARDS

Legal standards form the backbone of the environmental minimum: (1) where a specific risk is encountered and legal standards exist, a state is generally in compliance with the environmental minimum, to the extent that the relevant regulation was adopted in good faith and (2) is consistently enforced.

6.1.1 *Good Faith Regulation*

The older case law is ambiguous on the necessity of good faith regulation. The EComHR and the ECtHR are often dismissive of environmental harm, even though elements of their reasoning and the outcomes of cases support the environmental minimum. This older case law is well represented through the case of *Vearncombe*.[1] It involved a shooting range in close proximity to the applicants' homes. Although restrictions on its use were in place, the applicants attempted to have future, expanded use of the shooting range enjoined before both the High Court in London and the Berlin Administrative Court. The former struck out the action pursuant to the Crown Proceedings Act 1947, while the latter held the application inadmissible for lack of jurisdiction under legislation which prohibited German courts from hearing cases pertaining to the Allied forces. Having failed to prevail before these courts, the applicants engaged the EComHR and submitted expert testimony that the noise level was projected to rise to between 64 and 66 dbA, exceeding the permitted daytime value of 55 dbA. The Commission rejected the application on the grounds that a significant current noise emission had not been established by the applicants. Whether the future, more expansive use would lead to an infringement of Convention rights was left open.

On the one hand, the environmental minimum is compatible with this outcome, because as far as can be discerned the current use of the range was in line with existing legal standards. With respect to future use, the Court also applied a test akin to determining whether there was a specific risk, and the result likewise supports the environmental minimum. On the other hand, the reasoning following this test and the ultimate outcome are more problematic. A key argument was that the applicants had not shown that a significant noise emission arose from the shooting range. In the terminology of the environmental minimum, the Court rejected that there was a specific risk. However, this is difficult to reconcile with the facts. The applicants had gone to considerable lengths and submitted expert testimony that demonstrated a reasonable hypothesis that noise levels would exceed existing regulations, while the respondent government offered no rebuttal of this hypothesis.[2] Under the environmental minimum, the case would at least not have failed at this early stage.

The environmental minimum is noticeably more compatible with the modern case law: the Court often relies on existing good faith regulations to dismiss an application and has examined whether regulations adopted represent a fair balance between competing interests, not unlike the requirement of good faith under the environmental minimum. A typical example is found in

[1] *Vearncombe and others* v. *Germany and United Kingdom* App no 12816/87 (Commission, 18 January 1989).

[2] Ibid., 12.

the Grand Chamber ruling in *Hatton*.[3] Even though the case was widely seen as a setback for environmental protection under the Convention, the reasoning and outcome broadly support the environmental minimum.[4] The night-time flight policy at Heathrow was in line with domestic noise emission regulation, and the Court expressly rejected the substantial part of the application for that reason.[5] There is, however, a question as to whether this regulation was adopted in good faith. This depends significantly on the generally accepted guidelines in the relevant scientific community at the time. In this case, one could question whether a complete exclusion of individual challenges for noise emissions pursuant to the Noise Abatement Act 1960 is not as such indicative of bad faith. Indeed, it is perhaps not too surprising that the vast majority of aircraft noise cases were brought against the United Kingdom. However, this element of the complaint was upheld by the Grand Chamber as a violation of Article 6 ECHR. The environmental minimum is also, at least arguably, compatible with the outcome on the underlying policy. There at least appear to be no obvious and overwhelming reasons to assume the regulation permitting the night flights was adopted in bad faith, that is to say in departure from the standards generally accepted in the relevant scientific community.

It is noteworthy in this context that the Court examines regulations to determine whether they constitute a good faith balancing of competing interests in environmental cases. In the case of *Allen*, the applicants had complained of plans to add a further runway to Stansted Airport and the absence of compensation for diminished property values under the Home Owner Support Scheme.[6] The Court found that the regulatory approach adopted in a government white paper struck '(…) a fair balance between individuals whose property values may have been affected by airport expansion and the conflicting interests of BAA Stansted, the applicants and the community as a whole.'[7] Accordingly, the application was dismissed, as it constituted a good faith regulation in the eyes of the Court. The Court similarly dismissed

3 *Hatton and others* v. *United Kingdom* App no 36022/97 (Grand Chamber, 8 July 2003).

4 Jeremy Hyam, 'Hatton v United Kingdom in the Grand Chamber: One Step Forward, Two Steps Back?' [2003] *European Human Rights Law Review* 631; David Hart and Marina Wheeler, 'Night Flights and Strasbourg's Retreat from Environmental Human Rights' (2004) 16 *Journal of Environmental Law* 100; Shelton, 'Human Rights and the Environment', 155; Alan Boyle, 'Environment and Human Rights' in Rüdiger Wolfrum (ed.), *Max Planck Encyclopedia of Public International Law* (Oxford University Press 2015) [32].

5 *Hatton and others* v. *United Kingdom* App no 36022/97 (Grand Chamber, 8 July 2003) [120].

6 See *Allen and others* v. *United Kingdom* App no 5591/07 (ECtHR, 6 October 2009) [9] and [13]–[15].

7 Ibid. [68].

the application in *Kania*.[8] The applicants had complained of noise emissions from a neighbouring artisan cooperative. The cooperative had complied with domestic environmental regulations, and the applicants had failed to cast doubt on their adequacy, and were unable to provide a reasonable hypothesis of serious and long-term health problems.[9] Accordingly, the Court did not find a violation of Article 8 ECHR.

6.1.2 *Enforcement*

The second major claim with respect to legal standards is that a lack of consistent enforcement must result in a violation of human rights under the environmental minimum. In the overwhelming majority of cases, the Court was satisfied that a failure to enforce domestic regulations is sufficient on its own to find a violation of the Convention. Even the notorious margin of appreciation of states is limited to effective measures that achieve results in due course and further restricted whenever the state has contributed to the creation or development of environmental harm.[10]

The case of *Öneryildiz* provides a particularly strong example. The Court faulted the state for the lack of enforcement efforts to protect the right to life under Article 2 ECHR.[11] The Court held that states were under a positive duty to establish a legal and administrative framework that adequately protects against environmental risks, especially with regard to dangerous industrial activities.[12] The Court acknowledged that it would refrain from substituting its own judgment on policy for that of local authorities, unless the policy proved wholly inadequate.[13] This reasoning is very much in line with the environmental minimum: it primarily demands that states address environmental harm through appropriate regulatory and enforcement action, while not prescribing any specific enforcement measures. In many ways, the case of *Budayeva* confirms this reasoning.[14] The case concerned a town in an area

8 *Leon and Agnieszka Kania* v. *Poland* App no 12605/03 (ECtHR, 21 July 2009).

9 Ibid. [102].

10 *Mileva and others* v. *Bulgaria* App nos 43449/02; 21475/04 (ECtHR, 25 November 2010) [99]–[101]; *Oluic* v. *Croatia* App no 61260/08 (ECtHR, 20 May 2010) [64]–[66]; *Fadeyeva* v. *Russia* App 55723/00 (ECtHR, 9 June 2005) [132]; *Ledyayeva and others* v. *Russia* App nos 53157/99; 53247/99; 53695/00; 56850/00 (ECtHR, 26 October 2006) [110].

11 *Öneryildiz* v. *Turkey* App no 48939/99 (Grand Chamber, 30 November 2004) [71] and [89].

12 Ibid. [90]; confirmed in *Budayeva and others* v. *Russia* App nos 15339/02; 21166/02; 20058/02; 11673/02; 15343/02 (ECtHR, 20 March 2008) [128]–[130].

13 *Öneryildiz* v. *Turkey* App no 48939/99 (Grand Chamber, 30 November 2004) [107].

14 *Budayeva and others* v. *Russia* App nos 15339/02; 21166/02; 20058/02; 11673/02; 15343/02 (ECtHR, 20 March 2008).

known to be prone to landslides, where necessary repairs to a mud retention system had been neglected. Subsequently, a catastrophic mudslide killed eight people and left 19 missing. What separates *Budayeva* from typical Article 2 ECHR cases, is that the Court relied on the lack of preventative measures to find a violation. Had the authorities enforced environmental regulations, there would not have been inadequate preventative measures and in turn, the catastrophe may well have been avoided.[15]

The case of *Moreno Gomez* also presents facts where existing environmental regulations were not consistently enforced.[16] It concerned noise emission from a nightclub beyond the levels permissible under domestic regulations. The Spanish Constitutional Court had concluded that the applicant had not established a link between the noise and her health issues and dismissed the claim on that basis.[17] By contrast, the ECtHR found that, as the noise levels at night time exceeded the permitted levels, this alone sufficed to find a violation of Article 8 ECHR.[18] The Court stated that '[r]egulations to protect guaranteed rights serve little purpose if they are not duly enforced and the Court must reiterate that the Convention is intended to protect effective rights, not illusory ones.'[19]

In this sense, the case of *Di Sarno* falls into the same category, as a lack of enforcement measures directly led to the harm and a violation of the Convention.[20] The case involved problems with solid urban waste disposal in an Italian town. Waste eventually piled up in the streets close to the applicants' homes due to a complete breakdown of waste removal services.[21] The Court found that Italy had violated its obligations under Article 8 ECHR by not ensuring that domestic laws regulating waste removal and disposal were effectively enforced.[22]

A special category of inadequate enforcement measures of legal standards comes in the guise of a failure to act on fresh evidence. The case of *Giacomelli* concerned a dump site for hazardous waste, which was treated with chemicals under a licence from the Regional Council. This licence was subject to renewal every five years. A further extension of the operating licence in 1999 was made subject to a favourable environmental impact assessment.

15 Ibid. [141] and [157].
16 *Moreno Gomez* v. *Spain* App no 4143/02 (ECtHR, 16 November 2004).
17 Ibid. [37].
18 Ibid. [60].
19 Ibid. [61].
20 *Di Sarno and others* v. *Italy* App no 30765/08 (ECtHR, 10 January 2012).
21 Ibid. [94].
22 Ibid. [112].

An initially unfavourable assessment was set aside in court proceedings brought by the operator of the dump site. A second, fresh environmental impact assessment then supported the continued operation in light of further precautions promised by the operator. A separate investigation by the local authority concurred and the licence was duly extended. Despite assurances, subsequent inspections revealed serious shortcomings in terms of air pollution, accident control, and waste management, but the site remained in operation nonetheless. Before the ECtHR, the government did not dispute the existence of an infringement of Article 8, but instead argued that this was justified under Article 8 (2) ECHR: the public interest in adequate waste disposal allegedly outweighed the infringement.

The Court noted that the responsibility of a state may arise from a failure to properly regulate private sector activity.[23] The Court found that decisions with a significant environmental impact must be preceded by thorough risk assessments, and that under Article 8 ECHR the public must be given access to the records in order to form an opinion on the potential dangers.[24] Ultimately, the Court did not limit itself to this finding of a procedural violation. The wording of the judgment also points to a substantive violation of Article 8 ECHR, with the Court holding that the applicant had spent years impaired by the dangerous activities despite ample evidence suggesting an unlawful operation producing dangerous emissions.[25] The implication is that a failure to react to fresh evidence with enforcement action constituted a violation in itself.

Conversely, the case of *Hardy and Maile* demonstrates that consistent enforcement of good faith regulations effectively satisfies relevant human rights obligations.[26] The application was lodged in opposition to the construction and operation of two liquefied natural gas terminals. They alleged the authorities failed to assess the marine risks and failed to allow public access to all relevant information, in violation of Article 8 ECHR. The Court initially noted that environmental and economic policy making of states must '(. . .) involve appropriate investigations and studies so that the effects of activities that might damage the environment and infringe individuals' rights may be predicted and evaluated in advance and a fair balance may accordingly be struck between the various conflicting interests at stake.'[27] Against this backdrop, the Court examined the framework and its risk management system,

23 *Giacomelli* v. *Italy* 59909/00 (ECtHR, 2 November 2006) [78].
24 Ibid. [93]–[94].
25 Ibid. [95]–[97].
26 *Hardy and Malie* v. *United Kingdom* App no 31965/07 (ECtHR, 14 February 2012).
27 Ibid. [220].

concluding '(. . .) an extensive legislative and regulatory framework is in place in the United Kingdom, and more specifically at Milford Haven port, to promote safety and to limit the risks posed by the transfer and processing of LNG in the area.'[28] Relevant regulations were diligently followed, with the Court noting that

> It is clear that extensive reports and studies were carried out in respect of the proposed LNG terminals, by both HSE and MHPA, in cooperation with the developers. The planning and hazardous substances authorities as well as the domestic courts were satisfied with the advice provided by the relevant authorities. In the circumstances, it does not appear to the Court that there has been any manifest error of appreciation by the national authorities in striking a fair balance between the competing interests in the case.[29]

In terms of the disclosure of information, the applicants had failed to show that any substantive documents were being kept outside the public domain and scrutiny.[30] In addition, there were established procedures under which the applicants could obtain further disclosure as needed. Accordingly, the Court rejected a violation of Article 8 and dismissed the application.

We can leave to one side the problematic reasoning of the Court with respect to the existence and disclosure of documents in other cases as well as its faith in domestic tribunals to acquire all relevant evidence. This was already discussed at length with respect to the series of cases on *Test Veterans*.[31] The case of *Hardy and Maile* does, however, demonstrate that good faith regulations and effective enforcement are generally seen as sufficient to meeting the obligations under the ECHR. This general view is therefore compatible with the legal standards under the environmental minimum.

6.2 ESTABLISHED AND EMERGING STANDARDS

The notion of established and emerging standards is overall less compatible with the ECHR case law. There are many examples of the Court citing established standards that are relevant to the case,[32] in addition to frequent

[28] Ibid. [223].
[29] Ibid. [231].
[30] A failure that similarly defeated the application of *McGinley and Egan* v. *United Kingdom* App nos 21825/93; 23414/94 (ECtHR, 9 June 1998).
[31] See Chapter 5, Section 5.3.
[32] A prominent example are the frequent references to international standards in cases involving prison conditions, see, for instance, *Bragadireanu* v. *Romania* App no 22088/04 (ECtHR, 6 December 2007) [73].

references and reliance on the Aarhus Convention.[33] However, these standards rarely form the basis for a violation or the dismissal of an application. The modern case law is more promising, demonstrating a subtle shift in the approach of the Court over the past decade.

In the case of *Fägerskiöld*, the applicants complained of noise and light emission from wind turbines at their Swedish holiday home. At the relevant time, there were no regulations in Sweden on the maximum noise levels of wind turbines. In a move in line with the environmental minimum, the Court turned to the domestic recommendations of the Swedish Environmental Protection Agency and WHO guidelines.[34] Ultimately, the decision to reject the application was based on the fact that the WHO standards were upheld.[35] In two further cases, the Court based its finding of a violation in part on a breach of international guidelines, even though an infringement of domestic regulations was likewise established.[36]

The Court also makes use of international guidelines to evaluate the quality and scope of environmental regulations. In the case of *Tatar*, polluted water from mining activities contaminated waterways with cyanide, affecting numerous neighbouring countries. The Court relied on Principle 21 of the Stockholm Declaration and Principle 14 of the Rio Declaration to hold states under a duty to discourage and prevent transboundary environmental harm.[37] Remarkably, it held that '(...) beyond the national legislative framework established by the Environmental Protection Act, specific international standards existed which could have been applied by the Romanian authorities.'[38]

Even though the applicants ultimately did not provide sufficient evidence of noise pollution in the case of *Borysiewicz*, the Court made it clear that international environmental standards are relevant benchmarks to determine whether the pollution rises to the level of a violation.[39] In finding a violation in the case of *Brincat*, the Court relied on the absence of domestic regulation of asbestos exposure,[40] and then turned to the international scientific evidence

[33] See, for instance, *Taskin and others* v. *Turkey* App no 46117/99 (ECtHR, 10 November 2004) [99] and [119]; *Grimkovskaya* v. *Ukraine* App no 38182/03 (ECtHR, 21 July 2011) [39].

[34] *Fägerskiöld* v. *Sweden* App no 37664/04 (ECtHR, 26 February 2008) 9 citing Berglund and others, *Guidelines for Community Noise*.

[35] *Fägerskiöld* v. *Sweden* App no 37664/04 (ECtHR, 26 February 2008) 16.

[36] *Udovicic* v. *Croatia* App no 27310/09 (ECtHR, 24 April 2014) [111], [140] and [143]; *Oluic* v. *Croatia* App no 61260/08 (ECtHR, 20 May 2010) [60].

[37] *Tatar* v. *Romania* App no 67021/01 (ECtHR, 27 January 2009) [111].

[38] Ibid. [112], *my translation*.

[39] *Borysiewicz* v. *Poland* App no 71146/01 (ECtHR, 1 July 2008) [53].

[40] Ibid. [107].

available to decision makers at the time.[41] Thus in the terminology of the environmental minimum, the state failed to uphold the generally accepted standard and therefore violated the Convention.

With respect to emerging standards, there is little clear evidence that the Court has ever availed itself of them to decide a case. However, as the Court notes in general terms, comprehensive evidence is not required with respect to 'each and every aspect of the matter to be decided',[42] and the Court accepts that conflicting scientific evidence may suffice to engage the Convention rights.[43] This opens the door to emerging standards, which can provide a basis for action whenever there is a specific risk to human rights.

The Court has further accepted that states are under a duty to regulate dangerous environmental activities and that a failure to do so may lead to a violation of the Convention. This was established in the case of *Öneryildiz* for the purposes of positive obligations under Article 2 ECHR in the context of dangerous activities.[44] This position has been confirmed in subsequent cases,[45] and expressly expanded to regulatory measures to protect human health under Article 8 ECHR.[46] While this is evidently not analogous to finding a violation based on established and emerging standards, the obligation to regulate largely follows the same rationale. In many cases, there will not be much practical difference in finding a violation on the basis that environmental harm arising from dangerous activities is left unregulated by a state. Ultimately, however, the case law is not conclusive on the compatibility with established or emerging standards. At least on this point, it arguably contains the basic ingredients to support the environmental minimum, notwithstanding some significant challenges that we now turn to discuss.

41 Ibid. [37].

42 *Gaida* v. *Germany* App no 32015/02 (ECtHR, 3 July 2007) [1]; *Hatton and others* v. *United Kingdom* App no 36022/97 (Grand Chamber, 8 July 2003) [128]; *Giacomelli* v. *Italy* 59909/00 (ECtHR, 2 November 2006) [83].

43 *Gaida* v. *Germany* App no 32015/02 (ECtHR, 3 July 2007) [1].

44 *Öneryildiz* v. *Turkey* App no 48939/99 (Grand Chamber, 30 November 2004) [90].

45 *Budayeva and others* v. *Russia* App nos 15339/02; 21166/02; 20058/02; 11673/02; 15343/02 (ECtHR, 20 March 2008) [129]–[132]; *Brincat and others* v. *Malta* App no 60908/11; 62110/11; 62129/11; 62312/11; 62338/11 (ECtHR, 24 July 2014) [80]; *Kolyadenko and others* v. *Russia* App nos 17423/05; 20534/05; 20678/05; 23263/05; 24283/05; 35673/05 (ECtHR, 28 February 2012) [157]–[159].

46 *Tatar* v. *Romania* App no 67021/01 (ECtHR, 27 January 2009) [88]; *Bacila* v. *Romania* App no 19234/04 (ECtHR, 30 March 2010) [61]; *Brincat and others* v. *Malta* App no 60908/11; 62110/11; 62129/11; 62312/11; 62338/11 (ECtHR, 24 July 2014) [101]–[102]; *Di Sarno and others* v. *Italy* App no 30765/08 (ECtHR, 10 January 2012) [106].

6.3 LIMITATIONS

The modern case law generally does not challenge the claims of the environmental minimum. A minority of cases offer conclusions and arguments that would have to be revisited, and in part rejected. As the Court does not subscribe to a doctrine of *stare decisis*, this does not present much practical difficulty, if there is a willingness to amend and clarify doctrine. The challenging cases fall into two principal categories: first, cases where there were no legal standards covering a specific risk and yet the Court did not apply established standards in their place, and second, cases where legal standards existed, but the Court did not find a violation resulting from their inconsistent enforcement.

6.3.1 *Absence of Legal Standards*

The first significant modern case to expressly present an argument incompatible with the environmental minimum is the case of *Khatun*.[47] The case involved a government scheme to develop Canary Wharf and the special fast track dispensations from the Town and Country Planning Act 1971 that construction companies had been granted. The applicants complained of excessive noise and dust emissions which deprived them of the use of their gardens and forced them to keep windows shut during the daytime. While the dust nuisances fell below the statutory threshold of the Environmental Protection Act 1990, the Commission did not scrutinise whether this regulation was adopted in good faith in light of scientific evidence. Instead, it declined to find a violation following a balancing against the economic interests of the development.[48] Had the environmental minimum been applied to this case, a more thorough investigation of compliance with established standards would have been necessary, potentially finding the regulation was adopted in bad faith.

In this vein, the case of *Greenpeace e.V.* involved a complaint of four individuals and the German branch of Greenpeace against the refusal of German authorities to take specific measures against emissions from diesel engines.[49] While domestic courts accepted that the individuals could claim a right to be protected from these emissions, they declined to find a violation of fundamental rights. Domestic courts only required that some measures were

[47] *Khatun and 180 Others* v. *United Kingdom* App no 38387/97 (Commission, 1 July 1998).
[48] Ibid., 6.
[49] *Greenpeace e.V. and others* v. *Germany* App no 18215/06 (ECtHR, 12 May 2009).

taken against the emissions and that these were not entirely inadequate. When the case reached the ECtHR, it examined the case under Article 8 ECHR. However, instead of enforcing established standards, the ECtHR proceeded to recall the subsidiary role of the Convention, and relied on the margin of appreciation afforded to states to dismiss the application. This outcome is at odds with the requirements of the environmental minimum, especially in its emphasis on economic interests, an issue that we shall return to in Chapter 7.

At the relevant time, there were no binding threshold values in Germany with respect to air pollution. While EU Directive 2008/50/EC set out air pollution targets for member states, the ECJ had already clarified that the Directive only required governments to provide an action plan to ensure general compliance with these targets in due course.[50] To the extent that one may regard the Directive as setting an air pollution standard, it remains unclear whether it was adopted in good faith within the meaning of the environmental minimum.

The significant flexibility and dispensation the European Commission granted to member states on reaching the air pollution targets arguably points to bad faith. Air pollution constitutes a well-established specific risk, and regulations that merely require an action plan fail to tackle that specific risk. However, the targets of the Directive were at least based to a significant extent on WHO guidelines, which reflect the generally accepted recommendations of the relevant scientific community. The discrepancy and the timeline for meeting the targets is conceivably the result of a good faith balancing of environmental interests with competing considerations, for instance, with the time and effort necessary to adapt a wide range of conventional industrial activity, and offering the public alternatives to individual mobility. Depending on how one evaluates the divergence from the generally accepted standards, there is a reasonable case in support of both good and bad faith regulation. However, the various dispensations that individual member states have been afforded for highly polluted areas over the years by the European Commission,[51] in my view tilt the balance towards bad faith: for decades, the Commission was evidently not seriously interested in meaningful enforcement measures, making it doubtful that the Directive was ever intended to function as a legal standard for environmental regulation within the meaning of the framework.

50 Case C-237/07 *Dieter Janecek* v. *Freistaat Bayern* [2008] ECLI:EU:C:2008:447, see the more detailed discussion in Chapters 9 and 10.

51 'Commission Warns Germany, France, Spain, Italy, and the United Kingdom of Continued Air Pollution Breaches'.

In the absence of legal standards, assuming bad faith, the environmental minimum defaults to established standards, which plainly existed at the time of the ruling in the guise of the WHO air quality guidelines.[52] Under the environmental minimum, the ECtHR would then investigate whether the air pollution levels had exceeded these established standards. In the present case, this is somewhat difficult to discern, because the ECtHR made no such findings of fact. However, based on the domestic court proceedings and available data from the WHO database, we can conclude that the emissions considerably exceeded relevant guidelines at the time.[53] This in turn ought to have led the ECtHR to find a violation of the Convention.

Should one be inclined to disagree with that assessment, the decidedly mixed enforcement practice of the Directive through member states constitutes a separate concern. If we accept the Directive as a legal standard, then it was certainly not consistently enforced. Most regional action plans manifestly failed to live up to the requirements of the Directive, as evidenced by various domestic court rulings that struck down action plans.[54] The apparent lack of consistent enforcement is therefore a credible violation of the environmental minimum, at the time of the *Greenpeace e.V.* ruling of the ECtHR.

Overall, the case presents one of the strongest recent challenges to ECHR compatibility with the environmental minimum: either because established standards were not substituted for a bad faith legal standard, or because the inconsistent enforcement of a legal standard was not treated as a violation.

6.3.2 *Lack of Consistent Enforcement*

Cases where inconsistent enforcement is not found to violate the Convention are comparatively rare, but nonetheless raise serious compatibility concerns. The case of *Rupar* is one such example.[55] An illegally constructed apartment extension on a neighbouring property caused a significant fire and environmental hazard for the applicant. During the construction process, the applicant succeeded in obtaining injunctions, which enjoined the neighbour from

[52] *WHO Air Quality Guidelines.*

[53] 'WHO Global Urban Ambient Air Pollution Database', World Health Organization (2016) www.who.int/phe/health_topics/outdoorair/databases/cities/en/, accessed 1 April 2021; 'Ambient (Outdoor) Air Quality and Health'.

[54] In the United Kingdom, see *R (ClientEarth)* v. *Secretary of State for the Environment, Food and Rural Affairs* [2015] UKSC 28; *R (ClientEarth (No 2))* v. *Secretary of State for the Environment, Food and Rural Affairs* [2016] EWHC 2740 (Admin); for Germany, see Chapter 9.

[55] *Rupar* v. *Slovenia* App no 16480/02 (ECtHR, 18 May 2010).

continuing the construction.[56] However, due to a five-year lapse in enforcement, construction efforts were not halted by local authorities.[57] The ECtHR relied on the fact that the applicant had not demonstrated a loss in property value, nor shown an absence of appropriate compensation, to dismiss the application. The peculiarities of the Court's doctrinal approach to Article 1 AP-I notwithstanding, this is a result that does not square with the environmental minimum and its insistence on enforcing legal standards consistently.

The case of *Atanasov* provides a more recent example of this problematic approach of the ECtHR.[58] The applicant's home was situated one kilometre from a tailings pond and floatation plant of a former copper-ore mine. After the mine ceased operations, a reclamation scheme was introduced which called for the replanting of vegetation in the pond. After the scheme was abruptly cancelled, an alternative scheme proposed by a local business was implemented, even though authorities had rejected and questioned the viability of the proposed plan: it involved sludge, which was feared to contain high concentrations of heavy metals and thus constituted an environmental hazard. Following the results of an independent investigation the alternative scheme was eventually halted,[59] but not before, approximately 97 of 98.3 ha of the pond had already been covered in sludge.[60] Bulgarian law required a full environmental impact assessment for projects of this kind, which had not been completed by the time the alternative scheme commenced.[61] While the ECtHR noted that the protection of the environment was an increasingly important consideration, it ultimately did not find a violation of the Convention.[62]

The decision is problematic under the environmental minimum in two respects. For one, the Court relies on the lack of unequivocal information on the detrimental impact of the alternative reclamation scheme to conclude that there was no environmental harm, and hence no specific risk.[63] This runs counter to the environmental minimum, which requires that specific risks to human rights must suffice, provided they are based on a reasonable hypothesis. States could otherwise absolve themselves of responsibility simply by not properly investigating alleged environmental harms. Second, had the Court

56 Ibid. [8].
57 Ibid. [20].
58 *Atanasov* v. *Bulgaria* App no 12853/03 (ECtHR, 2 December 2010).
59 Ibid. [32].
60 Ibid. [40].
61 Ibid. [78] and [21].
62 Ibid. [66].
63 Ibid. [76]–[79].

applied the environmental minimum, it would have likely found a violation, as the state had not upheld and consistently enforced its own legal standards, namely by allowing the reclamation scheme to proceed without a full environmental impact assessment, as required by domestic law.

According to the environmental minimum, a failure to consistently enforce legal standards must lead to a violation, especially when findings to that effect are made by domestic courts. This must be the case regardless of the position one takes on the scope of the margin of appreciation afforded to states: that margin cannot extend to states violating their own domestic law, ignoring court rulings or their obligations under international law. This position is mostly, but as this section has shown, not invariably reflected in the doctrinal practice of the ECtHR.[64]

6.4 SUMMARY

As we have seen in this chapter there are some notable, but no systematic, challenges to the three main propositions of the environmental minimum with respect to minimum standards. The ECtHR will generally find a violation if the respondent state has failed to uphold legal standards, that is to say, requirements of its own domestic law, or clear international law obligations. There are a few unsystematic exceptions to this rule found in a smaller number of cases which would have to be consciously discarded under the environmental minimum. A failure to enforce legal standards was not invariably found to be a violation of the Convention, an absence of legal standards was not followed up based on established standards, and in some instances, domestic regulations were not examined as to whether they were adopted in good faith. On the latter, there is at least a reasonable argument in air pollution cases that relevant EU regulations were adopted in good faith. However, if that is the case, then their inconsistent enforcement remains a separate concern and cause for finding violations. Finally, recourse to emerging standards is rare and they ultimately do not appear to influence the outcome of a decision.

Apart from that, lapses in finding violations where domestic environmental regulations are not consistently enforced present the most serious area of dissonance between the environmental minimum and the doctrinal position under the ECHR. Some aspects of this ECHR case law would therefore have to be modified and consciously rejected to ensure a higher degree of

[64] Hilson, 'The Margin of Appreciation, Domestic Irregularity and Domestic Court Rulings in ECHR Environmental Jurisprudence', 285.

compatibility, but on the whole, it does not present an insurmountable challenge to the environmental minimum: the cases are isolated examples that are not indicative of the broader trends in the case law.

Chapter 7 considers the other side of the coin: the high-priority principle and the ECtHR approach to challenges directed against environmental regulations, as well as the overall balance of environmental protection with economic benefits and the margin of appreciation doctrine.

7

High Priority

This chapter explores the claim of the environmental minimum that environmental interests must be balanced with competing considerations, and that neither can be presumed to have priority from the outset. This becomes particularly relevant when considering cases where states justify infringements of human rights by invoking their duty to protect the environment and prevent environmental harms. Not every conceivable measure in furtherance of environmental protection is proportionate in light of its impact on other rights.

The question specifically for the ECHR case law is whether the accepted justifications of infringements demonstrate that environmental regulations are balanced with and can be outweighed by competing interests: in principle, the ECtHR must accept that environmental regulations can give rise to infringements of Convention rights and acknowledge that regulations must be proportional (Section 7.1). The chapter then discusses an enduring area of concern, namely the acceptance of vague economic justifications and the unprincipled application of the margin of appreciation doctrine to environmental harm (Section 7.2).

7.1 ENVIRONMENTAL CASES

7.1.1 *Data Analysis*

The range of Articles invoked to challenge environmental regulations (category two) generally spans the same spectrum as category one cases. A notable difference is the strong reliance on Article 1 AP-I (Right to property). Figure 7.1 demonstrates that the number of violations per invoked Article is markedly lower across the board than in category one cases.

The data shows that the practice and approach of the Court in category two cases has remained largely consistent throughout the decades. While the

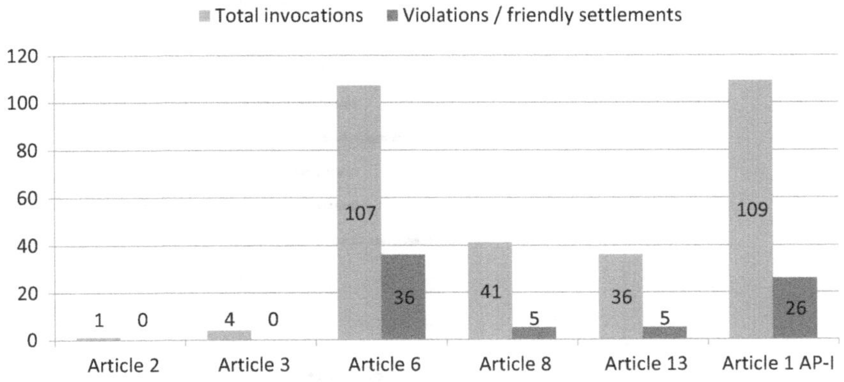

FIGURE 7.1 Number of successful invocations by Article in category 2

number of cases adjudicated in category two has fluctuated, the number of successful cases has remained strikingly consistent, as Figure 7.2 shows.

Although not all of the decisions surveyed exclusively involve environmental regulations, they all seek to regulate, at least partially, in the interests of environmental protection (even though there may be further policy considerations involved). The regulations can be broadly separated into the following sub-categories:

- *Planning and zoning*: this sub-category is closely related to expropriation. Applicants typically claim that a refusal to grant planning permission or a designation of their property as falling under environmental protection zoning regulations has violated their rights.[1] A prominent sub-category here is the string of cases involving the rights of 'gypsy'[2] applicants who were denied planning permission pursuant to so-called green belt policies in the United Kingdom.[3] Save for one, where a settlement was

[1] *Enzi* v. *Austria* App no 29268/95 (ECtHR, 8 February 2000); *Schädler-Eberle* v. *Liechtenstein* App no 56422/09 (ECtHR, 18 July 2013).

[2] I use the term solely within the confines of its legal meaning under the Caravan Sites and Control of Development Act 1960 in English Law and due to its prevalence in the judgments of the ECHR. It is my position that usage of the term is problematic considering the diversity of the groups and peoples thus categorised and its generally discriminatory and racist history; for further details, see, for instance, the statement from the Central Council of German Sinti and Roma: https://zentralrat.sintiundroma.de/sinti-und-roma-zigeuner/, accessed 1 April 2021.

[3] See, for instance, *Mabey* v. *United Kingdom* App no 28370/95 (Commission, 15 May 1996); *Woolhead* v. *United Kingdom* App no 31219/96 (Commission, 21 May 1997); *Eatson* v. *United Kingdom* App no 39664/98 (ECtHR, 30 January 2001); *Smith* v. *United Kingdom* App no 40435/98 (ECtHR, 30 January 2001).

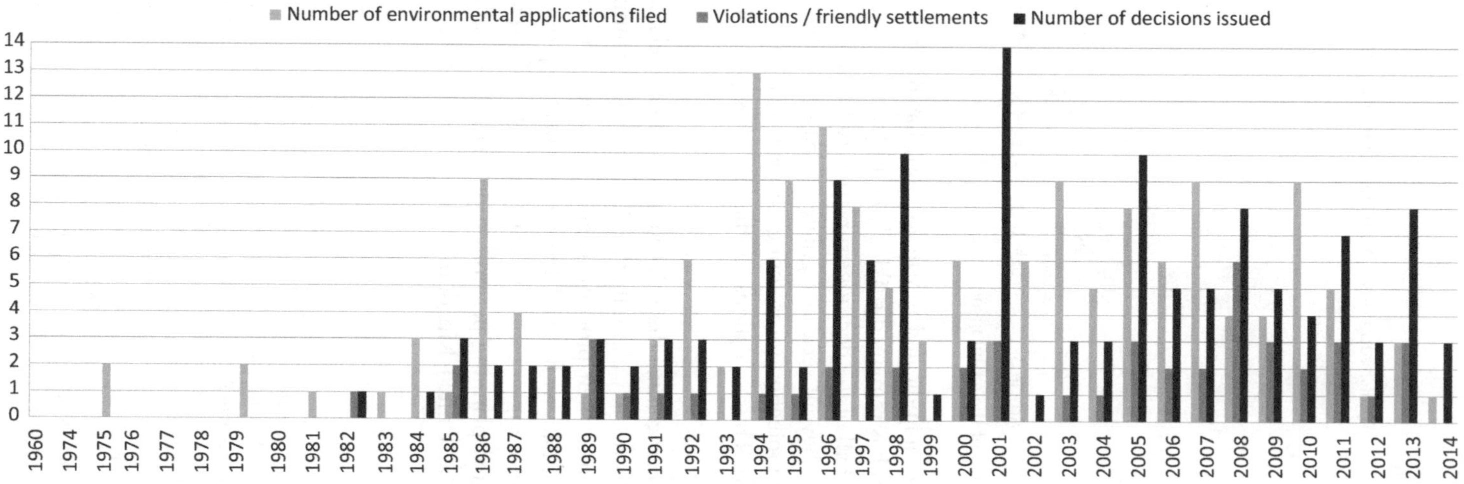

FIGURE 7.2 Number of environmental cases and violations per year in category 2 1960–2014

reached, all of the applications challenging this policy were ultimately dismissed or withdrawn.[4]

- *Expropriation*: cases in this sub-category deal with claims that the respondent state expropriated property either contrary to the law or failed to grant sufficient compensation. Representative examples centre on instances where authorities have expropriated property with a view to creating nature reserves,[5] at times with the motive of protecting specific endangered species,[6] or ensuring the environmental soundness of construction projects,[7] and in the interests of local disaster prevention measures, such as the reinforcement of dikes.[8] Decisions to expropriate have never amounted to a violation of the Convention in and of themselves, although applicants have been reasonably successful in claiming procedural violations or a lack of appropriate compensation.
- *Flora and fauna*: in this sub-category applicants challenge regulations that protect flora and fauna. Challenges directed against fauna regulations result in fewer violations (32 per cent) than those against flora regulations (53 per cent). The Court holds that states enjoy a wide margin of appreciation in taking measures to protect endangered species.[9]
- *Amenity*: this is a curious sub-category, covering regulations that are at least partially motivated by the preservation of a general environmental appeal. The ECtHR has recognised the interest in a pleasant environment as a key justification for infringements of Convention rights, chiefly Article 1 AP-I and Article 8 ECHR.[10] Indeed, whenever amenity was invoked as a justification, or implicitly motivated a regulation, violations were found in only 19 per cent of cases.
- *Industrial*: the industrial sub-category encompasses regulations that make some effort to provide environmental protection from industries and

4 *Varey* v. *United Kingdom* App no 26662/95 (ECtHR, 21 December 2000).

5 *Katte Klitsche de la Grange* v. *Italy* App no 12539/86 (ECtHR, 27 October 1994); *Turgut and Others* v. *Turkey* App no 1411/03 (ECtHR, 8 July 2008); *Eskelinen and 54 others* v. *Finland* App nos 19761/92; 21343/93 (Commission, 10 October 1994).

6 *Aarniosalo and others* v. *Finland* App no 39737/98 (ECtHR, 5 July 2005) and *Kukkola* v. *Finland* App no 26890/95 (ECtHR, 15 November 2005) [protection of threatened species of woodpeckers].

7 *Zumtobel* v. *Austria* App no 12235/86 (ECtHR, 21 September 1993); *Sporrong and Lönnroth* v. *Sweden* App nos 7151/75; 7152/75 (Plenary, 23 September 1982).

8 *Mens and Mens-Hoek* v. *Netherlands* App no 34325/96 (Commission, 20 May 1998).

9 *Jakobsson* v. *Sweden* App no 14459/88 (Commission, 19 February 1992) 6; *Günther* v. *Sweden* App no 26265/95 (Commission, 27 November 1996) 6; *Aarniosalo and others* v. *Finland* App no 39737/98 (ECtHR, 5 July 2005) 11.

10 *Herrick* v. *United Kingdom* App no 11185/84 (Commission, 11 March 1985); *Mabey* v. *United Kingdom* App no 28370/95 (Commission, 15 May 1996); *Kozubek* v. *Poland* App no 62040/00 (ECtHR, 11 December 2007).

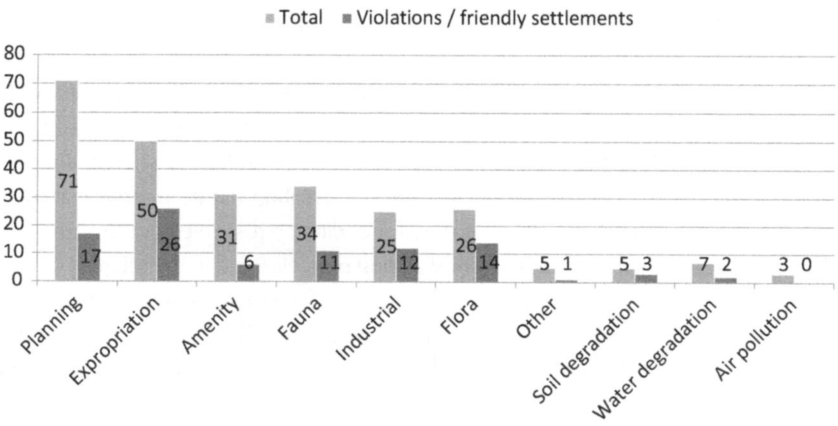

FIGURE 7.3 Nature of environmental regulations in category 2

business activities. Typical cases concern requirements for environmental impact assessments as well as the necessity and overall conditions attached to operating permits. However, there is an important caveat: all of the successful cases in this category were based either on Article 6 alone (11) or in conjunction with Articles 13 (1) and 1 AP-I (3). The cases therefore primarily relate to procedural violations, and do not draw into question the ability of states to regulate industry behaviour and business conduct as such.

- *Miscellaneous*: the remaining themes covered by environmental regulations are soil contamination (five cases), water pollution (seven cases), air pollution (three cases), and other (five cases). Owing to the comparatively low number of cases, it is difficult to identify any clear trends or doctrinal approaches towards such regulations.

Generally speaking, the Court grants states a wide margin of appreciation with respect to regulating environmental harms.[11] This is evidenced by the relatively low percentage of cases where a violation is found, on average only 33 per cent. Significantly increased violation rates appear in some categories, although this must be treated with a degree of caution, as the sample size varies significantly: soil contamination and water pollution are both examples where the low number of cases cautions against firm conclusions. Figure 7.3 provides a more detailed overview.

[11] *Fägerskiöld* v. *Sweden* App no 37664/04 (ECtHR, 26 February 2008) 19.

7.1.2 *Doctrinal Analysis*

In contrast to category one, there are no cases in category two that directly challenge the claims of the environmental minimum. The initial reluctance towards accommodating environmental harms which impeded the development of the case law in category one is absent from category two cases. The Court accepted from the beginning that environmental motives were suitable grounds for justifying infringements of Convention rights. The provisions most often engaged in category two cases are Article 1 AP-I (109) and Article 6 ECHR (107).[12] Under Article 1 AP-I, the measures and regulations challenged are too diverse to paint a clear picture of development akin to the case law in category one. The applicants challenge a plethora of regulations that at times only have incidental environmental motives. Much the same can be said of those applications that rely primarily on Article 6 ECHR. These cases typically revolve around the overall fairness of domestic proceedings, and hence are not primarily relevant to the environmental minimum.

7.1.2.1 Articles 2, 3, and 8

Alleged violations of both Articles 2 and 3 were consistently dismissed by the Court in all category two cases. Due to the severity of infringements of Articles 2 and 3 ECHR, it is unlikely that a state could justify any measure by reference to environmental protection.

Article 8 ECHR cases are also comparatively rare in category two. The provision is raised in only 41 cases, which is less than half the number of applications invoking Article 1 AP-I. The justification of infringements is handled flexibly by the Court, and environmental arguments are generally afforded the same weight as other considerations in justifying infringements. The case of *Herrick* provides an example where limitations placed on the use of property were deemed justified.[13] The applicant had bought a former bunker installation on the Isle of Jersey and converted it into a summer home. Local authorities initially prohibited its use, due to its location in an area of pristine natural beauty, but domestic courts partially overturned this decision. The EComHR did not rule whether this constituted a violation of Article 8 ECHR, as it would, in its view, be justified under paragraph 2 in any case: '(t)he existence and operation of planning controls which delimit areas where domestic development may be extended is a legitimate control measure to

[12] See the dataset gathered for this book.
[13] *Herrick* v. *United Kingdom* App no 11185/84 (Commission, 11 March 1985).

protect the amenity value of rural areas and thereby protect the rights of others.'[14] The EComHR thus held that environmentally damaging enjoyment of Article 8 ECHR rights may be justifiably limited.

Further examples are the previously mentioned cases brought by 'gypsy' applicants challenging the denial of planning permission.[15] Beginning in the 1990s and stretching into the early 2000s, several challenges were brought against refusals of planning permission for caravan sites pursuant to so-called green belt policies. Among the goals of the policies is the promotion of visual amenity and biodiversity,[16] as they are designed 'to check the unrestricted sprawl of large built-up areas; to prevent neighbouring towns merging into one another; to assist in safeguarding the countryside from encroachment; to preserve the setting and special character of historic towns; and to assist in urban regeneration, by encouraging the recycling of derelict and other urban land.'[17] According to the ECtHR, this interest in protecting the environment and visual amenity of the countryside was sufficient in all but one case to justify the infringements of Article 8 ECHR.[18]

7.1.2.2 Article 1 AP-I

The Court has consistently held that infringements of Article 1 AP-I may be justified through any number of competing interests. With respect to expropriation, environmental motives typically suffice as long as adequate compensation is ensured. This position has been clear in the case law since the earliest category two case of *Sporrong and Lönnroth*. In a narrow plenary decision (10–9) the Court found that Sweden had violated Article 1 AP-I and Article 6 ECHR (12–7) by failing for decades to resolve expropriation proceedings against the applicants. This left the applicants in legal limbo with a prohibition of all construction on the property, coupled with an inability to sell the property. However, at no point did the ECtHR judgment draw into question that Sweden was entitled to expropriate and limit the use of property for environmental motives: '(...) [the Court] finds it natural that, in an area as complex and difficult as that of the development of large cities, the

[14] Ibid., 280.
[15] I reiterate my previous reservations with respect to the use of this term (see n. 2).
[16] Louise Smith, *Green Belt – Briefing Paper 00934* (House of Commons Library, 2016).
[17] *National Planning Policy Framework* (Department for Communities and Local Government, 2012) 19.
[18] A settlement was reached in the case of *Varey v. United Kingdom* App no 26662/95 (ECtHR, 21 December 2000).

Contracting states should enjoy a wide margin of appreciation in order to implement their town planning policy.'[19]

Provided that sufficient compensation is afforded to the applicants and that the process is subject to review through a tribunal, no concerns typically arise under the ECHR.[20] Complete expropriation for environmental motives generally requires adequate and timely compensation.[21] Nonetheless, there are a few cases where no compensation was afforded, and yet no violation of the Convention was found. In *N.* v. *Austria*, local authorities had refused to allow industrial use of a property in an area designated for agricultural development, while *Simli* involved a complete prohibition on construction work on the applicant's property.[22] Even though no compensation was offered in either case, the EComHR rejected a violation of Article 1 AP-I, holding that there was a public interest in safeguarding the natural landscape. Similarly, demolishing an illegally constructed holiday home in a protected woodland area was considered proportional under Article 1 AP-I, even though no compensation was offered.[23]

Environmental concerns also play an important role in justifying infringements in the context of industrial activity. The case of *Fredin* involved the denial of a permit to extract gravel, justified chiefly on environmental grounds. The Court stated explicitly that environmental protection constitutes a legitimate interest limiting the protections under Article 1 AP-I.[24] Other measures for the protection of forests and the countryside,[25] as well as of coastal areas,[26] were likewise upheld. Owners may also be required to maintain their property

[19] *Sporrong and Lönnroth* v. *Sweden* App nos 7151/75; 7152/75 (Plenary, 23 September 1982) [27] and [69]; more recently, see *Varniene* v. *Lithuania* App no 42916/04 (ECtHR, 12 November 2013) [54].

[20] *Sporrong and Lönnroth* v. *Sweden* App nos 7151/75; 7152/75 (Plenary, 23 September 1982) [73] and [87].

[21] *Curmi* v. *Malta* App no 2243/10 (ECtHR, 22 November 2011); *Varniene* v. *Lithuania* App no 42916/04 (ECtHR, 12 November 2013) [54].

[22] *N* v. *Austria* App no 10395/83 (Commission, 18 July 1986); *Simli* v. *Belgium* App no 11965/86 (Commission, 12 December 1988).

[23] *Hamer* v. *Belgium* App no 21861/03 (ECtHR, 27 November 2007) [72].

[24] *Fredin* v. *Sweden* App no 12033/86 (ECtHR, 18 February 1991) [48]; for a more recent case see *Depalle* v. *France* App no 34044/02 (Grand Chamber, 29 March 2010) [81]; *Pine Valley Developments Ltd and others* v. *Ireland* App no 12742/87 (ECtHR, 29 November 1991) [maintenance of a green belt policy].

[25] *Turgut and Others* v. *Turkey* App no 1411/03 (ECtHR, 8 July 2008); *Köktepe* v. *Turkey* App no 35785/03 (ECtHR, 22 July 2008); *Şatır* v. *Turkey* App no 36192/03 (ECtHR, 10 March 2009).

[26] *Depalle* v. *France* App no 34044/02 (Grand Chamber, 29 March 2010); *NA and Others* v. *Turkey* App no 37451/97 (ECtHR, 11 October 2005).

in a certain way,[27] or denied specific uses of their property, without the need for compensation.[28]

7.2 LIMITATIONS

The tendency for the ECtHR to uphold existing environmental regulation in the face of rights-based challenges is encouraging with respect to the high-priority principle. There are no cases that suggest that environmental protection inevitably trumps conflicting rights, but neither are environmental interests automatically inferior. However, this promising picture is challenged somewhat through the acceptance of vague economic justifications and the at times unprincipled application of the margin of appreciation doctrine.

7.2.1 *Economic Benefits*

The ECtHR does not view environmental protection as superior to any other interest in the context of justifying infringements.[29] This is compatible with the high-priority principle, as it expressly does not require environmental considerations to be granted absolute priority. Rather, the principle is content with raising the importance of environmental protection to the same level as other interests based on human rights. Nonetheless, in some cases, the Court has accepted claims of vague economic benefits as justifications for environmental harm.[30] This is potentially problematic because an overly strong favouring of economic considerations raises the concern that specific risks under the environmental minimum are not adequately addressed.

The case of G *and* E involving indigenous reindeer herders in Finland is an early notable decision where a state succeeded in justifying environmental harm with economic benefits.[31] The applicants objected to the construction of a hydro-electric dam which threatened to flood an area of traditional grazing land. The Commission, while acknowledging the protection of minority rights through Article 8 ECHR in principle, rejected a violation under the specific

27 *Denev* v. *Sweden* App no 12570/86 (Commission, 18 January 1989) [on a domestic law requiring the planting of indigenous pine trees].

28 *Jakobsson* v. *Sweden* App no 14459/88 (Commission, 19 February 1992) [refusal of permission to hunt elk on the property].

29 *Fadeyeva* v. *Russia* App 55723/00 (ECtHR, 9 June 2005) [103]–[105].

30 *Noack and others* v. *Germany* App no 46346/99 (ECtHR, 25 May 2000); *Hatton and others* v. *United Kingdom* App no 36022/97 (Grand Chamber, 8 July 2003); on the ambiguous use of the margin of appreciation, *see* Letsas, 'Two Concepts of the Margin of Appreciation'.

31 G *and* E v. *Norway* App no 9278/81; 9415/81 (Commission, 3 October 1983) 36.

circumstances of the case. Even if the flooding constituted a violation, it would be justified under Article 8 (2) ECHR, as the production of energy constituted a legitimate economic goal. This is similar to the case of *Kjellberg* involving gravel extraction, where concerns over the environmental impact and noise were similarly dismissed by reference to economic benefits.[32]

The case of *Powell and Rayner* served to further bolster this trend of economic justifications.[33] The applicants claimed a violation of Article 8 ECHR from noise emissions from Heathrow Airport. While the Court held it possible that an infringement of Article 8 occurred, it deemed it justified as a legitimate economic goal was being pursued.[34] The case of *Noack* involved relocation proceedings to accommodate coal mining under the town of Horno in Germany.[35] While the Court agreed that Article 8 ECHR had been infringed, it deemed the violation justified. The state enjoyed a wide margin of appreciation, and the balancing of competing interests had been conducted fairly.

One particularly astonishing case is that of G. A. v. *Sweden*.[36] The applicant lived in close proximity to a dock used by a state-run ferry service, which caused low-frequency noise that the applicant found intolerable. An inquiry later determined that the noise levels reached 52–60 dB (IL) when the ferry passed and 52 dB (IL) when idling at the ferry berth. A series of complaints and appeals eventually led to only minor improvements for the applicant. The applicant claimed violations of Articles 3, 5, and 8 ECHR, as well as Article 1 AP-I. The Commission accepted and confirmed that in principle, noise emissions 'may constitute an interference with an individual's right to respect for his private life and home if the level of noise attains a certain level of intensity.'[37] However, it considered the infringement justified under Article 8 (2) ECHR as being in accordance with the relevant domestic legislation, which deemed noise levels of up to 90 dB acceptable. The case therefore raises serious questions about whether the relevant provision of Swedish law was adopted in good faith, as the noise levels are considerably outside the scope of generally accepted guidelines.[38] The Commission accepted the

32 *Kjellberg* v. *Sweden* App no 13724/88 Commission (Commission, 5 March 1991).

33 *Powell and Rayner* v. *United Kingdom* App no 9310/81 (ECtHR, 21 February 1990).

34 On airport noise justified by economic considerations, see also *Eckenbrecht and Ruhmer* v. *Germany* 25330/10 (ECtHR, 10 July 2014) [35]; *Flamenbaum and others* v. *France* App nos 3675/04; 23264/04 (ECtHR, 13 December 2012).

35 *Noack and others* v. *Germany* App no 46346/99 (ECtHR, 25 May 2000).

36 GA v. *Sweden* App no 12671/87 (Commission, 13 March 1989).

37 Ibid., 3.

38 Berglund and others, *Guidelines for Community Noise*, xi.

general contribution of the ferry service to the economic development as sufficient, instead of inquiring whether its specific noise emissions and operation was justified. It is difficult to imagine such a ruling would be issued by the Court under its modern doctrinal approach, but it does speak to the sway that economic justifications once held.

These cases leave the compatibility with the environmental minimum in some doubt. The environmental minimum insists that environmental interests, to the extent that they impact human rights, must be given a fair hearing, and must stand an even chance of prevailing over economic considerations in the course of human rights litigation. That being said, the high-priority principle does not require that environmental interests outweigh other considerations in every conceivable case, nor does it eliminate the need to balance environmental concerns with competing interests. However, unsubstantiated and general references to economic benefits should not suffice to justify specific environmental harm that impacts human rights. Generally speaking, that is also the doctrinal position under the ECHR, notwithstanding the line of cases we discussed. The more problematic elements of the doctrinal approach of the Court often rely on the margin of appreciation doctrine, which is deployed to justify outcomes that would otherwise constitute violations of the Convention.

7.2.2 *Margin of Appreciation Doctrine*

The ECHR Grand Chamber ruling in *Hatton* exemplifies how the margin of appreciation doctrine and economic justifications are deployed to reach conclusions that are problematic under the environmental minimum.[39] The applicants lived near Heathrow Airport and experienced significant daytime and night-time aircraft noise emissions.

The ECHR chamber judgment initially found a violation of Article 8 ECHR through the government's failure to strike a fair balance between the rights of the applicants and the economic interests of the state.[40] The chamber concluded that while it was likely that 'night flights contribute to a certain extent to the national economy as a whole, the importance of that contribution has never been assessed critically, whether by the government directly or by independent research on their behalf.'[41] Mere assertions of economic significance on their own were insufficient to justify the

[39] *Hatton and others* v. *United Kingdom* App no 36022/97 (ECtHR, 2 October 2001).
[40] Ibid. [107].
[41] Ibid. [102].

interference. A further violation of Article 13 was found due to the limited recourse of applicants to challenge government decisions in domestic courts.[42]

Citing the margin of appreciation, a majority of the Grand Chamber (12–5) overturned the finding of a violation under Article 8 ECHR, while upholding the ruling with respect to Article 13 ECHR. In part this decision was reached because the Grand Chamber found that the applicants were in a position to move without financial loss: the housing prices had apparently remained stable despite their proximity to the airport.[43] The Grand Chamber ruling was criticised as a major step backwards in a joint dissenting opinion and in legal scholarship.[44] From a modern doctrinal perspective, the discussion of the margin of appreciation appears puzzling.

The Grand Chamber invoked the case of *Dudgeon*,[45] the decision holding the criminalisation of homosexuality in Northern Ireland contrary to Article 8 ECHR, to muse on whether the 'intimacy' of night-time sleep called for a narrower margin of appreciation than environmental policy questions would ordinarily suggest.[46] It found that 'the sleep disturbances relied on by the applicants did not intrude into an aspect of private life in a manner comparable to that of the criminal measures considered in Dudgeon to call for an especially narrow scope for the State's margin of appreciation.'[47]

This comparison with *Dudgeon* was ostensibly relevant because the Grand Chamber viewed itself in a subsidiary role on matters of policy and found 'intimacy' of infringements a compelling standard. The Court decisively applied the margin of appreciation in favour of the state at the justification stage: at least this is a reasonable interpretation leaving to one side some contradictory statements earlier in the ruling.[48] Thus understood, the margin of appreciation grants states flexibility in justifying infringements even when the interests invoked, such as the economic benefits, remain general and unsubstantiated. This is problematic in at least three respects, which go some way to explaining the limited doctrinal impact of the margin of appreciation approach adopted in *Hatton* on subsequent environmental cases.

42 Ibid. [116].

43 *Hatton and others* v. *United Kingdom* App no 36022/97 (Grand Chamber, 8 July 2003) [127].

44 Hyam, 'Hatton v United Kingdom in the Grand Chamber'.

45 *Dudgeon* v. *United Kingdom* App no 7525/76 (ECtHR, 22 October 1981) [criminalisation of homosexuality].

46 *Hatton and others* v. *United Kingdom* App no 36022/97 (Grand Chamber, 8 July 2003) [102] and [103].

47 Ibid. [123].

48 Ibid. [99]–[100], where it is discussed together with the scope of Article 8 ECHR, thereby implying the margin of appreciation also applies to the scope of a Convention right.

First, there is a point on consistency. The margin of appreciation outlined by the Grand Chamber suggests a departure from ordinary rules of evidence. The Court typically requires substantiation and an evidentiary basis for assertions from both applicants and states, particularly with respect to infringements and grounds for justification.[49] In *Hatton*, the Grand Chamber instead appears to use the margin of appreciation to loosen the evidentiary requirements in favour of the state, to a point where it was content with mere general assertions as to the economic significance of this particular night-time flight policy.

Second, there is a point of principle connected to the shift in the rules of evidence. The margin of appreciation conceived by the Grand Chamber ultimately lowers the practical protection afforded to individuals: the greater the margin of appreciation, the freer states are to engage in actions that would otherwise constitute violations of Convention rights.[50] The Grand Chamber in *Hatton* expanded the range of possible justifications to include unsubstantiated claims of economic benefits that do not respond to the particular environmental harm challenged by the applicants. In part, this is because, as George Letsas has pointed out, the margin of appreciation remains ill-defined and nebulous in its normative implications in the ECtHR case law.[51] This makes it difficult to normatively assess the doctrine.[52] Some suggest that the margin of appreciation merely refers to and encapsulates the discretion enjoyed by states in fulfilling their obligations under the Convention, particularly with regard to qualified rights.[53] In this limited sense, however, the margin of appreciation is both not particularly controversial or useful as a concept.[54] States can of course fulfil their human rights obligations in a myriad of ways.

This is because, for instance with respect to qualified rights, not all infringements are invariably violations: states are required to balance the demands of competing interests (contrast with so-called absolute rights: Article 3 ECHR). The specific measures adopted by states remain subject to review, and it is ultimately for the ECtHR to determine whether states have struck a fair balance. States can only permissibly infringe rights if the conduct in question is justified under the invoked right, or where it forms part of a

49 Thienel, 'The Burden and Standard of Proof in the European Court of Human Rights', 563.

50 Letsas, 'Two Concepts of the Margin of Appreciation'.

51 George Letsas, 'The Margin of Appreciation Revisited: A Response to Føllesdal' in Adam Etinson (ed.), *Human Rights: Moral or Political?* (Oxford University Press 2018) 296.

52 Letsas, 'The Margin of Appreciation Revisited' 309.

53 A Føllesdal, 'Appreciating the Margin of Appreciation' in A Etinson (ed.), *Human Rights: Moral or Political?* (Oxford University Press 2018) 269.

54 Letsas, 'The Margin of Appreciation Revisited', 298.

permissible derogation under Article 15 ECHR. In those cases, the state, by definition, had 'discretion' to act in this manner. Conversely, should a right require regulation of an environmental harm, then the state has no 'discretion' to do otherwise. A 'discretion' concept of the margin of appreciation, therefore, adds little to the conventional analysis. The scope of Convention rights and the range of permissible justifications as interpreted by the Court determine the margin of appreciation for states, not the other way around.[55]

Letsas suggests a further possibility is to understand the margin of appreciation not as expressing normatively redundant 'discretion' of states in fulfilling their obligations, but instead as requiring a level of deference of the ECtHR *to* states.[56] The Court may determine that an unjustified infringement of the Convention has occurred but nonetheless declines to find against the state because of the weight it attaches to the state's domestic legal assessment. This clarifies the doctrine somewhat, but still lacks a clear normative basis and the conditions under which such deference is appropriate.[57] There is no need for our purposes to delve any deeper into this debate, save to note here that a 'deference' understanding of the margin of appreciation would have serious implications for the rule of law and the purpose of international treaty cooperation. Should those accused of infringing human rights law be empowered to make the final assessment of their compliance, then what is the point of domestic and international human rights protection regimes?[58]

It is not clear whether either the 'discretion' or 'deference' model was indeed the understanding of the ECtHR in *Hatton*. A key defect of the Grand Chamber decision is that it declines to define the margin of appreciation, and does not offer any specific guidelines for its application: the perplexed reader is left only with vague references to 'general policy' that supposedly justify a more generous margin, and the 'intimacy' of a protected right which requires a narrower margin of appreciation.[59] Thankfully, the peculiar and vague understanding of the margin of appreciation advanced by the Grand Chamber in *Hatton* has been largely ignored in the modern doctrinal approach, at least with respect to environmental cases.

As we have seen, in its modern doctrinal approach the Court does not primarily rely on the margin of appreciation doctrine when evaluating state action, especially with respect to Article 8 ECHR. Decisions still invoke the

55 Ibid., 299.
56 Ibid., 300.
57 Ibid.
58 Ibid., 301.
59 *Hatton and others* v. *United Kingdom* App no 36022/97 (Grand Chamber, 8 July 2003) [100]–[103].

concept, but it now primarily justifies the Court's reliance on the factual assessment of state entities (barring evidence to the contrary), especially with respect to compliance with domestic regulations and deference on appropriate remedies for violations. This still entails certain difficulties, as both of these questions can impact the scope of protection and hence the level to which the Court assigns high priority to environmental harm, but it is less overtly problematic. The contemporary approach is significantly less lenient on infringements of rights and justifications invoked by the state.

7.3 SUMMARY

Both the data and doctrinal analysis demonstrate that in contrast to category one cases, there are few cases in category two that directly challenge the compatibility of the ECHR with the environmental minimum. The initial reluctance towards accommodating environmental harms which impeded the development of the case law in category one is almost entirely absent from category two cases. The Court accepted early on that environmental motives were suitable grounds for justifying infringements of Convention rights, and it has frequently confirmed this doctrinal position. This is especially true with respect to property, which is frequently invoked but generally only found violated for procedural reasons or inadequate compensation. The ECHR does not draw into question the ability of states to limit Convention rights in the interests of environmental protection: that possibility has been almost universally affirmed across all category two cases.

The broader balance between environmental protection and competing human rights under the ECHR is somewhat upset through the strong emphasis on economic benefits. These are often successfully invoked as justifications for environmental harm, as the chapter demonstrated in a string of cases culminating in the Grand Chamber decision of G. A. v. *Sweden*. What makes that decision particularly troubling is that the ECtHR was prepared to accept blatantly inadequate environmental regulations relating to noise based on vague and general economic benefits that led to an intolerable state of affairs for the applicant.

Finally, the margin of appreciation doctrine, especially its vague nature and unclear conditions of application, constitutes a major concern for a consistent approach to environmental harm under the Convention. The Grand Chamber decision in *Hatton* put forward an understanding of the margin of appreciation that is difficult to justify on principled grounds and, at times, even lacks sufficient elaboration to determine basic coherence. On this basis, one can only guess what – if any – questions the Court might consider in order to determine whether a claim falls within its broad understanding of policy, what principles guide the intimacy of an interference, and why night-time sleep is less 'intimate' than sexual preferences. To an extent, the decision in *Hatton* can be seen as an outlier that does not comport with the modern doctrine on environmental cases, but it does raise concerns over the compatibility of the ECHR with the high-priority principle of the environmental minimum. The fact that the *Hatton* understanding of the margin of appreciation doctrine has not held sway, and that generally the role of the doctrine has been less prominent in subsequent environmental decisions, at least offers some hope.

PART III

Beyond the European Convention

Part III expands our perspective beyond the European Convention on Human Rights. It explores the doctrinal position in the other examined international human rights protection regimes as well as in the domestic jurisdictions of South Africa and Germany.

Chapter 8 considers the African Charter on Human and Peoples' Rights (ACHPR), the American Convention on Human Rights (ACHR), and the International Covenants on Civil and Political Rights (ICCPR), as well as Economic, Social and Cultural Rights (ICESCR). The position in these international human rights protection regimes is generally less encouraging than under the ECHR. Too few environmental cases have been decided to draw firm conclusions, but what has surfaced is far from unequivocally supportive of the environmental minimum framework.

In Chapter 9, we consider two case studies, namely of South Africa and Germany, which implemented a constitutional environmental right and a constitutional mission statement, respectively. Despite these provisions, there are doctrinal gaps that demonstrate that the environmental minimum can add something of value even to two comparatively sophisticated protection regimes. Both jurisdictions have thus far been unable to lend their respective constitutional provisions a decisive role in shaping environmental regulation and policy, a state of affairs the environmental minimum is well equipped to remedy.

8

International Protection Regimes

This chapter presents an overview of the case law of further international and regional human rights protection regimes examined for this book. The number of cases is generally limited and the doctrinal position on environmental human rights is therefore less clear when compared to the ECHR. Therefore, it is difficult to conclude with much certainty whether the environmental minimum is compatible with the prevailing doctrinal position under the UN Covenants, the African Charter on Human and Peoples' Rights, and the American Convention on Human Rights. Some aspects of the doctrinal approach are generally supportive of core normative claims and principles of the framework, while others are in tension with the environmental minimum. Nonetheless, if the framework were consistently applied to conceptualise environmental harm in these protection regimes, it would bring about significant improvements towards a more coherent and principled case law aligned with the normative claims of environmental human rights.

8.1 THE UN COVENANTS

The decisions of the HRC on the ICCPR contain fourteen broadly environmentally themed applications,[1] of which nine fall into category one (protective function), and five fall into category two (limitation function).[2] Overall, only four applications ultimately led to a violation, and the reasoning does not hold much encouragement for the proposition of environmental protection under the ICCPR. As a discrete issue, environmental harm was apparently not

[1] The Covenants employ the terminology of author and communication to describe, respectively, the individual bringing an application to the Human Rights Council and the application itself, the terminology in the book was standardised differently.

[2] See the dataset on the UN Covenants gathered for this book.

considered at the drafting stage,[3] although Alfred Rest is correct to point out that this alone does not close the door to future developments.[4]

In the matter of *Chief Bernard Ominayak* the HRC recognised a violation through the expropriation of traditional tribal lands for commercial interests.[5] However, this was expressly based on Article 27 of the ICCPR (Minority protection), and all other articles invoked by the applicants were dismissed. While the commercial development undoubtedly raised issues of environmental harm, it is not mentioned as a relevant factor in the decision. Indeed, the HRC avoided so much as mentioning the environment in the operative part of the decision.[6]

The HRC next had occasion to rule on the relevance of environmental harm in the matter of *Äärelä and Jouni*.[7] The application concerned a domestic court ruling allowing logging and road construction in an area vital to the Sami people for reindeer herding. The application before the HRC revolved primarily around procedural questions and the burden imposed on the applicants through an adverse domestic cost order. The applicants invoked Articles 2, 14, and 27 of the ICCPR. The HRC decision to dismiss Article 27 underscores the absence of environmental considerations.[8] It found that the award of costs and the inability of the applicants to challenge submissions before the domestic court constituted a violation of Articles 14 (Equality before the courts and tribunals) and 2 ICCPR.[9]

The matter of *Poma Poma* concerned diversion of a water source, which caused the drying out of wetlands and destruction of the local ecosystem as well as the collapse of the applicants' livestock herds, depriving the community of their livelihood.[10] The applicant invoked Articles 1, 2, 14, 17, 27 ICCPR. The HRC found violations of Articles 2 and 27 ICCPR, based on the loss of livelihood, the exclusion of the applicant from the decision-making process,

3 S. Hobe, 'Menschenrechte auf Umweltschutz? – Bestand und Begründungsmöglichkeiten' [1994] *Zeitschrift für Umweltrecht* 15, 16.

4 Alfred Rest, 'Europäischer Menschenrechtsschutz als Katalysator für ein verbessertes Umweltrecht' (1997) 19 *Natur und Recht* 209, 210.

5 *Chief Bernard Ominayak and the Lubicon Lake Band* v. *Canada* Communication No 167/1984 (UN Human Rights Committee, 26 March 1990).

6 Ibid. [33].

7 *Äärelä and Jouni* v. *Finland* Communication No 779/1997 (UN Human Rights Committee, 24 October 2001).

8 Ibid. [7.6].

9 Ibid. [7.2] and [7.4].

10 *Poma Poma* v. *Peru* Communication No 1357/2006; UN Doc CCPR/C/95/D/1457/2006 (2009) (UN Human Rights Committee, 27 March 2009).

and procedural shortcomings, including the absence of an economic impact assessment as well as the denial of an effective remedy.[11] The HRC again did not make any determination on the relevance of environmental harm under the ICCPR, even though the case clearly raised that issue. The environmental minimum emphasises the need to recognise environmental harm as a threat to human rights, regardless of traditionally accepted threat scenarios: all three cases appear at odds with this insight.

Significant changes to this dismissive attitude would take another decade. In General Comment No 36 the HRC first suggested that the right to life requires protections from environmental harm:

> Environmental degradation, climate change and unsustainable development constitute some of the most pressing and serious threats to the ability of present and future generations to enjoy the right to life. Obligations of States parties under international environmental law should thus inform the contents of article 6 of the Covenant, and the obligation of States parties to respect and ensure the right to life should also inform their relevant obligations under international environmental law.[12]

The case of *Cáceres* provided the first opportunity to flesh out new understanding. The widespread use of toxic agrochemicals had a devastating impact on the applicant's health, livelihood, and living conditions.[13] The use of the chemicals had been found illegal by domestic courts and their dangers as well as the absence of effective enforcement measures were openly acknowledged by the Ministry of the Environment. Based on these facts, the HRC accepted a threat to the right to life and found a violation of Articles 6 (Right to life), 17 (Freedom from arbitrary or unlawful interference) in conjunction with 2(3) ICCPR essentially due to the state's failure to enforce domestic court decisions.[14] Overall, the case is a significant step towards the normative framework of the environmental minimum. The HRC now appears to accept a specific risk as sufficient and was prepared to find a violation due to the lack of consistent enforcement of domestic legal standards. The unequivocal finding of the domestic court and the admission of deficient enforcement on the part of the state likely helped reach this conclusion. Nonetheless, it constitutes the

[11] Ibid. [7.7] and [7.8].

[12] *General Comment No. 36 (2018) on Article 6 of the International Covenant on Civil and Political Rights, on the Right to Life, Adopted by the Human Rights Committee on 30 October 2018* (CCPR/C/GC/36, 2000) [62].

[13] *Cáceres* v. *Paraguay* Communication No 2751/2016; UN Doc CCPR/C/126/D/2751/2016 (2019) (UN Human Rights Committee, 20 September 2019) [7.2].

[14] Ibid. [7.5].

first significant step towards recognising environmental human rights. Most recently, this approach was affirmed in principle in the case of *Teitiota*.[15]

The matter of *Haraldsson and Sveinsson* is the only case in category two where the HRC found a violation.[16] The case involved the repeated and ultimately unsuccessful attempts of the applicants to secure fishing quotas for their trawler. Requests for allocation were repeatedly denied by authorities and the applicants were forced to lease quotas from other fishermen at economically unsustainable rates. Faced with the prospect of bankruptcy, the applicants decided to catch fish without quotas, provoking legal action from the Fisheries Department. Before the HRC, the applicants relied on Article 26 ICCPR, claiming that current laws unduly privileged a group of fishermen entitled to free quotas solely on the basis that they were historic holders of such quotas in the 1980s. Fishermen that were not active in this time period were effectively forced to buy or lease quotas from this privileged group.[17] The HRC noted that the protection of fish stocks is a legitimate aim for states, thus at least indirectly recognising their ability to justify infringements of the ICCPR through environmental protection motives.[18] However, the difference in treatment between the groups of fishermen was deemed unreasonable and hence in violation of Article 26 ICCPR (Equality before the law). At the very least, the case recognised that environmental motives of states may justify infringements of the ICCPR, thus supporting the claim that the environment must be afforded a high priority.

The picture is entirely unclear with respect to environmental harm under the provisions of Articles 11 and 12 of the International Covenant on Economic, Social and Cultural Rights (ICESCR).[19] As the optional protocol to the ICESCR recognising the competence of the HRC to consider applications is a fairly recent creation, to date no pertinent environmental themed applications have been reported.

8.2 AFRICAN CHARTER ON HUMAN AND PEOPLES' RIGHTS

The case law of the African Commission of Human Rights (AfComHR) on environmental matters is limited when compared to other domestic and

[15] *Teitiota* v. *New Zealand* Communication No 2728/2016; UN Doc CCPR/C/127/D/2728/2016 (2020) (UN Human Rights Committee, 7 January 2020) [9.4].

[16] *Haraldsson and Sveinsson* v. *Iceland* Communication No 1306/2004; UN Doc CCCPR/C/91/D/1306/2004 (2004) (UN Human Rights Committee, 14 December 2007).

[17] Ibid. [10.3].

[18] Ibid. [10.4].

[19] International Covenant on Economic, Social and Cultural Rights.

international counterparts. From a total of six broadly environmentally themed cases, only three deal specifically with Article 24 ACHPR (Right to a general satisfactory environment).[20] There have been no pertinent environmentally themed rulings at all from the AfCtHR up to December 2019.

The case of *Social and Economic Rights Action Centre*[21] remains the only ruling to find a violation of Article 24 ACHPR. The case involved the exploitation of oil reserves in Ogoniland by the Shell Petroleum Development Corporation (SPDC), of which the Nigerian National Petroleum Company (NNPC) was a majority shareholder. The applicants alleged that exploitation efforts resulted in severe environmental harm and health problems among the Ogoni People, chiefly through numerous avoidable spills in the proximity of villages.[22] This led to the contamination of water, soil and air and skin infections, and gastrointestinal and respiratory illnesses, as well as an increased risk of cancer, neurological and reproductive conditions. Nigeria allegedly facilitated these actions through legal and military enforcement efforts in favour of the oil companies and through a string of consistent regulatory failures, including failures to monitor and enforce protective measures for the environment.[23]

The AfComHR held that Article 24 ACHPR required the state to take 'reasonable and other measures to prevent pollution and ecological harm, to promote conservation, and to secure an ecologically sustainable development and use of natural resources.'[24] This is said to include the scientific monitoring of environmental harm, requiring impact assessments, as well as providing information to affected communities and integrating them into the decision-making process.[25] These obligations stretch beyond immediate state action towards a requirement to protect citizens from the conduct of third parties.[26] The Commission held that Nigeria had failed to take appropriate measures to protect the Ogoni communities from environmental harm, and had not involved them in the decision-making process, in violation of Articles 16 (Right to health), 21 (Peoples' rights), and 24 ACHPR.[27] This is a result that

[20] See the dataset gathered on the African Charter system for this book.

[21] *Social and Economic Rights Action Centre and another* v. *Nigeria* App no 155/96 (Commission, 27 October 2001).

[22] Ibid. [1].

[23] Ibid. [3]–[6].

[24] Ibid. [52].

[25] Ibid. [53].

[26] Ibid. [57]; *Commission nationale des droits de l'Homme et des libertés* v. *Chad* App no 74/92 (Commission, 11 October 1995) [20].

[27] *Social and Economic Rights Action Centre and another* v. *Nigeria* App no 155/96 (Commission, 27 October 2001) [55] and [58].

is generally compatible with the environmental minimum, as it endorses environmental harm as a credible threat to human rights. Moreover, states are under an obligation to take measures directed at preventing environmental harm through regulation and consistent enforcement.

A further opportunity to rule on the substance of Article 24 ACHPR arose in the case of *Kevin Mgwanga Gunme*.[28] The case revolved around the impact of the 1961 plebiscite for southern Cameroonians, where they were presented with alternative options to join Nigeria or Cameroon, but not with the option of independent statehood.[29] The AfComHR declined to investigate the alleged violation of Article 24 ACHPR, as it deemed to have been presented with insufficient evidence.[30] This case is harder to reconcile with the environmental minimum as it remains unclear whether the applicants failed to demonstrate a specific risk, or whether the Commission placed an excessive burden on the applicants. In a further case, *Socio-Economic Rights and Accountability Project (SERAP)*, an Article 24 ACHPR claim was dismissed in its entirety due to non-exhaustion of domestic remedies.[31] Beyond these cases, there are two rulings that are environmentally themed but do not engage with Article 24 ACHPR.

In *Sudan Human Rights Organisation* the AfComHR dealt with allegations of large-scale killings and ethnic cleansing in the Darfur region.[32] The complaints alleged that Sudan had been complicit '(...) in looting and destroying foodstuffs, crops and livestock as well as poisoning wells and denying access to water sources.'[33] The AfComHR held that Article 16 ACHPR (Right to health) can be violated through state conduct or insufficient regulation of third parties.[34] States are obligated to provide basic services such as safe drinking water,[35] and must prevent the destruction of homes, livestock, farms and the poisoning of water sources, as it exposed victims to serious health risks.[36] This outcome can be reconciled with the environmental minimum,

28 *Kevin Mgwanga Gunme et al.* v. *Cameroon* App no 266/03 (Commission, 27 May 2009).
29 Ibid. [3].
30 Ibid. [208].
31 *Socio-Economic Rights and Accountability Project (SERAP)* v. *Nigeria* App no 338/07 (Commission, 24 November 2010) [69] and [71].
32 *Sudan Human Rights Organisation & Centre on Housing Rights and Evictions (COHRE)* v. *Sudan* App nos 279/03-296/05 (Commission, 27 May 2009).
33 Ibid. [207].
34 Ibid. [210].
35 *Free Legal Assistance Group and others* v. *Democratic Republic of the Congo* App nos 25/89; 47/90; 56/91; 100/93 (Commission, 4 April 1996).
36 *Sudan Human Rights Organisation & Centre on Housing Rights and Evictions (COHRE)* v. *Sudan* App nos 279/03-296/05 (Commission, 27 May 2009) [212].

notwithstanding the question of whether the Commission declined to investigate Article 24 ACHPR because it did not conceptualise the harm as environmental, which would be problematic, or because it applied Article 16 ACHPR as a *lex specialis* under the circumstances of the case.

Finally, in *Centre for Minority Rights Development*, the AfComHR dealt with violations resulting from the displacement of the Endorois community through the designation of a game reserve on their lands by the Kenyan state.[37] The applicants sought the return of their ancestral lands, relying on Article 14 ACHPR (Right to property),[38] which the Commission judged to be violated.[39] However, no reference is made to Article 24 ACHPR when discussing whether the infringement was justified; it is merely invoked to support the argument that Article 14 ACHPR may be claimed collectively.[40] The environmental nature of this case is somewhat doubtful, and so a reasonable argument could be made that it raised no relevant environmental harm and therefore did not necessitate separate discussion of Article 24.

Overall, there are too few cases to draw firm conclusions on the prevailing doctrine with respect to environmental harm. Some cases and approaches are encouraging, but they are too isolated and limited in their scope to pronounce the ACHPR definitively supportive of the environmental minimum.

8.3 AMERICAN CONVENTION ON HUMAN RIGHTS

The Inter-American system of human rights protection rests on two pillars: The American Declaration on the Rights and Duties of Man (American Declaration),[41] as well as the ACHR.[42] The ACHR has largely superseded the American Declaration, with the notable exceptions of the United States, Canada, and Cuba, which have all not ratified the ACHR and hence remain subject to the American Declaration and the jurisdiction of the Inter-American Commission on Human Rights (IAComHR) only.[43] Both mechanisms were jointly examined for the presence of environmentally themed petitions and evaluated as a whole for their doctrinal support of the

[37] *Centre for Minority Rights Development (Kenya) and Minority Rights Group (on behalf of Endorois Welfare Council)* v. *Kenya* App no 276/03 (Commission, 25 November 2009) [3].

[38] Ibid. [174].

[39] Ibid. [238].

[40] Ibid. [151].

[41] American Declaration of the Rights and Duties of Man (Bogota Declaration) (adopted 2 May 1948), OAS/Ser.L/V/I.4 Rev. 9 (2003) (American Declaration).

[42] American Convention on Human Rights.

[43] Beyerlin and Marauhn, *International Environmental Law*, 396.

environmental minimum: overall, 10 environmentally themed cases have reached the Inter-American system, all falling in category one (protection function).[44]

Generally speaking, the doctrinal position adopted by the IAComHR and the Inter-IACtHR does not engage with environmental harm directly.[45] Although specific environmental rights were adopted (see Article 11 ACHR, Right to a healthy environment), Article 19 (6) of the Protocol to the American Convention on Human Rights (Protocol of San Salvador) expressly prevents individuals from relying on Article 11 ACHR directly.[46] This results in an indirect consideration of environmental harm under other provisions, often by emphasising the importance of the environment for indigenous cultures and communities.

The early case of *Yanomani Indians* provides a good illustration of this approach.[47] The applicants complained that the construction of a road and mining licences granted on indigenous land led to environmental harm and an influx of outsiders, bringing with them epidemics and diseases, as well as leading to the forced displacement, and death of hundreds of members of an indigenous community. The IAComHR held the state responsible for failing to protect the native population, finding violations of Articles 1 (Right to life, liberty, and personal security), 8 (Right to residence and movement), and 11 (Right to the preservation of health and to well-being) of the American Declaration. Environmental harm as such was not conceptualised as a separate threat or potential source of infringements.

In the more recent case of *Kichwa Indigenous People*, the IACtHR was called upon to rule on oil exploration rights granted for lands traditionally occupied by the indigenous group.[48] It found that Ecuador had violated Articles 1, 2, 4, 5, 8, 21, and 25 of the ACHR. The facts of the case included grave examples of environmental harm, but the decision centred on its cultural and subsistence impact.[49] Ultimately, all violations rested on the fact that the indigenous group had not been adequately consulted during the

44 See the dataset gathered on the American Convention system for this book.

45 Lucas Lixinski, 'Treaty Interpretation by the Inter-American Court of Human Rights: Expansionism at the Service of the Unity of International Law' (2010) 21 *European Journal of International Law* 585, 594.

46 Additional Protocol to the American Convention on Human Rights in the Area of Economic, Social and Cultural Rights (Protocol of San Salvador) (entered into force 16 November 1999) OAS Treaty Series No 69 (1988).

47 *Yanomani Indians* v. *Brazil* Resolution No 12/85 (IAComHR, 5 March 1985).

48 *Kichwa Indigenous People of Sarayaku* v. *Ecuador* Series C No. 245 (2012) (IACtHR, 27 July 2012).

49 Ibid. [105] and [219].

decision-making process,[50] and on the absence of an effective domestic remedy.[51] While procedural environmental rights can form an important part of an effective protection regime, it is a core claim of the environmental minimum that specific risks arising from environmental harm must in principle give rise to a separate infringement of substantive rights. No amount of consultation and procedural safeguards can absolve states from their obligation to provide a minimally sound environment under general commitments to human rights. Nonetheless, an emphasis on consultation is typical of environmentally themed cases in the Inter-American system.[52] A smaller number of cases stand apart from this general trend and therefore merit closer attention.

The case of *Mossville Environmental Action Now* involved allegations that permits for industrial complexes had a discriminatory effect on the African-American community of Mossville.[53] The IAComHR admitted the application under Article 2 (Equality before the law),[54] and Article 5 (Right to private and family life) of the American Declaration.[55] The admissibility decision illustrates a trend that one would perhaps recognise from the ECHR. The use and reliance on Article 5 of the American Declaration in particular are reminiscent of the well-established ECHR case law on Article 8 ECHR. However, according to a report published in 2018, the *Mossville* case has been archived by the Court: in other words it has been struck from the list of pending cases without a final decision on the merits.[56] This could be due to inactivity or a formal withdrawal of the petition, the registry was unable to provide further details on request.[57] There is some scattered anecdotal evidence that the applicants exhausted their financial resources and were forced to abandon the case.

50 Ibid. [211] and [220].

51 Ibid. [278].

52 See, for instance, *Ngobe Indigenous Communities* v. *Panama* Report No. 75/09 (IAComHR, 5 August 2009) [failure to consult before construction of hydroelectric dam]; *Maya Indigenous Community of the Toledo District* v. *Belize* Report No. 40/04 (IAComHR, 12 October 2004) [155] [failure to recognise an indigenous group and consult them before granting logging concessions].

53 *Mossville Environmental Action Now* v. *United States* Report No. 43/10 (IAComHR, 17 March 2010).

54 Ibid. [42].

55 Ibid. [43].

56 *Annual Report 2018* (Inter-American Commission on Human Rights, 2018) www.oas.org/en/iachr/docs/annual/2018/docs/IA2018cap.2-en.pdf, accessed 1 April 2021.

57 The author contacted the registry of the commission and was directed to the legal basis for archiving decisions and otherwise encouraged to use a publicly available search engine for further information; Email on file with author.

In a further decision reminiscent of the ECHR case law, the IAComHR admitted the application in *Community of La Oroya* chiefly on the grounds that Peru had failed to enforce a final domestic ruling against a metallurgical complex.[58] A failure to enforce domestic decisions was likewise at the heart of the admissibility decision in *Community of San Mateo de Huanchor.*[59] However, it is too early to draw any firm conclusions. Both admissibility decisions certainly have the potential to validate a core claim of the environmental minimum in the doctrinal approach of the IAComHR: the idea that a failure to consistently enforce legal standards must invariably result in a violation.

The most recent statement on the level of environmental protection available under the ACHR comes in the guise of an advisory opinion concerning state obligations in relation to the environment, with a particular focus on the right to life.[60] The opinion acknowledged a core claim of the environmental minimum, namely the close link between the enjoyment of human rights and the protection of the environment. The advisory opinion bases state duties on a right to a healthy environment derived from Article 26 ACHR (Progressive development) in conjunction with Article 11 of the Protocol of San Salvador.[61] The provisions encompass both substantive protections of the environment for its own sake as well as procedural guarantees.[62] Notably, this approach sidesteps the otherwise non-justiciable Article 11 ACHR to grant individuals an enforceable right. However, as it is derived from Article 26 ACHR, this implies only a duty towards progressive realisation. The autonomous right to environmental protection is therefore closer conceptually to the rights found in the ICESCR.

Although the court emphasises the environmental dimensions of rights that go beyond aspirational obligations, including Articles 1 (Right to life) and 5 ACHR (Personal integrity), the discussion focuses almost entirely on 'due diligence' requirements, that is to say, a progressive realisation and discrimination free access.[63] Indeed, the court notes that 'most environmental obligations are based on this duty of due diligence.'[64] The Court determines compliance with this duty through well-established principles of international

[58] *Community of La Oroya* v. *Peru* Report No. 76/09 (IAComHR, 5 August 2009) [75].
[59] *Community of San Mateo de Huanchor and Its Members* v. *Peru* Report No. 69/04 (IAComHR, 15 October 2004) [48]–[64].
[60] *Advisory Opinion: Environment and Human Rights* OC-23/18 (IACtHR, 15 November 2017).
[61] Ibid. [57]–[58].
[62] Ibid. [62]–[63].
[63] Ibid. [111]–[114].
[64] Ibid. [124].

environmental law, including the duty to prevent transboundary environmental harm, the precautionary principle, and the duty to cooperate.[65]

More recently, in the case of *Indigenous Communities of the Lhaka Honhat Association*, the Court considered state failure to prevent illegal logging as well as environmentally and culturally harmful activities on indigenous land. Ultimately, the court found a violation of Article 21 ACHR (Right to property), but also took the opportunity to develop its doctrinal position on Article 26 ACHR. The Court affirmed that states have an obligation to prevent and mitigate environmental harms, including ensuring access to water.[66] Given the environmentally harmful activities, a slim majority (4–3) concluded that Argentina had failed to guarantee the right to a healthy environment and associated rights under Article 26 ACHR.[67]

Overall, it remains to be seen whether the advisory opinion and the more recent rulings are the starting point towards the type of substantive environmental protections advocated by the environmental minimum. Equally, it might be the less ambitious restatement of procedural protections and aspirational rights that generally only require due diligence in responding to environmental harm. The case in *Lhaka Honhat Association* provides some indication that the Court may wish to go further, but it was a slim majority that found a violation of Article 26 ACHR and it was ultimately not decisive to the outcome of the case. Violations of the right to property featured prominently and also constituted the primary focus of the earlier ruling from the Commission.

8.4 SUMMARY

The position in the examined regional human rights protection regimes is overall less encouraging for the environmental minimum than under the ECHR. In many respects, the protection regimes are hampered by the limited number of environmental cases. What small number of cases have been decided is far from a doctrinal position unequivocally supportive of the environmental minimum. The level of support appears to correlate with the presence and enforceability of a right to an environment of a certain quality. Most support comes from the ACHPR, where decisions based on the enforceable right under Article 24 and related environmental rulings approach a

65 Ibid. [125]–[126].
66 *Indigenous Communities of the Lhaka Honhat Association (Our Land)* v. *Argentina* Series C No 400 (IACtHR, 6 February 2020) [208], [229]–[230].
67 Ibid. [289].

doctrinal position broadly in line with core claims of the environmental minimum. However, the overall small number of cases makes this doctrinal approach less stable and settled: especially considering that the AfCtHR has not yet had a meaningful opportunity to comment.

Less support comes from the Inter-American system. While there is a provision that specifically provides a right to a healthy environment, its practical impact is significantly reduced because applicants cannot invoke it against states. Despite the recent advisory opinion which certainly seems to point in a promising direction, the Inter-American system therefore tends to rely on indigenous rights and emphasises consultation and procedural safeguards to address environmental harm. Most of the more recent rulings were on admissibility only, and it is unclear whether and when decisions on merits will signal further meaningful doctrinal shifts in line with the important advisory opinion.

Finally, environmental protection is weakest under the UN Covenants, but that doctrinal position is currently undergoing a shift. Historically, the HRC failed to unequivocally endorse a fundamental claim of the environmental minimum: the notion that environmental harm presents a serious threat to human rights that cannot be ignored on principled grounds, even if it falls outside the scope of traditionally accepted threat scenarios. General Comment No. 36 suggested a change in that position, which was confirmed in the case of *Cáceres*. Despite the fact that a domestic court ruling finding a violation of legal standards was ultimately decisive, it nonetheless represents an encouraging development that signals a broader embracing of environmental human rights.

9

Environmental Constitutionalism

This chapter discusses the constitutional protections of the environment in two sophisticated protection regimes. South Africa and Germany represent two broad approaches in environmental constitutionalism that go beyond the greener interpretation of existing rights through courts. Both jurisdictions have adopted express environmental provisions into their constitutional frameworks: South Africa codified an enforceable, individual right to an environment of a certain quality and Germany introduced a general constitutional mission statement requiring environmental protection and sustainable development.

However, as the case studies demonstrate, both approaches have been limited in their practical impact: key conceptual and doctrinal questions remain unresolved and the provisions have arguably failed to significantly impact environmental regulations. In South Africa, the environmental minimum could contribute towards the development of a principled doctrinal approach to section 24 of the constitution and clarify its relationship to dominant environmental framework legislation. With respect to Germany, the chief contribution of the environmental minimum lies in limiting the deference granted to regulators and stimulating the development of clear positive obligations under fundamental rights. Further improvements could be achieved on established and emerging standards, but the numbers of cases in both jurisdictions are too limited for meaningful comments on the compatibility.

Apart from its ability to contribute to the development of doctrine, the environmental minimum also addresses a common concern in environmental constitutionalism, namely translating general and abstract constitutional norms into specific and practically enforceable safeguards for the environment. While this chapter is modelled on key claims of the environmental minimum, it focuses on the extent to which the environmental provisions provide a constitutional obligation to regulate environmental harm.

9.1 SOUTH AFRICA

Section 24 of the Bill of Rights of the Constitution incorporates the principle of environmental protection and sustainable development into the constitutional law of South Africa, and generally recognises the fundamental importance of the environment.[1] The provision states:

Everyone has the right

a. to an environment that is not harmful to their health or well-being; and
b. to have the environment protected, for the benefit of present and future generations, through reasonable legislative and other measures that
   i. prevent pollution and ecological degradation;
   ii. promote conservation; and
   iii. secure ecologically sustainable development and use of natural resources while promoting justifiable economic and social development.

It is a central provision that underpins the environmental governance regime in South Africa.[2] As we shall see, while there is some uncertainty surrounding the provision, section 24 is generally thought to contain a fundamental individual right (paragraph a), as well as a source of positive obligations for the state (paragraph b).[3] These constitutional provisions are flanked by a comprehensive environmental framework statute, the National Environmental Management Act (NEMA).[4]

1 This section draws in large parts on my earlier article, 'The Problem with the Normative Content of Section 24 of the Constitution of South Africa' (2019) 37 *Nordic Journal of Human Rights* 105.

2 *Fuel Retailers Association of SA (Pty) Ltd* v. *Director General, Environmental Management Mpumalanga and others* [2007] ZACC 13 [102]; Louis J. Kotzé, 'The Judiciary, the Environmental Right and the Quest for Sustainability in South Africa: A Critical Reflection' (2007) 16 *Review of European Community & International Environmental Law* 298, 299; Loretta Feris, 'Constitutional Environmental Rights: An Under-utilised Resource' (2008) 24 *South African Journal on Human Rights* 29, 49; Rachel Wynberg and David Fig, 'Realising Environmental Rights' in Malcolm Langford and others (eds.), *Socio-Economic Rights in South Africa: Symbols or Substance?* (Cambridge University Press 2014) 310; Jan Glazewski, *Environmental Law in South Africa* (Butterworths 2000) 78; May and Daly, *Global Environmental Constitutionalism*, 50.

3 *HTF Developers* v. *Minister of Environmental Affairs and Tourism and Others* [2006] ZAGPHC 132 [17]; Michael Kidd, 'Environment' in Iain Currie and Johan de Waal (eds.), *The Bill of Rights Handbook* (6th ed., Juta 2016) 518; Feris, 'Constitutional Environmental Rights', 36; Glazewski, *Environmental Law in South Africa*, 85.

4 National Environmental Management Act 107 of 1998, as amended 18 December 2014; Environment Conservation Act 73 of 1989, as amended 18 September 2009.

An absence of legal standards is generally not a concern in South Africa. The few cases that engage with substantive environmental protection suggest some issues with the consistent enforcement of regulations, but they do not offer much evidence of a systemic problem. Instead, the case law of the Constitutional Court of South Africa (ZACC) leaves three core areas of concern with respect to the doctrinal position on section 24 of the Constitution.

First, the doctrine requires further definition and clarification: courts primarily conceptualise key terminology such as 'environment', 'health', and 'well-being' through framework legislation, and not on the basis of the Constitution. An independent constitutional understanding is however crucial to determining the constitutionality of framework legislation but also to ascertain whether section 24 supports core normative claims of environmental human rights.

Second, it is unclear whether section 24 is self-executing or requires 'reasonable legislative and other measures' (paragraph b) to give effect to the safeguards it outlines. The answer turns largely on whether we view section 24 as encompassing one individual right (paragraph a) and a separate positive obligation (paragraph b) or whether it constitutes a composite right. In turn, this influences whether section 24 chimes with core claims of the environmental minimum: that certain minimum standards of environmental protection must be upheld, even when these are balanced with competing interests.

Finally, assuming that framework legislation at least partially expresses the normative requirements of section 24 of the Constitution in a permissible manner, it remains unclear to what extent applicants must first invoke the framework legislation. Generally, the South African principle of constitutional subsidiarity applies to such cases, requiring applicants to rely first and foremost on the framework legislation in bringing their claims.[5] Apart from a practical concern over how one can determine whether ordinary legislation gives effect to a constitutional right without a clear notion of its content,[6] the principle of

5 The general existence of this principle is settled case law, see *Member of the Executive Council for Development Planning and Local Government, Gauteng* v. *Democratic Party and Others* [1998] ZACC 9; 1998 (4) SA 1157 [62]; *My Vote Counts NPC* v. *Speaker of the National Assembly and Others* [2015] ZACC 31 [54]; on the principle in general, see Lourens Du Plessis, 'Subsidiarity: What's in the Name for Constitutional Interpretation and Adjudication' (2006) 17 *Stellenbosch Law Review* 207.

6 A similar point is made by Anel Du Plessis, 'South Africa's Constitutional Environmental Right (Generously) Interpreted: What Is in It for Poverty?' (2011) 27 *South African Journal on Human Rights* 279, 303.

subsidiarity is problematic to the extent that it hinders the enforcement of minimum standards, as advocated by the environmental minimum.

As Section 9.1.1 demonstrates, courts have provided some tentative responses to these areas of concern that could form part of a more comprehensive doctrinal approach based on the environmental minimum. However, currently the doctrine is too underdeveloped and there are significant limitations which we explore in Section 9.1.2.

9.1.1 *Scope of Protection*

Most cases engaging with section 24 revolve around the interpretation of procedural protections under statutory provisions. The case of *Bengwenyama* dealt with the fairness requirement for administrative proceedings in the allocation of prospecting rights,[7] *National Society for the Prevention of Cruelty to Animals* decided whether the NSPCA was authorised under the SPCA Act to institute private prosecution against individuals,[8] and the crux of the case in *HTF Developers* was whether the observance of a 30-day notice and commentary period was necessary for environmental protection directions under the Environmental Conservation Act.[9] Section 24 was also discussed in a case regarding the delineation of provincial and national spheres of competence in zoning law.[10] Apart from these procedurally focused cases, there is only a limited number of substantive environmental decisions under section 24 (a) and (b) of the Constitution.

9.1.1.1 Subsection a

The ZACC has clarified that section 24 (a) contains an individual right to a healthy environment, and that it forms the basis for legislation that guarantees the protection of the environment.[11] The term 'environment' is not specified

7 *Bengwenyama (Pty) Ltd and Others* v. *Genorah (Pty) Ltd and Others* [2010] ZACC 26.

8 *National Society for the Prevention of Cruelty to Animals* v. *Minister of Justice and Constitutional Development and others* [2016] ZACC 46.

9 *MEC Department of Agriculture, Conservation and Environment and Another* v. *HTF Developers (Pty) Limited* [2007] ZACC 25; see also *Earthlife Africa Johannesburg* v. *Minister of Environmental Affairs and Others* [2017] ZAGPPHC 58 [failure to consider the climate change impacts of coal fire power station].

10 *Maccsand (Pty) Ltd* v. *City of Cape Town and Others* [2012] ZACC 7.

11 *Fuel Retailers Association of SA (Pty) Ltd* v. *Director General, Environmental Management Mpumalanga and others* [2007] ZACC 13 [40]; Feris, 'Constitutional Environmental Rights', 39.

in the constitution or the case law of the ZACC, and it has been suggested that it is 'open-ended and manifestly (...) incapable of precise definition.'[12] The Supreme Court of Appeal suggests only that the terminology of 'health' and 'well-being' must be interpreted in the sense that 'well-being' overlaps, but goes beyond 'health' to encompass a more holistic sense of the environment.[13] Section 24 can thus be read as incorporating interests of people in the environment that are rooted in utility, as well as intrinsic and aesthetic value.[14] Cases commonly proceed on the basis of the definition of the environment given by the NEMA framework legislation:

> (...) the surroundings within which humans exist and that are made up of— (i) the land, water and atmosphere of the earth; (ii) micro-organisms, plant and animal life; (iii) any part or combination of (i) and (ii) and the interrelationships among and between them; and (iv) the physical, chemical, aesthetic and cultural properties and conditions of the foregoing that influence human health and well-being.[15]

The definition is broad enough to encompass the claims of the environmental minimum, but it is problematic that it is not expressly tied to section 24, the primary expression of environmental human rights in the Constitution.

Judicial pronouncements offer little detail on the scope of protection. In a series of rulings, the High Court has opined that section 24 contains a positive obligation to protect individuals from the effects of environmental pollution and degradation.[16] In the case of *Hichange Investments*, the applicant complained over chemical waste originating from a tanning business on

[12] *HTF Developers* v. *Minister of Environmental Affairs and Tourism and Others* [2006] ZAGPHC 132 [18]; for some preliminary definitions of key terms, see Du Plessis, 'South Africa's Constitutional Environmental Right (Generously) Interpreted', 292.

[13] *HTF Developers* v. *Minister of Environmental Affairs and Tourism and Others* [2006] ZAGPHC 132 [18]; Kidd, 'Environment', 522; Louis J. Kotzé and Anél du Plessis, 'Some Brief Observations on Fifteen Years of Environmental Rights Jurisprudence in South Africa' (2010) 3 *Journal of Court Innovation* 157, 166–7.

[14] *Hichange Investments (Pty) Ltd* v. *Cape Produce Co (Pty) Ltd t/a Pelts Products and Others* [2004] 2 SA 393 (E) [which seemed to endorse this broader reading]; *Paola* v. *Jeeva NO and Others* [2003] ZASCA 100 [on the aesthetic value of an ocean view]; Kotzé, 'The Judiciary, the Environmental Right and the Quest for Sustainability in South Africa', 300; Glazewski, *Environmental Law in South Africa*, 77.

[15] See section 1 of the National Environmental Management Act 1998.

[16] *Hichange Investments (Pty) Ltd* v. *Cape Produce Co (Pty) Ltd t/a Pelts Products and Others* [2004] 2 SA 393 (E) 16; *Minister of Water Affairs and Forestry* v. *Stilfontein Gold Mining Company Ltd and Others* [2006] ZAGPHC 47 [16.9]; *HTF Developers* v. *Minister of Environmental Affairs and Tourism and Others* [2006] ZAGPHC 132 [16].

neighbouring property.[17] The applicant argued that noxious gases and odour constituted a threat to health and well-being within the meaning of section 24 (a). The High Court accepted an infringement of the applicant's well-being,[18] and proceeded to examine whether the state had discharged its positive obligation, determined by the severity of the pollution and based on scientific evidence.[19] In determining what constitutes sufficiently severe pollution for the purposes of section 28 NEMA, the High Court referred to section 24 of the Constitution, noting that the broad definition of health and well-being suggested a low threshold.[20] However, the High Court was ultimately not convinced that the pollution cleared this threshold and accordingly dismissed the claim.[21] This case is partially compatible with the environmental minimum because it recognises the threats that environmental harm presents to human rights. However, there is some doubt as to whether the High Court considered and applied established standards to arrive at this conclusion, as the environmental minimum would have required.

The case of *Harmony Gold* was not directly concerned with section 24 of the Constitution,[22] but established that private actors could be lawfully required to contribute to the costs of preventing pollution. The case concerned a government order under section 19(3) of the National Water Act that sought to prevent water pollution by requiring the appellant to contribute towards the costs of on-going pumping operations in a defunct mine. The Court of Appeal found that the appellant was lawfully required to contribute to the costs, even though they were not formally and directly responsible for the defunct mine or the intruding water.[23] Section 24 of the Constitution is mentioned only in passing as providing the basis for the interpretation of section 19 of the National Water Act.[24] While the decision is therefore ultimately based on the interpretation of the National Water Act, it does point to an important dimension of section 24 (a) of the Constitution. Insofar as the economic activity of companies threatens the environment, individual rights

[17] *Hichange Investments (Pty) Ltd* v. *Cape Produce Co (Pty) Ltd t/a Pelts Products and Others* [2004] 2 SA 393 (E).
[18] Ibid., 21.
[19] Ibid., 16.
[20] Ibid., 16–17.
[21] Ibid.
[22] *Harmony Gold Mining Company (Ltd)* v. *Regional Director Free State Department of Water Affairs and Forestry and Another* [2006] ZASCA 66.
[23] Ibid. [35].
[24] Ibid. [17].

can be justifiably infringed to prevent and alleviate environmental harm. This is a position broadly supportive of the environmental minimum.

More recently, the High Court decision in *Stern* epitomises the lack of consistent engagement with section 24 of the Constitution.[25] The applicants specifically challenged the constitutionality of a decision by the Minister of Mineral Resources to create regulations for petroleum exploration. Nonetheless, the High Court does not mention section 24 at all and instead engages exclusively with provisions of ordinary legislation, notably NEMA.

9.1.1.2 Subsection b

Section 24 (b) is generally believed to oblige the state to (i) prevent pollution and ecological degradation, (ii) promote conservation, and (iii) secure ecologically sustainable development and use of natural resources while promoting justifiable economic and social development.[26] A recent ruling in the Supreme Court of Appeal hints at this interpretation when it held that despite the provision of section 24 (b) the customary rights of access to marine resources constituted a defence against unlawful conduct under the Marine Living Resources Act.[27]

Only one case before the ZACC has dealt with the interpretation of section 24 (b) in any meaningful way: *Fuel Retailers*.[28] The case involved an application for the construction of a petrol station, which was granted by local environmental authorities. The case afforded the ZACC the opportunity to discuss the nature of environmental protection under section 24 and its relationship with socio-economic and sustainable development:

> Economic and social development is essential to the well-being of human beings. (. . .) But development cannot subsist upon a deteriorating environmental base. Unlimited development is detrimental to the environment and the destruction of the environment is detrimental to development. Promotion of development requires the protection of the environment.[29]

[25] *Stern NO and Others* v. *Minister of Mineral Resources* [2017] ZAECGHC 109.

[26] Kidd, 'Environment', 528; Du Plessis, 'South Africa's Constitutional Environmental Right (Generously) Interpreted', 299.

[27] *Gongqose and Others* v. *Minister of Agriculture, Forestry and Others, Gongqose and S* [2018] ZASCA 87 [2].

[28] *Fuel Retailers Association of SA (Pty) Ltd* v. *Director General, Environmental Management Mpumalanga and others* [2007] ZACC 13.

[29] Ibid. [44] (footnotes omitted).

The ZACC criticised the local environmental department for failing to consider the environmental impact of the petrol station. Environmental officials were held under a duty to conduct their own independent review, instead of delegating this task to town planning authorities.[30] The case presents the Constitution as the crucial arbiter and guideline to balancing competing considerations of development and environmental protection.[31] This stance is compatible with the environmental minimum, particularly the high-priority principle which requires the balancing of environmental protection with competing interests. However, the decision is limited to procedural shortcomings of the administrative process as opposed to the normative claims of section 24 of the Constitution.

The High Court has expanded on this understanding of section 24 to include a requirement that legislative measures must be capable of realising environmental protection.[32] Again, the decision is compatible with the environmental minimum to the extent that it confirms that the government must regulate environmental harm. However, it is less helpful with respect to the specific standard of review. The High Court requires only some reasonable legislative measures, regardless of whether these are in line with generally accepted guidelines of the relevant scientific community as the environmental minimum would require.

Crucially, the High Court and the Supreme Court of Appeal have affirmed that economic development and environmental protection must be considered jointly and are of equal importance within the range of constitutional interests.[33] Overall, the normative scope and implications of section 24 (b) remain largely unclear on some key questions that we turn to discuss in Section 9.1.2.[34] Primarily, this uncertainty is owed to the limited number of pertinent cases, and the apparent reluctance of the ZACC and lower courts to move beyond a primarily procedural review of environmental regulation. It is

30 Ibid. [92].

31 Ibid. [93].

32 *BP Southern Africa (Pty) Limited* v. *Mec for Agriculture, Conservation, Environment & Land Affair* [2004] ZAGPHC 18 [22]; *MEC for Agriculture Conservation, Environment and Land Affairs Gauteng* v. *Sasol Oil (Pty) Ltd and another* [2005] ZASCA 76 [14]

33 *BP Southern Africa (Pty) Limited* v. *Mec for Agriculture, Conservation, Environment & Land Affair* [2004] ZAGPHC 18 22; *All the Best Trading CC and Others* v. *S N Nayagar Property Development and Construction CC and Others* [2005] ZAGPHC 17 [4]; *MEC for Agriculture Conservation, Environment and Land Affairs Gauteng* v. *Sasol Oil (Pty) Ltd and another* [2005] ZASCA 76 [16]; Eric C. Christiansen, 'Empowerment, Fairness, Integration: South African Answers to the Question of Constitutional Environmental Rights' (2013) 32 *Stanford Environmental Law Journal* 215, 257.

34 May and Daly, *Global Environmental Constitutionalism*, 265.

correct that section 24 is supplemented and indirectly discussed in the context of some associated specified rights, such as the access to sufficient water and healthcare (section 27), right to life (section 11), equality (section 9), and dignity (section 10).[35] However, these occasional and limited references serve only to highlight the absence of clear constitutional doctrine and the resulting limitations for compatibility with the environmental minimum.

9.1.2 *Limitations*

The limitations of compatibility are rooted in the uncertain doctrinal scope of section 24: it is difficult to assess whether the doctrine supports the environmental minimum framework if the constitutional provision is rarely discussed and clarified. In the parlance of the environmental minimum, this raises the question of whether section 24 sufficiently recognises the relevance of environmental threats to constitutional rights and accepts that specific risks are sufficient to engage safeguards.

This concern is connected to and compounded by the crucial role that NEMA often plays in environmental cases. To the extent that the provisions of NEMA are in harmony with constitutional requirements, there is nothing objectionable in principle about specifying constitutional safeguards in this manner. However, doctrine on section 24 of the Constitution is not in a sufficient state of maturity to provide a clear normative standard to determine whether ordinary legislation comports with constitutional requirements. Even solely on the terms of South African constitutional law, NEMA is filling the numerous normative gaps of section 24 in a questionable manner. Especially the formal normative relationship of section 24 with framework legislation and the related principle of constitutional subsidiarity is largely unsettled: it is unclear whether framework legislation is required to give effect to the right and whether applicants must primarily rely on ordinary legislation when bringing environmental cases.

For the purposes of the environmental minimum, this also challenges the idea that constitutional environmental rights provide a consistent legal standard of review. Should framework legislation be required to give effect to section 24, and should applicants be effectively barred from relying on constitutional protections of the environment, then most environmental regulations and enforcement actions are likewise insulated from constitutional challenges. The specialised South African environmental right would then

[35] *Environmental Law and Local Government in South Africa* (Anél du Plessis ed., Juta 2016).

provide little beyond the protections available under a 'green' interpretation of other constitutional rights.

9.1.2.1 Framework Legislation

The case law is limited on the relevance of framework legislation for the purposes of section 24, but a few passages in the case of *Fuel Retailers* hint at a crucial role for ordinary legislation:

> Section 24 of the Constitution guarantees to everyone the right to a healthy environment and contemplates that legislation will be enacted for the protection of the environment. ECA and NEMA are legislation which give effect to this provision of the Constitution.' (...) '[The principles of NEMA] provide guidance for the interpretation and implementation not only of NEMA but any other legislation that is concerned with the protection and management of the environment. It is therefore plain that these principles must be observed as they are of considerable importance to the protection and management of the environment.[36]

The Court suggests that NEMA does indeed give effect to the environmental right, a view shared in parts of the scholarship. Loretta Feris argues that issues over the vagueness of section 24 are addressed through legislation, and that the judiciary is primarily tasked with enforcing these provisions.[37] This is echoed by Rachel Wynberg and David Fig, who emphasise that enabling legislation gives 'effect to the constitutional right to an environmentally sustainable society, as well as sector-specific legislation that can be used to uphold environmental rights.'[38] NEMA is thus presented as a quasi-constitutional statute, which takes precedence over other legislation on environmental matters.[39] This position is challenged somewhat in *HTF Developers*, where the Constitutional Court held that

> NEMA is a legislative measure contemplated by section 24(b) of the Constitution. Section 2 of NEMA sets out a series of principles that give effect to the understanding that 'the environment is a composite right, which

[36] *Fuel Retailers Association of SA (Pty) Ltd* v. *Director General, Environmental Management Mpumalanga and others* [2007] ZACC 13[40] and [67] (footnotes omitted).

[37] Feris, 'Constitutional Environmental Rights', 36–38 and 49; this emphasis on enforcement is also expressed by Kidd, 'Environment', 529; and Christiansen, 'Empowerment, Fairness, Integration', 254.

[38] Wynberg and Fig, 'Realising Environmental Rights', 316, see also Christiansen, 'Empowerment, Fairness, Integration', 254.

[39] Christiansen, 'Empowerment, Fairness, Integration', 270; see also Kidd, 'Environment', 527.

> includes social, economic and cultural considerations in order to ultimately result in a balanced environment. The composite nature of this right is captured in the principle that [d]evelopment must be socially, environmentally and economically sustainable.[40]

Because the passage refers to section 24 (b) alone, it is unclear whether the ZACC endorses the idea that framework legislation is also necessary to give effect to section 24 (a). The confusion is compounded by the fact that the ZACC does not consistently distinguish in its case law between these two subsections. NEMA itself conceptualises section 24 of the Constitution as a 'composite' right, which ostensibly supports the argument that ordinary legislation is required to give effect to both section 24 (a) and (b): a view also endorsed in parts of the scholarship.[41] This assertion admittedly has some basis in references to section 2 of NEMA in *HTF Developers*,[42] and a further reference in the context of section 1 ECA in the case of *BP Southern Africa*.[43] In both decisions, the ZACC ostensibly based its reasoning on the goals and principles set out in NEMA and the ECA, most notably the stated aim to 'give effect' to section 24. However, because it is precisely the *constitutional* significance of ordinary legislation that we are probing here, references to the provisions and aims of NEMA and the ECA have little persuasive force. Should ordinary legislation not be constitutionally required to give effect to section 24, then it is constitutionally irrelevant how NEMA conceptualises and purports to delineate constitutional safeguards: ultimately, only the Constitution can make such determinations.

It is therefore worth comparing section 24 to other constitutional rights that unequivocally require ordinary legislation.[44] Both sections 32 and 33 of the Constitution state that 'national legislation must be enacted to give effect to this right.' It is debatable whether the absence of a similar provision from section 24 is relevant: the text clearly matters to constitutional interpretation, but it is not necessarily conclusive absent further judicial clarification.[45] Even without going so far as asserting a categorical difference between sections

40 *MEC Department of Agriculture, Conservation and Environment and Another* v. *HTF Developers (Pty) Limited* [2007] ZACC 25 [24], footnotes omitted.

41 Christiansen, 'Empowerment, Fairness, Integration', 271.

42 *MEC Department of Agriculture, Conservation and Environment and Another* v. *HTF Developers (Pty) Limited* [2007] ZACC 25 [24].

43 *BP Southern Africa (Pty) Limited* v. *Mec for Agriculture, Conservation, Environment & Land Affair* [2004] ZAGPHC 18, 26.

44 Heinz Klug, *The Constitution of South Africa – A Contextual Analysis* (Hart 2010) 115–16.

45 Kate O'Regan, 'Text Matters: Some Reflections on the Forging of a New Constitutional Jurisprudence in South Africa' (2012) 75 *Modern Law Review* 1, 10–11.

32 and 24, there is room for reasonable doubt whether ordinary legislation alone can ever comprehensively define the protection afforded by constitutional rights, and whether this is in any case desirable.[46]

It seems that all constitutional rights must retain at least some independent normative content, because otherwise their normative claims over time become synonymous with the protection afforded under easily amended framework legislation. It would then be impossible to distinguish between a permissible specification of constitutional provisions through ordinary legislation, and questionable attempts to limit constitutional protections through ordinary legislation that does not meet the formal requirements for constitutional amendment.[47]

Applicants would be effectively limited to protections granted through ordinary legislation and subject to entrenched regulatory practices largely insulated from constitutional review.[48] Constitutional rights then no longer provide a robust standard of review for environmental regulation. This would lead to tensions with the general duty posited in section 39 (2) of the Constitution, which states that '[w]hen interpreting any legislation, and when developing the common law or customary law, every court, tribunal or forum must promote the spirit, purport and objects of the Bill of Rights.'[49]

There is therefore sufficient scope for the environmental minimum to provide helpful guidelines as to this independent normative content of section 24, both through the normative claims of environmental human rights and their application outlined by the specific risk, minimum standards, and high-priority principle. These would give at least some notion of independent constitutional environmental protection and some metric by which to evaluate the state and future development of dominant framework legislation like NEMA.

Perhaps the best reading of the limited case law, scholarship, and broader constitutional context is that, while legislation is not strictly required, it may nonetheless specify some of the normative claims of section 24. If that is the case, then legislation such as NEMA takes on a less problematic role, because

46 *Carmichele* v. *Minister of Safety and Security* [2001] ZACC 22 [33]– [34]; Klug, *The Constitution of South Africa*, 126; Kotzé, 'The Judiciary, the Environmental Right and the Quest for Sustainability in South Africa', 309–10.

47 See section 74 Constitution of the Republic of South Africa, 1996 [No. 108 of 1996] – G17678 (adopted 18 December 1996, in force 4 February 1997).

48 On the problems with the practice of 'constitutional avoidance' by the US Supreme Court, see Neal Kumar Katyal and Thomas P. Schmidt, 'Active Avoidance: The Modern Supreme Court and Legal Change' (2014) 128 *Harvard Law Review* 2109, 2165.

49 Section 39 (2) Constitution of the Republic of South Africa, 1996.

it does not specify protections to the exclusion of broader safeguards mandated independently by section 24, particularly those posited by the environmental minimum. However, in that case, the question remains whether applicants are nonetheless required to primarily rely on NEMA before invoking constitutional rights pursuant to the principle of constitutional subsidiarity.

9.1.2.2 Constitutional Subsidiarity

The principle of constitutional subsidiarity is crucial to understanding South African constitutional law.[50] It requires that an applicant relies '(. . .) on legislation when enforcing constitutional rights rather than circumventing the legislation in favour of direct application of a constitutional provision.'[51] Subsidiarity thus ideally provides a principled basis for determining when a constitutional right applies directly to a case, or whether it instead finds expression in legislation designed to give effect to the constitutional right.[52]

9.1.2.2.1 GENERAL PRINCIPLES. The ZACC articulated the modern principle in the case of *South African National Defence Union*,[53] and more recently confirmed it in *My Vote Counts*.[54]

My Vote Counts involved a constitutional claim from a non-profit organisation seeking disclosure of information regarding private funding of political parties. The issue was whether the Constitution afforded such a right, and whether the applicant was required, under the principle of constitutional subsidiarity, to first rely on the statutory provisions of the Promotion of Access to Information Act (PAIA).[55] The ZACC confirmed the subsidiarity principle and stated that '[o]nce legislation to fulfil a constitutional right exists, the Constitution's embodiment of that right is no longer the prime mechanism for its enforcement.'[56] Crucially, the ZACC applies this principle more generally to any 'legislation Parliament adopts with the clear design of

[50] The general existence of this principle is settled case law, see *Member of the Executive Council for Development Planning and Local Government, Gauteng* v. *Democratic Party and Others* [1998] ZACC 9; 1998 (4) SA 1157 [62]; *My Vote Counts NPC* v. *Speaker of the National Assembly and Others* [2015] ZACC 31 [54].

[51] Du Plessis, 'Subsidiarity', 47.

[52] Melanie Murcott and Werner Van der Westhuizen, 'The Ebb and Flow of the Application of the Principle of Subsidiarity-Critical Reflections on Motau and My Vote Counts' (2015) 7 *Constitutional Court Review* 43, 62.

[53] *South African National Defence Union* v. *Minister of Defence and Others* [2007] ZACC 10.

[54] *My Vote Counts NPC* v. *Speaker of the National Assembly and Others* [2015] ZACC 31.

[55] Promotion of Access to Information Act 2 of 2000, as amended 22 January 2014.

[56] *My Vote Counts NPC* v. *Speaker of the National Assembly and Others* [2015] ZACC 31 [53].

codifying a right afforded by the Bill of Rights',[57] and to those provisions of the Bill of Rights that specifically require the enactment of legislation, such as sections 9 (4), 25 (9), 32 (3), and 33 (3) of the Constitution.[58] Until the decision in *My Vote Counts*, there was some uncertainty over whether legislation gave effect only if it met some substantive requirements, or whether purporting to give effect to a right was sufficient.[59] Subsequently, it was clear that legislation enacted for the purposes of codifying any right will lead to the application of the principle of subsidiarity in order to avoid the development of parallel systems of law.[60]

In line with some earlier rulings,[61] the ZACC attached three conditions to the application of the subsidiarity principle: (1) it applies to legislation that claims to give effect to provisions of the Bill of Rights, primarily but not exclusively legislation enacted pursuant to an express constitutional requirement; (2) the constitutional validity of the statute must not be at stake; and (3) there must be an attempt to bypass the provisions of the statute by invoking constitutional rights.[62] It of course remains open to applicants to challenge legislation on the basis that it does not in fact give effect to a right, or where it does, to challenge the constitutionality of the legislation.[63]

9.1.2.2.2 ENVIRONMENTAL CASES. The subsidiarity principle has not been applied to section 24 of the Constitution. It therefore remains unclear whether there is a difference in the application of the principle depending on whether

57 Ibid. [55]; *National Education Health & Allied Workers Union (NEHAWU)* v. *University of Cape Town and Others* [2002] ZACC 27; 2003 (2) BCLR 154 [14]; *South African National Defence Union* v. *Minister of Defence and Others* [2007] ZACC 10; 2007 (5) SA 400 [52].

58 *My Vote Counts NPC* v. *Speaker of the National Assembly and Others* [2015] ZACC 31 [57]; *Minister of Health and Another* v. *New Clicks South Africa (Pty) Ltd and Others* [2005] ZACC 14; 2006 (2) SA 311 (CC) [96] and [437].

59 The minority opinion operates on this understanding, see *My Vote Counts NPC* v. *Speaker of the National Assembly and Others* [2015] ZACC 31 [67], but is expressly rejected by the majority [166].

60 Ibid. [63]; Murcott and Van der Westhuizen, 'The Ebb and Flow', 62.

61 *Zantsi* v. *Council of State, Ciskei and Others* [1995] ZACC 9 [4]; *Pharmaceutical Manufacturers Association of South Africa and Another* v. *President of the Republic of South Africa and Others* [2000] ZACC 1 [33] [rejecting the notion that the review of administrative action remained determined by common law, as opposed to the Constitution]; *Harksen* v. *Lane NO and Others* [1997] ZACC 12 [26] [rejecting the need to exhaust non-constitutional remedies before claiming constitutional remedies].

62 *My Vote Counts NPC* v. *Speaker of the National Assembly and Others* [2015] ZACC 31 [68]; while expressed in the minority opinion, the conditions were expressly adopted by the majority, see [121]–[122] and [161]–[162].

63 Ibid. [155].

ordinary legislation is envisioned to 'give effect' to the right or not.[64] The general doctrinal position we discussed above suggests that there may be no such difference, but the cases considered do not engage with section 24: they revolve mostly around rights that contain an express provision requiring legislation and where the legislation in question expressly purported to give effect to the right.[65]

For instance, the ZACC considered applying the principle of constitutional subsidiarity in the environmentally themed case of *Mazibuko*.[66] The case concerned a pilot scheme run by the City of Johannesburg to address water loss and outstanding payments for water services in Soweto: it involved the installation of prepaid metres which were designed to ensure payment once households had exceeded their six thousand litre per month water allowance. The applicants challenged this allowance and the lawfulness of the prepaid metres on the basis of section 27 (1) (b) of the Constitution (Access to Water). The previous ruling in the High Court and Supreme Court of Appeal held that subsidiarity did not apply, because the legislation did not intend to comprehensively give effect to the constitutional right.[67] The ZACC left open the question whether the subsidiarity principle precluded scrutiny of the policy on constitutional grounds, because it ultimately upheld the city policy.[68]

In reaching this conclusion, the ZACC engaged with the constitutional aspects of the case and fleshed out the normative content of the right to water: the court qualified section 27 (1) (b) as a positive obligation on the government to take reasonable legislative and other measures to progressively realise the right to access to water.[69] Ultimately, the ZACC declined to determine what constitutes 'sufficient water' under the constitutional right, deeming it a matter best addressed through legislation by policymakers.

64 Kidd, 'Environment', 528.

65 See, for instance, *Bato Star Fishing (Pty) Ltd* v. *Minister of Environmental Affairs and Tourism and Others* [2004] ZACC 15 [25]–[26] [ordinary legislation giving effect to Section 33 (administrative justice) of the Constitution]; *MEC for Education Kwazulu-Natal and others* v. *Pillay* [2007] ZACC 21 [40] [ordinary legislation giving effect to Section 9 (Equality) of the Constitution]; *South African National Defence Union* v. *Minister of Defence and Others* [2007] ZACC 10 [51]–[52] [ordinary legislation to give effect to labour rights protected in Section 23 of the Constitution]; *PFE International Inc (BVI) and Others* v. *Industrial Development Corporation of South Africa Ltd* [2012] ZACC 21 [4] [ordinary legislation giving effect to Section 32 of the Constitution (access to information)]; see also Murcott and Van der Westhuizen, 'The Ebb and Flow', 65.

66 *Mazibuko and Others* v. *City of Johannesburg and Others* [2009] ZACC 28.

67 Ibid. [75]– [76].

68 Ibid. [97], [102].

69 Ibid. [94].

9.1.2.2.3 CONCLUSIONS. There are a number of problems that render it difficult to draw any firm conclusions from the case law for the purposes of section 24 of the Constitution.

First, the ZACC considered, but ultimately did not state whether the principle of subsidiarity precluded a review of the reasonableness of the water policy in *Mazibuko* on constitutional grounds. It did not follow or reject the position of the previous judgments in the High Court and Supreme Court of Appeal, which both held that the legislation in question was never intended to comprehensively give effect to section 27 (1) (b).

Second, the general applicability of the principle of constitutional subsidiarity to the Bill of Rights appears well established. However, it is not clear how the principle applies where a right is concerned that does not unequivocally require legislation to give effect to its safeguards, as is the case with section 24. Even with broad legislation such as NEMA, there is scope for legislative blind spots, quite apart from whether it meets the minimum standards and consistent enforcement requirements of the environmental minimum.

The environmental minimum would at least ask whether the regulation in question presented a good faith attempt to regulate environmental harm and was consistently enforced. This is an analytical step that the ZACC hints at in *Mazibuko*, when expressing doubts as to whether mere compliance with legislation is sufficient to demonstrate that an administrative decision was reasonable, and hence legal.[70] The compatibility of the principle of subsidiarity with the environmental minimum at least for the purposes of section 24 therefore appears in some doubt, but this is largely due to the doctrinally underdeveloped case law on this constitutional provision. Certainly, there is no insurmountable barrier to integrating the normative claims and principles of the environmental minimum into the evolving doctrine. However, such a step requires a shift in focus onto constitutional rights and their normative content independent from currently dominant framework legislation.

As we explore in the following section, in Germany there is a similar issue with deference to the legislature on fulfilling positive obligations arising from constitutional rights. It is arguably based on the limited impact of the constitutional mission statement in the broader constitutional framework.

9.2 GERMANY

As with South Africa, the enforcement of environmental regulations is generally not a source of systemic concern in Germany. Once an environmental

[70] Ibid. [74].

regulation is established, there are few recent examples of inadequate enforcement. The German *Rechtsstaatsprinzip* implies a strong and largely unqualified duty on all branches of government to implement regulations in good faith.[71] The problems in Germany are more closely related to the limited practical relevance and impact of the constitutional mission statement requiring environmental safeguards. While the constitutional mission statement influences the interpretation of constitutional rights, there are few clear obligations to regulate, mostly due to the deferential approach of the Constitutional Court to positive obligations. Before turning to the level of deference, it is useful to briefly consider the scope of duties under Article 20a of the German Basic Law (GG) and how they might influence constitutional rights.

9.2.1 *Scope of Protection*

9.2.1.1 Article 20a

The German Constitution refers to the environment in the constitutional mission statement of Article 20a GG:

> Mindful also of its responsibility toward future generations, the state shall protect the natural foundations of life and animals by legislation and, in accordance with law and justice, by executive and judicial action, all within the framework of the constitutional order.

The provision was introduced in 1994 and initially concerned itself only with the 'natural foundations of life', while the reference to the protection of animals was added in 2002. It was intended to provide a binding constitutional mission statement, and is generally seen as having a direct legal effect, acting as a guideline to the interpretation of constitutional provisions and the application of ordinary legislation.[72] Article 20a GG is generally not thought to contain an individual right, even though this has not yet been unequivocally confirmed by the Constitutional Court.[73]

There is overall little Constitutional Court case law on Article 20a GG, making lower court rulings and scholarship the primary source for the scope of

[71] Jarass and Pieroth, *Grundgesetz für die Bundesrepublik Deutschland*, 532–9.

[72] *Bericht der Gemeinsamen Verfassungskommission* (Bundestag Drucksache 12/6000, 1993) 67.

[73] Federal Administrative Court, 6 C 996, 27 August 1997, NJW 1998, 843; Rupert Scholz, 'Artikel 20a' in Theodor Maunz and others (eds.), *Grundgesetz Loseblatt Kommentar* (76th ed., C. H. Beck 2016) [33]; Helmuth Schulze-Fielitz, 'Artikel 20a' in Horst Dreier (ed.), *Grundgesetz Kommentar*, vol 2 (3rd ed., Mohr Siebeck 2015) [82].

obligations. According to the prevailing view, the state must refrain from taking measures that harm the environment,[74] and take positive steps towards conservation and repair of environmental harm.[75] This requires regulation of the conduct of private entities and justifies burdening the responsible actors with the associated costs pursuant to the Polluter Pays principle.[76] Article 20a GG limits state incentives for actions that entail environmental harm and mandates a friendly disposition towards private organisations that seek environmental protection.[77] The 'responsibility toward future generations' obliges the state to acknowledge and account for long-term risks,[78] including preventative measures and the principle of sustainability.[79] Natural resources may only be exploited to the extent that this does not defeat the possibility of future generations benefitting from them.[80]

The doctrine of the Constitutional Court is more developed with respect to the protection of animals. The case of *Legehennenhaltung* involved a challenge against delegated legislation regulating the treatment of hens. Under § 16b of the Federal Animal Protection Act, the animal welfare committee had to be consulted before any changes to subordinate legislation could be enacted. However, in the case at hand, the committee was consulted too late to effect any real changes to the proposed amendment: state entities had no real interest or practical ability to make further changes based on the committee's views, effectively rendering the consultation pointless.[81] The Federal Constitutional Court struck down the amended subordinate legislation for this reason, holding that the Animal Protection Act contributed towards fulfilling Article 20a GG, and thus the amendment was inconsistent with

[74] Norbert Bernsdorff, 'Artikel 20a' in Thomas Clemens and Dieter C. Umbach (eds.), *Grundgesetz Kommentar*, vol 1 (C. F. Müller 2002) [31]; Astrid Epiney, 'Artikel 20a' in Hermann von Mangoldt, Friedrich Klein and Christian Stark (eds.), *Kommentar zum Grundgesetz*, vol 2 (6th ed., Franz Vahlen 2010) [57].

[75] Dietrich Murswiek, 'Artikel 20a' in Michael Sachs (ed.), *Grundgesetz Kommentar* (7th ed., C. H. Beck 2014) [33].

[76] Jarass and Pieroth, *Grundgesetz für die Bundesrepublik Deutschland*, Article 20a, 576 [9]; Murswiek, 'Artikel 20a' [35]; Epiney, 'Artikel 20a' [74].

[77] Jarass and Pieroth, *Grundgesetz für die Bundesrepublik Deutschland*, Article 20a, 577 [5]; Murswiek, 'Artikel 20a' [34]; Simone Westphal, 'Artikel 20a GG – Staatsziel "Umweltschutz"' [2000] *Juristische Schulung* 339, 340.

[78] Epiney, 'Artikel 20a' [31].

[79] Federal Administrative Court, 8 B 25597, 22 December 1997, NJW 1998, 2301; Epiney, 'Artikel 20a' [73]; Schulze-Fielitz, 'Artikel 20a' [69]; Murswiek, 'Artikel 20a' [37]; Stefan Huster and Johannes Rux, 'Artikel 20a ' in Volker Epping and Christian Hillgruber (eds.), *Grundgesetz Kommentar* (2nd ed., CH Beck 2013) [16].

[80] Schulze-Fielitz, 'Artikel 20a' [40].

[81] Federal Constitutional Court, 2 BvF 1/07, 12 October 2010, BVerfGE 127, 293 [105] and [119].

the authorising legislation as interpreted through the mission statement.[82] The case also afforded the Constitutional Court an opportunity to make some general comments on the nature of protections under the constitutional mission statement.

Article 20a GG protects all animals to the extent that they are sentient,[83] but only applies to individual specimens, rather than the ecosystem or the species as a whole.[84] The state must further refrain from harming animals and must regulate the conduct of private actors appropriately.[85] Notably, animal welfare and environmental protection can justify limitations of fundamental rights and other constitutionally sanctioned interests, without necessarily taking priority in every case.[86] The state enjoys broad discretion in balancing competing interests when drafting animal welfare and environmental protection legislation.[87] Apart from the constitutional mission statement, the Constitutional Court has interpreted several constitutional rights to require protections from environmental harm, most prominently among them the right to life and health.

9.2.1.2 Constitutional Rights

The right to life (Article 2 (2) GG) constitutes a key provision in the rationale of the German Basic Law with strong connections to the fundamental commitment to human dignity, under Article 1(1) GG.[88] The right to health encompasses human health in a physiological sense,[89] and bodily integrity,[90] while human well-being is protected at least to the extent that it is disrupted by physical pain.[91] The protections are primarily directed against state actions that infringe on these rights. These are not of primary relevance here, as the

82 Ibid. [102] and [120].
83 Ibid. [121]; 914; Schulze-Fielitz, 'Artikel 20a' [55]; Murswiek, 'Artikel 20a' [31b].
84 Jarass and Pieroth, *Grundgesetz für die Bundesrepublik Deutschland*, Article 20a, 578 [13]; Epiney, 'Artikel 20a' [88]; Murswiek, 'Artikel 20a' [51b].
85 Schulze-Fielitz, 'Artikel 20a' [59].
86 Federal Constitutional Court, BVerfGE 127, 293 [121].
87 Ibid. [122].
88 Federal Constitutional Court, 1 BvR 357/05, 15 February 2006, BVerfGE 115, 118 (139); Federal Labour Court, 2 AZR 638/99, 8 June 2000, BAGE 95, 78.
89 Federal Constitutional Court, 1 BvR 612/72, 14 January 1981, BVerfGE 56, 54 (74).
90 Federal Constitutional Court, 2 BvR 882/09, 23 March 2011, BVerfGE 128, 282 (302) [emphasising self-determination and consent with regard to surgical procedures].
91 Federal Constitutional Court, BVerfGE 56, 54(75) [left open with respect to psychological pain].

state is rarely directly responsible for environmental harm.[92] Major environmental regulations are found in the *Bundesnaturschutzgesetz* (Federal Nature Conservation Act),[93] and further provisions adopted by the German *Länder* (the constituent federal states of Germany). Most environmentally themed cases litigated under constitutional rights involve nuclear power production and the allegation that the state has failed to protect life and health through appropriate regulation of third-party activity.[94]

The case of *Mühlheim-Kärlich* concerned the construction of the nuclear power plant of that name.[95] The Constitutional Court heard a procedural challenge against the order of the Federal Administrative Court that allowed construction to go ahead notwithstanding an on-going legal challenge concerning the legality of deviations from the original planning permission.[96] The case occasioned the Constitutional Court to make some remarks about the relationship between constitutional rights and environmental regulation. The state typically satisfies its positive obligations under constitutional rights through appropriate regulations,[97] and fair administrative proceedings for issuing relevant permits.[98] These measures must be adequate and effective,[99] and a threat to life and health may be sufficient on its own to find a violation.[100] The requirements are not any less exacting simply because the state is not typically directly responsible for environmentally risky ventures.[101] Generally, these regulations permit the state to balance competing interests of individuals and the public. However, in the specific case of nuclear energy, the interests of protecting life, health, and integrity of neighbouring property

92 For a rare exception, *see* Federal Administrative Court, 7 C 3387, 29 April 1988, BVerwGE 79, 254 (257) [state run business causing noise pollution].

93 Bundesnaturschutzgesetz – German Federal Nature Conservation Act.

94 Federal Constitutional Court, BVerfGE 56, 54 (73); Federal Administrative Court, 4 C 995, 21 March 1996, BVerwGE 101, 1 (10); Federal Constitutional Court, 1 BvR 518/02, 4 April 2006, BVerfGE 115, 320 (346); Federal Constitutional Court, 1 BvQ 5/77, 16 October 1977, BVerfGE 46, 160 (164).

95 Federal Constitutional Court, 1 BvR 385/77, 20 December 1979, BVerfGE 53, 30.

96 Ibid. [14].

97 Federal Administrative Court, 7 C 6981, 18 March 1982, BVerwGE 65, 157 (160).

98 Federal Constitutional Court, BVerfGE 53, 30 [53]; Federal Administrative Court, 7 C 10178, 17 July 1980, BVerwGE 60, 297 (305).

99 Federal Constitutional Court, 2 BvF 2/90; 2 BvF 4/90; 2 BvF 5/92, 28 May 1993, BVerfGE 88, 203 (254); Federal Constitutional Court, BVerfGE 56, 54 (78); Federal Constitutional Court, 1 BvR 1025/82, 1 BvL 16/831, 1 BvL 10/91, 28 January 1992, BVerfGE 85, 191 (212).

100 Federal Constitutional Court, 2 BvR 1060/78, 19 June 1979, BVerfGE 51, 324 (346); Federal Constitutional Court, 2 BvR 1160/83; 2 BvR 1565/83; 2 BvR 1714/83, 16 December 1983, BVerfGE 66, 39 (58).

101 Federal Constitutional Court, BVerfGE 53, 30 [53].

take precedence over the promotion of nuclear energy production.[102] The Constitutional Court found no reason to hold the general regulatory regimes for nuclear power plants inadequate, and held the Federal Administrative Court had not misunderstood the gravity and implications of fundamental rights. The constitutional challenge was therefore dismissed.

The case of *Kalkar I* clarified that future environmental harm could constitute a violation of constitutional rights. The case concerned a referral to the Constitutional Court regarding the validity of a provision of the Atomic Energy Act that allowed for the construction of a certain type of experimental power plant.[103] The Constitutional Court held that constitutional duties to protect can require the state to take measures to limit potential future violations. Notably, the nature, immediacy, and gravity of a risk, as well as the reversibility of infringements, are important considerations.[104] However, complete certainty can never be achieved in risk assessments and the state was not under a duty to eliminate the possibility of all conceivable harms.[105] Crucially, the Act granted administrative bodies discretion to reject planning permissions for nuclear power plants on the basis of best practices and scientific evidence not available when the Act was originally drafted.[106] The Constitutional Court therefore found that the regulatory regime of the Atomic Energy Act fully lived up to the requirements of constitutional rights.

The importance of updating regulations in line with scientific evidence was confirmed and expanded in the *Fluglärm* decision.[107] The Constitutional Court found that aircraft noise engages the right to health and derived a duty to adjust regulatory regimes in light of fresh evidence, even if the health risks could not have been foreseen at the time of drafting.[108] This is crucial where the state is indirectly responsible for the infringements of rights through the regulatory framework it adopts and the issuing of permits.[109] The generic provision in the environmental regulation limiting permissible noise to what is 'technically unavoidable' did not suffice to discharge this duty, because feasibility is not the sole determinant of environmental protection standards.[110] If technical solutions do not suffice to meet the required safeguards, then

[102] Ibid. [54].
[103] Federal Constitutional Court, 2 BvL 8/77, 8 August 1978, BVerfGE 49, 89.
[104] Ibid. [115].
[105] Ibid. [117].
[106] Ibid. [122].
[107] Federal Constitutional Court, BVerfGE 56, 54.
[108] Ibid. [81].
[109] Ibid. [82].
[110] Ibid. [83].

additional mitigating action must be considered to achieve a balance between the competing interests. The state may thus be required to prohibit harmful behaviour entirely, if it cannot be conducted in a manner that satisfies fundamental rights. However, the Constitutional Court ultimately found that the state had discharged its positive obligations on the facts of the case, citing the margin of appreciation that applies to implementing positive constitutional rights.

Similarly to the South African case study, the German Constitutional Court at times affords the state a problematic level of deference which sanctions at least some inadequate regulations and inconsistent enforcement.

9.2.2 *Limitations*

The German Constitutional Court invokes judicial self-restraint to justify discretion of state entities in balancing competing interests, especially with respect to the scope and implications of positive obligations. This is problematic under the environmental minimum because it allows state entities to systematically ignore specific risks to fundamental rights, despite the notional inclusion of an environmental mission statement in the Constitution. The Constitutional Court generally resists the idea that the right to life mandates any specific welfare obligations.[111] Under most circumstances, the state has broad discretion in determining appropriate regulatory actions.[112]

This was made clear by the Constitutional Court in the case of *Lagerung chemischer Waffen* concerning the storage of chemical weapons on German territory in US military depots.[113] The applicants alleged that Germany had failed to live up to its obligations to protect their rights by permitting the storage of chemical weapons near their homes and businesses as well as failing to ensure adequate safety standards and precautionary measures.[114] Crucially for our purposes, the Constitutional Court held that there was a positive obligation, in principle, to safeguard the right to life and that the state enjoyed considerable discretion in determining the correct balance with the public interest. To the extent that the applicants demanded the removal of chemical

[111] A notable exception is the existential minimum, where Article 2 (2) functions in conjunction with Article 1 (1) (human dignity), see Federal Constitutional Court, 1 BvL 1/09; 1 BvL 3/09; 1 BvL 4/09, 9 February 2010, BVerfGE 125, 175 (223).

[112] Federal Constitutional Court, 2 BvR 624/83; 2 BvR 1080/83; 2 BvR 2029/83, 29 October 1987, BVerfGE 77, 170 (214); Federal Constitutional Court, 1 BvR 1301/84, 30 November 1988, BVerfGE 79, 174 (202); Federal Constitutional Court, BVerfGE 85, 191 (212).

[113] Federal Constitutional Court, BVerfGE 77, 170.

[114] Ibid. [10]–[15].

weapons due to their (potentially) hazardous nature, they had failed to establish that this was the only means of living up to the positive obligation.[115] Because the Basic Law expressly permits a military-based defence in the event of an attack, on balance the Constitutional Court held that this entails certain limitations on fundamental rights.[116] The court required only that the measures are not entirely unsuitable or inadequate.[117] Barring such rare cases, it is unlikely that the protection of life and health will mandate any specific regulatory action to prevent environmental harm. Positive obligations therefore do not currently provide a firm standard of review for environmental regulations based on the normative claims of environmental human rights.

The mission statement of Article 20a GG for its part does not play a significant role in the interpretation of fundamental rights because it is itself limited through these rights. For instance, the kosher slaughter of animals is permissible in light of the protection of religious freedoms,[118] and limitations have been placed on the protection of animals in the interests of public safety. For instance, the Constitutional Court found in *Kampfhunde* that it is constitutionally permissible to criminalise the import of so-called aggressive dog breeds, provided that the state reviewed the evidence for the underlying assumption that they pose a higher risk to humans.[119] A prohibition on breeding these dogs was struck down, but this was primarily because the federal legislature lacked the necessary competence to enact the provisions. On both issues, the constitutional mission statement did not substantially alter the analysis: neither the import ban nor the prohibition on breeding was found to impact on the protection of animals, and even if it did, the law was deemed an appropriate limitation of the duties arising from the constitutional mission statement.[120]

Therefore, both the discretion afforded and the limited constitutional mission statement ensure that the doctrinal protection of the environment through constitutional rights falls considerably short of the environmental minimum framework. In striking a balance with other constitutional rights and values, the Constitutional Court affords regulators considerable discretion,[121] which limits the scope for positive obligations that may arise in

[115] Ibid. [112].
[116] Ibid. [126].
[117] Ibid. [112].
[118] Federal Administrative Court, 3 C 3005, 23 November 2006, BVerwGE 127, 183 [12].
[119] Federal Constitutional Court, 1 BvR 1778/01, 16 March 2004, BVerfGE 110, 141 [88].
[120] Ibid. [87].
[121] This is settled case law, see Federal Constitutional Court, BVerfGE 127, 293 [122]; Federal Constitutional Court, 1 BvF 1/05, 13.03.2007, BVerfGE 118, 79 [111].

conjunction with Article 20a GG. This becomes particularly problematic with respect to inadequate enforcement measures, for instance against air pollution in major cities, which have become the subject of several legal challenges over the past decade.

9.2.3 *Air Pollution Litigation*

The leading case challenged the lack of air pollution reduction measures in Munich. The applicant lived along a busy road and the closest monitoring station showed that the threshold value for PM10 of 50 μg/m³ was breached over 100 times in the year 2005.[122] Bavarian regulators had drawn up an action plan for Munich as a whole, but not specifically for the applicant's area.[123] Before the lower Bavarian Administrative Court, the applicant sought specific measures and commitments to reduce the pollution levels in his area. The ultimately engaged Federal Administrative Court requested a preliminary ruling from the ECJ on the question of whether Directive 96/62/EC (later consolidated in EU Directive 2008/50/EC) confers individual rights and hence whether the applicant could claim a violation of EU law.[124]

The ECJ determined that Directive 96/62/EC contains an individual right, but only to require the implementation of action plans that can conceivably achieve compliance with air quality threshold values in due course.[125] Applicants have no recourse against existing breaches of air pollution guidelines, and the Directive does not permit them to seek any specific measures to promote clean air. Member states are primarily required to take measures that are in principle capable of reducing the risk of exceeding thresholds, and are expressly permitted to take opposing interests, such as business opportunities, into account. The subsequent judgment of the German Federal Administrative Court hence focused on the question of whether the applicant could force the city of Munich to adopt measures to ensure that the threshold values for PM10 were not breached.[126]

The Court ruled that even in the absence of a comprehensive action plan, local government was not excused from taking appropriate measures to combat air pollution, as mandated by federal laws regulating emissions and

[122] Federal Administrative Court, BVerwGE 129, 296 [2].
[123] Ibid. [4].
[124] Federal Administrative Court, BVerwG 7 C 906, 27 March 2007, BVerwGE 128, 278.
[125] Case C-237/07 *Dieter Janecek* v. *Freistaat Bayern* [2008] ECLI:EU:C:2008:447 [42] and [47].
[126] Federal Administrative Court, BVerwGE 129, 296.

vehicular traffic.[127] The refusal of the city of Munich to take any measures constituted a violation of the applicant's right to health,[128] and gave rise to an individual claim towards drawing up a comprehensive action plan as well as suitable measures that can address the level of air pollution.[129] However, the measures must usually be directed to target those responsible for the emissions (vehicular traffic and industries) by the percentage of their contribution towards the overall air pollution.[130] Additionally, the local government has discretion in determining which measures it deems most appropriate and it must weigh the rights of the individual with the interests of broader society.[131] The Court indicated that any such evaluation could scarcely avoid concluding that a ban of LGV and HGV traffic from the affected area was necessary,[132] and that in principle the applicant could also demand measures to curb emissions from other sources.[133]

With that, the Federal Administrative Court remanded the case back to the Bavarian Administrative Court for further factual determinations, especially as to the level of air pollution in the immediate vicinity of the applicant's home. However, both the applicant and the respondent city of Munich issued declarations that they no longer required adjudication of the case (*Erledigungserklärung*) and the proceedings were terminated without a final ruling.[134] Why Janecek ultimately, after all the effort and time invested into litigation, withdrew his application remains a mystery that only those involved can resolve. From an outsider's perspective, that procedural decision appears to have little merit, as it exposes Janecek to carrying a significant burden of the litigation costs (50 per cent), which would have been borne primarily, or perhaps even in their entirety, by the defeated city had the judgment been in Janecek's favour, which seemed a likely outcome.

The city of Munich reacted to the litigation with ever more ambitious action plans to bring air pollution into line with regulatory requirements. While air quality improved primarily with respect to regulated indicators, on

[127] In this case namely §§ 17, 41, 45 *Bundesemissionsschutzgesetz* and § 4 of *Verordnung über Immissionswerte für Schadstoffe in der Luft*.
[128] Federal Administrative Court, BVerwGE 129, 296 [21]–[22].
[129] Ibid. [25].
[130] Ibid. [27].
[131] Ibid. [28]; Federal Administrative Court, BVerwG 7 C 7684, 4 June 1986, BVerwGE 74, 234, 239.
[132] Federal Administrative Court, BVerwGE 129, 296 [34].
[133] Ibid. [40].
[134] Bavarian Administrative Court, 22 BV 072587, 21 December 2007, unpublished.

other dangerous pollutants progress has been slow, even considering that the thresholds set by EU law are more generous than WHO guidelines.

The ruling in *Janecek* inspired a number of further legal challenges in other major German cities. The administrative court of Wiesbaden required the State of Hessia to amend the air quality legislation for the city of Darmstadt so as to ensure that the yearly threshold values for nitrous dioxide were met.[135] Almost identical decisions were reached with respect to the action plans for air quality in the cities of Limburg,[136] Hamburg,[137] and Reutlingen.[138]

The nonetheless persistently sluggish pace of improvement is a symptom of inconsistent and inadequate enforcement measures, at times overtly flouting legal obligations and ignoring final court decisions. Through the course of further and separate litigation to *Janecek*, the city of Munich was in 2018 required by a final court ruling to consider – and justify any decision to refrain from issuing – a prohibition on older diesel vehicles driving in the city in the interest of reducing NO_2 emissions.[139] However, local and state officials openly refused over many months to even contemplate such a ban, much less provide legally relevant reasons for their refusal.[140] Applications brought to enforce the ruling were stalled in no small part by the peculiarities of the German Code of Administrative Enforcement.[141]

The code generally provides for a fine of up to 25,000 Euros for failure to perform a task required by a court.[142] While the imposed fines were promptly paid by the government, they have had no discernible success inducing the desired behaviour: this is because the fines are payable from one government entity of the Bavarian state into the general Bavarian state public budget, thus limiting its deterrent effect on officials.

The alternative to a fine is imprisonment of the responsible officials for up to two weeks at the court's discretion until they comply with the court order.[143] At the time, it was unclear whether EU law required the use of such heavy-handed enforcement measures. For these reasons, the Bavarian Administrative Court asked the ECJ to clarify the position on imprisonment as a sanction for

[135] Wiesbaden Administrative Court, 4 K 165/12WI, 16 August 2012, unpublished.
[136] Wiesbaden Administrative Court, 4 K 97/15WI(2), 30 June 2015, unpublished.
[137] Hamburg Administrative Court, 9 K 1280/13, 5 November 2014, unpublished.
[138] Sigmaringen Administrative Court, 1 K 154/12, 22 October 2014, unpublished.
[139] Bavarian Administrative Court, 22 C 181718, 9 November 2018, unpublished [1]–[7].
[140] Ibid. [26].
[141] Verwaltungs-Vollstreckungsgesetz of 27 April 1953, last amended through Article 42 of the Ordinance of 19 June 2020 (BGBl. I S. 2490).
[142] See § 11 Verwaltungs-Vollstreckungsgesetz.
[143] See § 16 Verwaltungs-Vollstreckungsgesetz.

violations of EU law.[144] The ECJ ruled in the *Deutsche Umwelthilfe* decision that direct coercive measures, such as imprisonment, were possible if these were provided by German law and where such measures are sufficiently precise and foreseeable, taking into account the right to liberty guaranteed in Article 6 of the Charter of Fundamental Rights.[145]

In the interim, Munich passed the seventh amendment of the air pollution control plan in October 2019, which boasted significant reductions of many pollution indicators such as PM10 and PM2.5, while remaining in breach of threshold values for NO_2.[146] Crucially, the amendment expressly contemplated a ban on diesel vehicles, but perhaps predictably, concluded that such a ban was not necessary anywhere in the city despite persistently high levels of pollution. On this basis, the Bavarian Administrative Court declined to impose further sanctions over the failure to implement its ruling. It reasoned that the Bavarian government had only been required to contemplate and justify the absence of a ban on diesel vehicles. The 2018 ruling never imposed specific outcome requirements and hence the government had complied with its legal obligations in respect of the judgment.[147] Although there were strong suggestions in the 2018 ruling that diesel vehicle bans would have to be part of the overall package of measures,[148] the Court declined to rule on the matter pointing to separate on-going litigation.[149]

As things stand, some core claims of the environmental minimum framework are not vindicated through the prevailing deferential approach to positive obligations under constitutional rights: environmental harm must be actionable under fundamental rights, and both good faith environmental regulation and consistent enforcement are required to meet these obligations. In the air pollution saga, Germany has revealed a significant environmental blind spot in its otherwise detailed constitutional law and fundamental rights protection regime.

9.3 SUMMARY

The purpose of this chapter was to demonstrate some of the environmental gaps that persist in two sophisticated human rights regimes with different

144 Case C-752/18 *Deutsche Umwelthilfe* v. *Freistaat Bayern* [2019] ECLI:EU:C:2019:1114.

145 Ibid. [57].

146 *Luftreinhalteplan für die Stadt München* (Bayerisches Staatsministerium für Umwelt und Verbraucherschutz, 2019) www.muenchen.de/rathaus/Stadtverwaltung/Referat-fuer-Gesundheit-und-Umwelt/Luft_und_Strahlung/Luftreinhalteplan.html, accessed 1 April 2021, 13.

147 Bavarian Administrative Court, 22 C 2044, 27 August 2020, ZUR 2020, 689 [48] and [50].

148 Bavarian Administrative Court, 22 C 181718, 9 November 2018, unpublished [107] and [139].

149 See Bavarian Administrative Court, 22 B 18.1952 (case pending).

constitutional commitments to environmental protection. In both jurisdictions, it is no longer in serious doubt that the Constitution imposes a duty to regulate environmental harm that constitutes a specific risk. This normative question has been settled in the case law of the South African and German Constitutional Courts. Hence, regulators must address environmental harm, but the manner in which it is addressed, and the level of scrutiny exercised through constitutional rights review, remains unsatisfactory. Both courts grant a level of deference that is problematic under the environmental minimum.

With respect to section 24 of the Constitution of South Africa, the doctrine is underdeveloped and in practice, decisions often rely exclusively on framework legislation. This is due to doctrinal uncertainty over the necessity of ordinary legislation to give effect to constitutional rights and the applicability of the constitutional subsidiarity doctrine. Crucially, the normative claims and principles of section 24 are underdeveloped, thereby making it difficult to challenge framework legislation as unconstitutional. The resulting reliance on NEMA in the vast majority of environmental cases serves only to deepen the doctrinal marginalisation of section 24 of the Constitution: as things stand, section 24 cannot provide a suitable standard of review for environmental regulations. This renders it difficult to say anything definitive about the compatibility of South African constitutional law and fundamental rights with the environmental minimum, other than to note that it is certainly possible to develop a more robust doctrine based on the framework.

The German Constitutional Court for its part recognises positive obligations to regulate environmental threats to human life and health, chiefly in the context of nuclear energy production. The Court has voiced support for the idea that regulatory frameworks can be unconstitutional if they are not updated to reflect modern scientific evidence, a key claim that connects with the bad faith requirement of the environmental minimum. The primary cause for concern is undue deference on positive obligations. This is particularly noticeable with respect to air pollution regulation in Munich. Legislatures and administrative bodies are generally afforded broad discretion, especially when deciding whether and how to regulate environmental harm. The constitutional mission statement of Article 20a GG has been unable to significantly shift this doctrinal approach and fails to meaningfully impact practical environmental regulation standards. So marginal is the impact that it was not of particular note that none of the court decisions regarding air pollution relied on the constitutional mission statement in any meaningful sense, and often did not even so much as mention it during the proceedings. The scope and requirements of Article 20a GG might be doctrinally clearer in some respects than section 24 of the South African Constitution, but the provision is nonetheless virtually irrelevant for the practical business of air pollution regulation.

Regulations remain inadequate and inconsistently enforced, at times even in open defiance of court rulings. The focus on air pollution action plans and the reluctance to deploy fundamental rights guarantees against existing environmental harm are difficult to reconcile with the environmental minimum. At least some of the most serious environmental harms covered by established standards, and certainly domestic legal standards, should not be open to any meaningful discretion on compliance. Under any serious commitment to human rights and the rule of law, they require nothing short of swift and full compliance as well as consistent enforcement.

PART IV

Beyond Human Rights Law

Part IV of the book explores some of the contributions that the environmental minimum can make to environmental regulation under domestic and international law.

Chapter 10 focuses on domestic and international efforts to regulate environmental harm and suggests that three pathologies have historically hampered their success: political lag, which describes the gap between the best available scientific evidence and regulatory efforts to address environmental harm; industry resistance, which arises from the fruitful ground of political lag, permitting vested interests to entrench harmful, but profitable business models and practices; and finally, regulatory inertia, which means that regulators are less likely to burden vested economic interests with effective and consistently enforced regulations. Historic atmospheric ozone and asbestos regulations provide contrasting examples of regulatory success. Meanwhile, pervasive contemporary failures in addressing air pollution demonstrate that the pathologies remain salient concerns. The chapter argues that conventional regulatory approaches can be an effective tool to address environmental harm, provided there is a close relationship between scientific evidence and regulatory action.

Chapter 11 then considers two key implications of the environmental minimum for international environmental law. For one, there is considerable difficulty in applying the human rights-based framework of the book to issues surrounding transboundary environmental harm, particularly the overarching global issue of man-made climate change. Due to enduring temporal and geographic concerns as well as the at most lacklustre commitment of many states to upholding them as practical safeguards, human rights are unlikely to contribute decisively to the current policy debate. The environmental minimum presupposes a robust commitment to human rights, environmental protection, and the rule of law that is questionable with respect to many of the most notorious global polluters. The framework is more suited to specifying the scope, context, and relationship of competing environmental obligations, notably among the Sustainable Development Goals, and can provide a viable framework to enforce otherwise non-binding or underenforced treaty obligations under international environmental law.

10

Environmental Regulation

The history of environmental regulation abounds with examples of missed opportunities, failures to prevent harm, undue deference to industry interests, and politically expedient ignorance. In some parts of the environmental scholarship, past regulatory failures ground a general scepticism towards the enduring relevance and effectiveness of conventional human-centred regulations. Suggested alternatives focus on reconceptualising regulations from an ecological perspective and re-examining foundational principles of law across all fields,[1] and less commonly, to rejecting regulations altogether in favour of market-based mechanisms.[2] Despite their mixed track record, conventional regulatory approaches have in the past achieved some notable success, securing tangible improvements at both the domestic and international level.

This chapter first explains that conventional regulatory approaches can be an effective tool to address environmental harm, provided that there is a close relationship between scientific evidence and regulatory action. This chapter then argues that many concerns expressed in scholarship can be traced back to three broadly categorised pathologies, which feature in historic and contemporary regulatory responses:

- political lag, which describes the gap between the best available scientific evidence and regulatory efforts to address environmental harm;

[1] Montini, 'The Double Failure of Environmental Regulation and Deregulation'; Redgwell, 'Life, The Universe and Everything'; Grear, 'The Vulnerable Living Order'; Fitz-Henry, 'Decolonizing Personhood', 133; Grant, 'Re-imagining Adjudication'; Pierre- Dupuy and Viñuales, *International Environmental Law*, 300; Francioni, 'International Human Rights in an Environmental Horizon'; Viñuales, 'Law and the Anthropocence', 26; Kotzé, *Global Environmental Constitutionalism in the Anthropocene*.

[2] Klaus Bosselmann and B. J. Richardson, 'Introduction: New Challenges for Environmental Law and Policy' in Klaus Bosselmann and B. J. Richardson (eds.), *Environmental Justice and Market Mechanisms* (Kluwer Law International 1999) 1.

- industry resistance, which arises from the fruitful ground of political lag, permitting vested interests to entrench harmful, but profitable business models and practices; and finally
- regulatory inertia, the reluctance of regulators, resulting from political lag and industry resistance, to burden vested economic interests with effective regulations or enforce them consistently, even where they produce serious environmental harm.

Contemporary air pollution is a good example of how these three pathologies remain salient and the chapter closes by explaining what improvements the environmental minimum can achieve.

10.1 REGULATIONS TO PROTECT THE OZONE LAYER

There are many reasons to doubt the assertion that environmental regulations are 'inherently and inevitably'[3] unable to address environmental harm appropriately. Certainly, there have been prominent failures in the past and present, but there are equally examples of successful interventions, where conventional regulation achieved its stated goals and brought about tangible improvements.

Consider, for instance, regulations designed to protect the ozone layer. Early signs of significant damage to the ozone layer, along with allegations of a causal link to certain substances, were first raised in 1974.[4] The allegations were followed up by a comprehensive study and formulation of a scientific consensus at the World Meteorological Organization in 1985,[5] which culminated in the framework Vienna Convention for the Protection of the Ozone Layer (1985),[6] and the phasing out of harmful substances through the Montreal Protocol (1987).[7] Both are among the most widely adopted international treaties, ratified by over 197 parties. Overall, it took a global community of scientists and regulators less than 15 years to devise comprehensive international regulations that ultimately eliminated the emission of chlorofluorocarbons (CFCs) and other ozone-depleting substances. Had decisive regulatory action not been taken, 67 per cent of globally averaged ozone would have been destroyed by 2065 resulting in dramatic changes in global

[3] Montini, 'The Double Failure of Environmental Regulation and Deregulation', 12.

[4] Mario J. Molina and F. Sherwood Rowland, 'Stratospheric Sink for Chlorofluoromethanes: Chlorine Atom-Catalysed Destruction of Ozone' (1974) 249 *Nature* 810.

[5] *Atmospheric Ozone, 1985: Assessment of Our Understanding of the Processes Controlling Its Present Distribution and Change* (World Meteorological Organization, 1985).

[6] Vienna Convention for the Protection of the Ozone Layer.

[7] Montreal Protocol.

climate and temperatures.[8] More recent studies now show that the Antarctic hole in the ozone layer is gradually closing, thus reversing one of the most prominent global environmental harms of the 1990s.[9]

Given the history of successful regulations and failures, the question appears less as one of their general suitability in the abstract, but rather what factors hinder effective regulatory intervention, specifically with respect to their drafting and enforcement. The claim defended in this chapter is that regulations can be a powerful and effective tool to further environmental goals if they are adopted in good faith and consistently enforced. The example of atmospheric ozone demonstrates that environmental regulations tend to be most effective when tied closely to the best available scientific evidence. Provided that regulators are willing, institutionally empowered, and adequately resourced to enforce restrictions, there is no reason to doubt the efficacy of environmental regulations. Indeed, examples of regulatory failures like asbestos are notable for the absence of these requirements and for the pathologies of effective intervention they exhibit.

10.2 ASBESTOS AND THE THREE PATHOLOGIES

Asbestos is the commercial name given to a range of mineral fibres whose modern exploitation began in the late nineteenth century but whose use can be traced back to antiquity.[10] Exposure to asbestos is widely recognised as a significant risk to health, with early anecdotal evidence coming from factory inspectors and doctors at the turn of the twentieth century,[11] and the first major

8 On the catastrophic consequences that were avoided, see P. A. Newman and others, 'What Would Have Happened to the Ozone Layer if Chlorofluorocarbons (CFCs) Had Not Been Regulated?' (2009) 9 *Atmospheric Chemistry and Physics* 2113.

9 Susan Solomon and others, 'Emergence of Healing in the Antarctic Ozone Layer' (2016) 353 *Science* 269.

10 Robert Murray, 'Asbestos: A Chronology of Its Origins and Health Effects' (1990) 47 *British Journal of Industrial Medicine* 361.

11 Lucy Deane, *Report on the Health of Workers in Asbestos and Other Dusty Trades* (HM Chief Inspector of Factories and Workshops, 1899, 1899); H. M. Murray, *In Departmental Committee on Compensation for Industrial Diseases, 1907, Minutes of Evidence* (1906) 127 [4076–4104]; M. Auribault, *Note sur L'hygiène et la Sécurité des Ouvriers Dans les Filatures et Tissages D'amiante* (Bulletin de l'inspection du travail, 1906); Frederick Ludwig Hoffman, *Mortality from Respiratory Diseases in Dusty Trades (Inorganic Dusts)* (1918); as early as 1914 in Germany, see Xaver Baur, 'Asbestos-Related Disorders in Germany: Background, Politics, Incidence, Diagnostics and Compensation' (2018) 15 *International Journal of Environmental Research and Public Health* 143.

scientific studies that followed.[12] The history of asbestos regulation is synonymous with decades of regulatory and enforcement deficits: solid scientific evidence was available and effective measures could have been taken at an early stage to prevent harm to countless individuals. Instead, regulators failed to decisively address asbestos well into the 1980s. The disconnect between decades of compelling evidence and regulatory action can ultimately be accounted for through three pathologies that to a greater or lesser degree feature in most modern environmental regulatory failures.

First, political lag, which describes the gap between the best available scientific evidence and regulatory efforts to address environmental harm. Regulation, to the extent that it exists, does not necessarily reflect the best practices and procedures that scientific evidence would support. To an extent, political lag is to be expected and is at times a desirable element in a prudent regulatory strategy. Not every new study and not every piece of evidence requires regulatory attention, nor must it necessarily prompt wholesale reform. However, an overly cautious approach becomes problematic when evidence of harm is ignored on a wider scale, and when inaction jeopardises the rights (and health) of individuals. As asbestos demonstrates, this can have devastating consequences, especially when harmful business models and industry practices flourish in a regulatory atmosphere of laissez-faire indifference.

Before the 1980s, asbestos regulation in the United Kingdom largely focused on managing certain aspects of the risks posed to workers but stopped well short of comprehensive protection of the public.[13] Modern authors generally agree that early regulations such as the Asbestos Regulation 1931 and its subsequent updates were lagging well behind the historically available evidence.[14] Asbestos apologists such as Robert Murray, and to a lesser

[12] W. E. Cooke, 'Fibrosis of the Lungs due to the Inhalation of Asbestos Dust' (1924) 2 *British Medical Journal* 147; Edward Rowland Alworth Merewether and C. W. Price, *Report on Effects of Asbestos Dust on the Lungs and Dust Suppression in the Asbestos Industry. Part II. Processes Giving Rise to Dust and Methods for Its Suppression* (HMSO, 1930); see especially the landmark study by Y. Baris, 'The Clinical and Radiological Aspects of 185 Cases of Malignant Pleural Mesothelioma' [1979] IARC Scientific Publications 937, which continues to inform much of modern mesothelioma understanding.

[13] Asbestos Regulation of 1931; Factories Act 1937; The Asbestos Regulation 1969, passed under the authority of the Factories Act 1961; Health and Safety at Work etc. Act 1974.

[14] Nick Wikeley, 'The Asbestos Regulations 1931: A Licence to Kill?' (1992) 19 *Journal of Law and Society* 365; Morris Greenberg, 'Knowledge of the Health Hazard of Asbestos Prior to the Merewether and Price Report of 1930' (1994) 7 *The Society for the Social History of Medicine* 493; David J. Jeremy, 'Corporate Responses to the Emergent Recognition of a Health Hazard in the UK Asbestos Industry: The Case of Turner & Newall, 1920–1960' (1995) 24 *Business and Economic History* 254; Tweedale, *Magic Mineral to Killer Dust*; David Gee and Morris Greenberg, 'Asbestos: From "Magic" to Malevolent Mineral' in Paul Harremoes and others

extent Peter Bartrip, have argued that contemporary criticisms of regulators and the asbestos industry are largely the product of hindsight.[15] The regulations introduced, they suggest, were generally keeping pace with scientific evidence. Murray's ultimate claim that asbestos throughout the history of its use has saved more lives than it claimed is at best misleading,[16] as the true human cost of asbestos is still far from clear. Estimates suggest that as many as 40 per cent of workers were employed in industries that entailed some level of asbestos exposure.[17] For Germany alone, the statutory accident insurance database contains a record of over 500,000 workers exposed to asbestos who received some level of compensation.[18] As Geoffry Tweedale points out, there is no need to resort to hindsight to cast the actions of the asbestos industry and regulators in an unflattering light: indifference, delays, and suppression of scientific evidence as well as wilful exposure of countless workers, their families, and the public at large were the hallmarks of widespread industry resistance to regulation.[19]

Industry resistance arises from the fruitful ground of political lag, permitting vested interests to entrench harmful, but profitable, business models and practices. When subsequent attempts at regulation are made, industries will often resist any meaningful changes, especially where they threaten their business model, established practices, and wider economic interests.[20] Established industries are generally hostile towards regulatory interventions that reduce autonomy and perceived competitiveness, and entail high compliance costs.[21] Regulatory attempts are commonly opposed and resisted, including through attempts to downplay, discredit, or otherwise suppress scientific evidence that point to harm;[22] lobbying of regulators, especially

(eds.), *The Precautionary Principle in the 20th Century: Late Lessons from Early Warnings* (Routledge 2002) 3.

15 Murray, 'Asbestos', 364; Peter Bartrip, 'Too Little, Too Late? The Home Office and the Asbestos Industry Regulations, 1931' (1998) 42 *Medical History* 421, 422.

16 Murray, 'Asbestos', 362, citing J. C. Wagner, *Asbestos and Cancer* (Abbot 1968).

17 Baur, 'Asbestos-Related Disorders in Germany', 2; Antti Tossavainen, 'Asbestos, Asbestosis, and Cancer: The Helsinki Criteria for Diagnosis and Attribution' (2010) 23 *Scandinavian Journal of Work, Environment & Health* 311.

18 *National Asbestos Profile for Germany* (Federal Institute for Occupational Safety and Health, 2014) 32.

19 Tweedale, *Magic Mineral to Killer Dust*, 278.

20 Wansley, 'Regulation of Emerging Risks', 412.

21 Dorothy Thornton, Robert A. Kagan, and Neil Gunningham, 'When Social Norms and Pressures Are Not Enough: Environmental Performance in the Trucking Industry' (2009) 43 *Law & Society Review* 405, 406.

22 David Kotelchuck, 'Asbestos: The Funeral Dress of Kings' and Others' in D. Rosner and G. Markowitz (eds.), *Dying for Work: Workers' Safety and Health in Twentieth-Century*

through public participation processes and strategic litigation;[23] influencing public opinion through disinformation campaigns,[24] and justifying harm with countervailing economic interests.[25] Industry resistance of this kind has been a common feature, for instance, in the cases of tobacco,[26] climate change,[27] and lead.[28]

The infamous campaign of denial propagated by the tobacco industry has become synonymous with profit-seeking behaviour at the expense of public health,[29] and contemporary climate change denialism expierienced a resurgence under the previous US administration.[30] Resistance by the lead industry is particularly disturbing as the substantial health detriments, especially for children, were well understood even by the ancient Romans.[31] However, even with the benefit of this knowledge, the industry had no qualms about

America (Indiana University Press 1987); David E. Lilienfeld, 'The Silence: The Asbestos Industry and Early Occupational Cancer Research – A Case Study' (1991) 81 *American Journal of Public Health* 791; D. Ozonoff, 'Failed Warnings: Asbestos-Related Disease and Industrial Medicine' in R. Bayer (ed.), *Case Studies in the Politics of Professional Responsibility* (Oxford University Press 1988) 139; Gerald Markowitz, 'Cater to the Children: The Role of the Lead Industry in a Public Health Tragedy, 1900–1955' (2000) 90 *American Journal of Public Health* 36; Yussuf Saloojee and Elif Dagil, 'Tobacco Industry Tactics for Resisting Public Policy on Health' (2000) 78 *Bulletin of the World Health Organization* 902.

23 For instance the attempts of the tobacco industry to challenge so-called plain packaging legislation, see *British American Tobacco Australasia Limited and Ors* v. *The Commonwealth of Australia* [2012] HCA 30 [plain packaging law ruled constitutional]; *Philip Morris Asia Limited (Hong Kong)* v. *The Commonwealth of Australia* PCA Case No 2012-12 [ruling held it had no jurisdiction to hear the case]; *Australia – Certain Measures Concerning Trademarks, Geographical Indications and Other Plain Packaging Requirements Applicable to Tobacco Products and Packaging*.

24 Carter and Chapman, 'Smoking, Disease, and Obdurate Denial'; Robert N. Proctor, 'The History of the Discovery of the Cigarette–Lung Cancer Link: Evidentiary Traditions, Corporate Denial, Global Toll' (2012) 21 *Tobacco Control* 87; Riley E. Dunlap and Aaron M. McRight, 'Organized Climate Change Denial' in John S. Dryzek, Richard B. Norgaard, and David Schlosberg (eds.), *The Oxford Handbook of Climate Change and Society* (Oxford University Press 2011) 144; Riley E. Dunlap and Peter J. Jacques, 'Climate Change Denial Books and Conservative Think Tanks Exploring the Connection' (2013) 57 *American Behavioral Scientist* 699.

25 Wikeley, 'The Asbestos Regulations 1931', 373.

26 Saloojee and Dagil, 'Tobacco Industry Tactics for Resisting Public Policy on Health'.

27 Jorge E. Viñuales, 'Legal Techniques for Dealing with Scientific Uncertainty in Environmental Law' (2010) 43 *Vanderbilt Journal of Transnational Law* 437, 439.

28 Markowitz, 'Cater to the Children'.

29 Proctor, 'The History of the Discovery of the Cigarette–Lung Cancer Link'.

30 'President Trump Announces U.S. Withdrawal from the Paris Climate Accord' *The White House* (Washington, DC, 1 June 2017) www.whitehouse.gov/blog/2017/06/01/president-donald-j-trump-announces-us-withdrawal-paris-climate-accord, accessed 1 April 2021.

31 Jack Lewis, 'Lead Poisoning: A Historical Perspective' *EPA Journal* www.epa.gov/aboutepa/lead-poisoning-historical-perspective, accessed 1 April 2021.

specifically marketing their products to children and categorically denying any adverse health impacts.[32]

Industry resistance emerges and stiffens the longer industries are permitted to consolidate and engage in their business model free from meaningful regulation.[33] Despite mounting evidence of irreparable harm to humans emerging throughout the twentieth century, the attractive properties of asbestos meant its use continued largely unabated, most notably to provide insulation and as a fire retardant, but also in a wide range of consumer products.[34] Regulators took notice of scientific evidence but were exceedingly slow and cautious in translating this knowledge into law, much less enforcing it against a highly profitable industry and major employer.

The asbestos industry systematically suppressed the publication of independent scientific evidence on harmfulness, financed highly dubious studies expounding the safety of the mineral, and consistently obfuscated their lobbying efforts through intermediaries, all in an attempt to resist even basic regulations and litigation to protect workers and the public.[35] The result was widespread regulatory inertia.

Regulatory inertia means that regulators are less likely to burden vested economic interests with effective regulations or enforce them consistently. This behaviour points to a greater problem in contemporary environmental regulation, which presents a fragmented approach to the relationship between environmental harm and human well-being. Regulations often fail to address established harms and do not pursue a comprehensive strategy, especially within a general political atmosphere of economic deregulation.[36]

With respect to asbestos, a significant problem with passing comprehensive and effective regulation in the United Kingdom was the fact that regulators were encouraged to see themselves as mediators between organised labour

32 Markowitz, 'Cater to the Children', 42.

33 Wansley, 'Regulation of Emerging Risks', 414, refers to this as 'interest group entrenchment' in the US regulatory context.

34 For a list of former and current consumer products still in circulation in the United States, see 'Consumer's Guide to Asbestos', *Asbestos Nation* www.asbestosnation.org/facts/consumers-guide-to-asbestos/, accessed 1 April 2021.

35 David Michaels, *Doubt Is Their Product* (Oxford University Press 2008) chapter 2, James Rice, 'The Global Reorganization and Revitalization of the Asbestos Industry, 1970–2007' (2011) 41 *International Journal of Health Services* 239, 243, for a concise overview of industry resistance from German asbestos producers, see Baur, 'Asbestos-Related Disorders in Germany', 9–10.

36 Bosselmann and Richardson, 'Introduction', 3; E. Rehbinder, 'States between Economic Deregulation and Environmental Responsibility' in Klaus Bosselmann and B. J. Richardson (eds.), *Environmental Justice and Market Mechanisms* (Kluwer Law International 1999) 93; Montini and Volpe, 'Rethinking Regulation for Promoting an Ecologically Based Approach to Sustainability', 22.

and the asbestos industry. Many labour unions remained conspicuously ambivalent on asbestos regulation as they feared a significant threat to the livelihood of employees.[37]

Even though the hazards were well documented, there is little doubt that regulators failed to consistently enforce even limited safety regulations on an industry where both employers and organised labour were seemingly content with the status quo.[38] The regulators saw themselves as building a consensus on the issue of worker safety, rather than acting as a watchdog in the interests of the broader public. Factory owners were seldom, if ever, taken to court over violations of asbestos regulations,[39] and existing regulations in any case only afforded protection to employees working directly in factories and only for the duration of their employment.[40] They did not protect the families of workers, or so much as consider safeguards for the general public from the dangers of exposure. Everyday products containing the mineral were abundant, and ventilation of factories contaminated the surrounding neighbourhoods,[41] while asbestos-laden building materials were used in countless buildings and residential homes throughout the twentieth century.[42]

It was not until well into the 1980s that sufficient public pressure mounted to fill these regulatory gaps. As is often the case in the United Kingdom, it was a commission of inquiry, the Simpson Committee, which prompted important reforms.[43] Across Europe, this timeline was widely mirrored as asbestos regulations remained deficient for most of the twentieth century. Modern regulations and enforcement in the European Union have since been significantly tightened. Directive 1999/77/EC of 26 July 1999[44] added asbestos to

37 Jonathan Morgan, 'Technological Change and the Development of Liability for Fault in England and Wales' in Miquel Martin-Casals (ed.), *The Development of Liability in Relation to Technological Change*, vol 4 (Cambridge University Press 2010) 70; Baur, 'Asbestos-Related Disorders in Germany', 10.

38 For a concise chronology of events, see Gee and Greenberg, 'Asbestos', 53.

39 A. Dalton, *Asbestos Killer Dust* (British Society for Social Responsibility in Science 1979); Wikeley, 'The Asbestos Regulations 1931', 373.

40 Wikeley, 'The Asbestos Regulations 1931', 373.

41 Lorna Arblaster and others, 'Occupational and Environmental Links to Mesothelioma Deaths Occurring in Leeds during 1971–1987' (1995) 17 *Journal of Public Health Medicine* 297, 298.

42 'Where Can You Find Asbestos?' *Health and Safety Executive* (2017) www.hse.gov.uk/asbestos/building.htm, accessed 1 April 2021.

43 Gee and Greenberg, 'Asbestos', 57–8; *Asbestos – Final Report of the Advisory Committee*, vol 1 (HMSO 1979).

44 Commission Directive 1999/77/EC of 26 July 1999 Adapting to Technical Progress for the Sixth Time Annex I to Council Directive 76/769/EEC on the Approximation of the Laws, Regulations and Administrative Provisions of the Member States Relating to Restrictions on the Marketing and use of Certain Dangerous Substances and Preparations (Asbestos).

Annex 1 of Directive 76/769/EEC,[45] thereby effectively banning its use in any products throughout the EU.[46] Nonetheless, and even with the benefit of modern scientific evidence, many countries including the United States continue the use of asbestos in various forms.[47] Asbestos exposure, chiefly of those individuals historically and currently tasked to work with the mineral, remains a significant cause for litigation.[48] In 2015, around 2,542 deaths in the United Kingdom were attributable to mesothelioma, the cancer caused by asbestos exposure.[49] In addition to the human tragedies, there were also substantial economic detriments as a direct result of inadequate regulation.[50] As the following section explores, contemporary air pollution regulation in Europe suffers from the same three pathologies that allowed asbestos and other environmental harms to persist for decades as a public health crisis.

10.3 CONTEMPORARY AIR POLLUTION

As we have seen in Chapter 3, air quality standards are provided throughout the EU under Directive 2008/50/EC,[51] which bases its threshold values on WHO guidelines and their subsequent updates.[52] Among the harmful health effects of air pollution is an increase in instances of lung cancer,[53] cardiovascular mortality, chronic illnesses such as diabetes, links to atherosclerosis, birth defects, and childhood respiratory diseases, alongside detriments to neurodevelopment and cognitive functions, including links to Alzheimer's disease.[54] While air quality improved with respect to some regulated pollutants, progress has been marked by regulatory shortcomings and an absence of effective enforcement.

45 Directive 76/769/EEC.

46 There is no longer any significant asbestos use throughout the EU, *see* Daniel M. Flanagan, 2015 *Minerals Yearbook – Asbestos* (United States Geological Survey, 2016) 8.7.

47 Arthur L. Frank and T. K. Joshi, 'The Global Spread of Asbestos' (2014) 80 *Annals of Global Health* 257, 261; Gee and Greenberg, 'Asbestos', 52.

48 See, for instance, *McDonald (Deceased) (Represented by Mrs Edna McDonald) (Respondent)* v. *National Grid Electricity Transmission Plc (Appellant)* [2014] UKSC 53.

49 Andrew Darnton, *Mesothelioma in Great Britain – Mesothelioma Mortality in Great Britain 1968–2015* (Health and Safety Executive, 2016) 4.

50 Gee and Greenberg, 'Asbestos', 58; Baur, 'Asbestos-Related Disorders in Germany', 8–9.

51 They are based on the *WHO Air Quality Guidelines*.

52 *Review of Evidence on Health Aspects of Air Pollution*.

53 Raaschou-Nielsen and others, 'Air Pollution and Lung Cancer Incidence in 17 European Cohorts'.

54 Maher and others, 'Magnetite Pollution Nanoparticles in the Human Brain'; *WHO Air Quality Guidelines*; *Review of Evidence on Health Aspects of Air Pollution*, 6.

The focus of the Directive 2008/50/EC on government action plans is deficient in providing effective and consistent protection from air pollution. As we have seen in the case of *Janecek*, the ECJ determined that Directive 2008/50/EC contains an individual right only to the implementation of action plans that can conceivably achieve compliance with air quality threshold values in due course.[55] EU law does not grant individuals any rights to enforce compliance with threshold values or seek any specific regulatory actions. Admittedly, this has some procedural benefits for applicants.

Evidentiary burdens and scientific evidence are seldom problematic when challenging action plans, as the standard of review arises directly from the Directive. Environmental NGOs and individuals with limited resources can thus challenge inadequate action plans on the basis of the projections and assumptions that underpin the measures adopted by regulators. This eliminates the need to collect and provide independent scientific evidence beyond what regulators already gather and are legally required to publish. Action plans have also admittedly achieved some tangible improvements, even if these come with significant limitations. Directive 2008/50/EC is a prospective regulation, seeking compliance with air pollution standards at some unspecified point in the future, and in practice it has been subject to a striking level of regulatory inertia and industry resistance at the domestic and EU levels.

10.3.1 *Regulatory Inertia*

We observed in Chapters 3 and 9 that both the United Kingdom and Germany remain in breach of their EU law obligations: the root cause is regulatory inertia, an unwillingness to take the measures necessary to reduce air pollution to acceptable (and legally mandated) levels. In Germany, this has taken the shape of open defiance by Bavarian officials. As we saw, the case of *Deutsche Umwelthilfe* before the ECJ did little to address the openly hostile resistance to a final court ruling.[56] However, regulatory inertia has also taken more subtle forms in the EU.

To the surprise of many, the Brussels capital region long reported air quality measurements of NO_2 below the maximum threshold values permitted under EU law.[57] This made Brussels an outlier amongst major European cities who routinely breach their air quality targets. It was long suspected that this was

[55] Case C-237/07 *Dieter Janecek* v. *Freistaat Bayern* [2008] ECLI:EU:C:2008:447 [42] and [47].

[56] See the more detailed discussion of Germany in Chapter 9.

[57] 'Brussels Ordered to Get Real on Its Pollution Problem' *Politico* (27 June 2019) www.politico.eu/article/brussels-ordered-to-get-real-on-its-pollution-problem/, accessed 1 April 2021.

due to officials excluding data from monitoring stations covering the most polluted areas of the city in official reports and instead opting to measure primarily in convenient areas experiencing low pollution levels.[58] The placement of some monitoring stations in Brussels was expressly questioned by the European Commission in its formal notice of infringement proceedings to Belgium over failures to fully implement Directive 2008/50/EC.[59]

The applicants in *Craeynest* alleged before domestic Belgian courts that the location of the monitoring stations in Brussels had not followed the requirements of Directive 2008/50/EC and hence misleadingly gave the impression that the city remained compliant with NO_2 threshold values. Article 7(1) of Directive 2008/50, read in conjunction with Annex III, section B, paragraph 1(a) requires amongst other things that data is gathered in the 'areas within zones and agglomerations' with the 'highest concentrations' of pollutants.[60] The domestic Belgian court found that the Directive gave significant discretion to authorities and asked the ECJ whether it was required to scrutinise the monitoring station locations. The ECJ responded that Article 7 required the gathering of representative data and that the purpose of the Directive would be undermined if this provision was not enforced and subject to strict judicial review.[61] Crucially, EU law prevented the city of Brussels from averaging out the pollution levels across its monitoring stations as it had done in the past. Instead, any breach of the threshold values at any monitoring station required regulatory action under the Directive.[62]

The case demonstrates a more subtle form of regulatory inertia that attempts to avoid the difficult policy trade-offs and regulations of environmental harm through selective monitoring and misleading averaging of pollution levels. Beyond this general enforcement deficit on air quality action plans, there is also related sector-specific industry resistance.

10.3.2 *Industry Resistance and Regulatory Complicity*

The car industry responses to EU emission standards, particularly the diesel emission scandal and the ensuing domestic and EU regulatory response,

58 'New Brussels Air Pollution Figures Mask Health Risk Reality' *ClientEarth* (3 October 2019) www.clientearth.org/press/new-brussels-air-pollution-figures-mask-health-risk-reality/, accessed 1 April 2021.

59 'Commission Calls on Belgium to Comply with the EU Legislation on Air Quality' (8 November 2018) https://ec.europa.eu/commission/presscorner/detail/en/MEMO_18_6247, accessed 1 April 2021.

60 Case C-723/17 *Craeynest and Others* v. *Brussels* [2019] ECLI:EU:C:2019:533 [20].

61 Ibid. [45] and [49].

62 Ibid. [68].

provide a textbook example of industry resistance. Initially, voluntary agreements on emission standards were struck between the European Commission and the car industry in 1998, named for the car industry association involved. The initial European Automobile Manufacturers Association (ACEA) agreement had three main objectives:

- the introduction of car models that emit 120 g of CO_2 per kilometre or less by 2000;
- an emission target of 140 g of CO_2 per kilometre by 2008, with an intermediate target range of 165–170 g of CO_2 per kilometre by 2003; and
- a target of 120 g of CO_2 per kilometre by 2012.

By 2007, the agreement was widely considered a failure by the European Commission, the European Parliament, and advocacy groups, as manufacturers failed to deliver on their promises.[63] Annual reports indicated that only two car manufacturers reached the 2008 target by the intended deadline, and only one reached the 2012 target. Across all car fleets, the 140 g per km target for 2008 was achieved neither in the ACEA nor in the identical agreements struck with other car manufacturer organisations (Japanese Automobile Manufacturers Association – JAMA and Korean Automobile Manufacturers Association – KAMA). For this reason, the EU codified mandatory emission targets in Regulation (EC) 443/2009.[64]

As we know today, the car industry systematically undercut these legally mandated emission targets for years, as was prominently exposed in the diesel car emissions scandal involving the world's largest car manufacturer Volkswagen.[65] The ECJ has since ruled that the widespread use of defeat devices was illegal under EU law, specifically Article 5 of Regulation No 715/2007, which deals with the testing requirements for EURO 5 and 6 emission

[63] 'Proposal for a Regulation of the European Parliament and of the Council setting emission performance standards for new passenger cars as part of the Community's integrated approach to reduce CO2 emissions from light-duty vehicles', *European Commission Communication COM/2007/0856* (19 December 2007); 'Legislative resolution on the proposal for a regulation of the European Parliament and of the Council setting emission performance standards for new passenger cars as part of the Community's integrated approach to reduce CO_2 emissions from light-duty vehicles', *European Parliament COM(2007)0856-C6-0022/2008 -2007/0297 (COD)* (17 December 2008); Paul Rincon, 'Car Makers "Failing" on Climate' *BBC News* (20 April 2006) http://news.bbc.co.uk/1/hi/sci/tech/4922614.stm, accessed 1 April 2021.

[64] Regulation (EC) No 443/2009 of the European Parliament and of the Council of 23 April 2009 setting emission performance standards for new passenger cars as part of the Community's integrated approach to reduce CO_2 emissions from light-duty vehicles.

[65] 'Volkswagen: The Scandal Explained' *BBC News* (10 December 2015) www.bbc.co.uk/news/business-34324772, accessed 1 April 2021.

standards.[66] However, domestic German and European authorities failed to uncover and react to clear violations of regulations, at times actively shielding car manufacturers from meaningful consequences.

In Germany, a commission of inquiry of the Bundestag concluded that the Federal Motor Transport Authority (FMTA),[67] the domestic regulator responsible for licensing new car models, failed to uncover the use of defeat devices, despite ample grounds for suspicion.[68] Up to sevenfold increases in emissions from benchmarking scenarios to realistic driving conditions were common knowledge, and experts at the FMTA lacked the 'imagination' to suspect a defeat device, instead blaming the legally prescribed benchmarking scenario for the discrepancy.[69] Further suspicions were prominently raised by associated German government regulators. From 1993 the Federal Environment Agency conducted annual assessments into the road condition emissions of licensed car models, and the resulting public reports consistently showed significant discrepancies from advertised figures.[70] For instance, in its final report for the year 2006, the Agency concluded that '(. . .) the results of the measurements of diesel cars are alarming. Meeting the threshold value for nitrogen oxides is proving particularly difficult for these vehicles.'[71] Despite these early warnings, the FMTA continued with established benchmarking procedures and licensed all affected car models.

Even though the commission of inquiry ultimately found no illegal behaviour, it nonetheless noted structural deficiencies that have since been partially addressed to increase the rigour of future testing regimes.[72] It is worth noting that the commission's report was passed with the majority of the coalition government votes and has been criticised in academic scholarship. Some commentators see the licensing behaviour and absence of investigation on the part of the FMTA as clear violations of both domestic and EU law.[73] The argument is that, at the very least, the use of defeat devices was illegal, invalidating the resulting licences for affected car models.[74] The persistent

66 Case C-693/18 CLCV and others [2020] ECLI:EU:C:2020:1040.

67 The Federal Motor Transport Authority (Kraftfahrtbundesamt).

68 *Beschlussempfehlung und Bericht – 5. Untersuchungsausschuss der 18. Wahlperiode* (Deutscher Bundestag, 2017) 526–33.

69 Ibid., 527–8.

70 Martin Führ, *Gutachterliche Stellungnahme* (Öffentliche Sachverständigenanhörung des 5 Untersuchungsausschusses der 18 Wahlperiode, 2016) 15.

71 Cited per ibid., 11, my translation.

72 *Beschlussempfehlung und Bericht* – 5, 533.

73 Führ, *Gutachterliche Stellungnahme*, 1.

74 Remo Klinger, 'Dieselgate öffentlich-rechtlich – EG-Typengenehmigung, Übereinstimmungsbescheinigung und Konsequenzen für deutsche Pkw-Halter' (2017) 22

failure of the FMTA and the competent Ministry to act on this information constitutes a strong contemporary example of regulatory inertia, which also raises questions over the rule of law and the political priority afforded to the car industry.[75]

At the EU level, the revelations of the diesel emissions scandal exposed the complicity of the European Commission in safeguarding and prioritising industry interests over public health concerns. In March 2016, the European Commission initially reacted to the diesel scandal with Commission Regulation 2016/427,[76] which amended the benchmarking system to require measuring real driving emissions for the licensing of car models for the EURO-6 emission standard.[77] This laudable initiative was, however, undercut through the subsequent Commission Regulation 2016/646, which permitted car manufacturers to exceed the EURO-6 emission standard by more than double the threshold value until the year 2019 without risking the licensing of car models.[78] This move led the city of Paris to successfully seek the annulment of the EU Regulation before the General Court for lack of competence.[79] Against the backdrop of serious regulatory inertia, it is perhaps not surprising that leading car manufacturers felt emboldened to circumvent environmental regulations and long denied responsibility and compensation to consumers. Overall, a clear specific risk to human rights is currently permitted to remain in major European cities, and action plans will at best resolve these concerns at some future point in time, provided that widespread industry resistance and regulatory inertia is indeed more robustly challenged.

Zeitschrift für Umweltrecht 131; Martin Führ, 'Der Dieselskandal und das Recht – Ein Lehrstück zum technischen Sicherheitsrecht' (2017) 36 *Neue Zeitschrift für Verwaltungsrecht* 265.

75 Führ, 'Der Dieselskandal und das Recht', 273; Klinger, 'Dieselgate öffentlich-rechtlich'; Kurt Faßbender, 'Der Dieselskandal und der Gesundheitsschutz' [2017] Neue Jurisitische Wochenschrift 1995, 1999.

76 Commission Regulation (EU) 2016/427 amending Regulation (EC) No 692/2008 as regards emissions from light passenger and commercial vehicles (Euro 6) of 10 March 2016.

77 The framework is provided by Commission Regulation (EC) No 692/2008 implementing and amending Regulation (EC) No 715/2007 of the European Parliament and of the Council on type-approval of motor vehicles with respect to emissions from light passenger and commercial vehicles (Euro 5 and Euro 6) and on access to vehicle repair and maintenance information of 18 July 2008.

78 See the amendments to Annex 2 of Regulation (EC) No 692/2008, in Commission Regulation (EU) 2016/646 amending Regulation (EC) No 692/2008 as regards emissions from light passenger and commercial vehicles (Euro 6) of 20 April 2016; Faßbender, 'Der Dieselskandal und der Gesundheitsschutz', 2000; Christoph Sobotta, 'EuGH: Neue Verfahren im Umweltrecht' (2017) 22 *Zeitschrift für Umweltrecht* 241, 243–4.

79 Case T-339/16 *Ville de Paris* v. *Commission* [2018] ECLI:EU:T:2018:927; an appeal before the ECJ is pending, see Case C-179/19 P *Commission* v. *Ville de Paris* [2019] ECLI:EU:C:2019:836.

10.4 SUMMARY

Regulations adopted to protect the ozone layer offer an encouraging example that a close relationship between scientific evidence and assertive regulatory action can bring about tangible environmental improvements. The environmental minimum fosters precisely this type of engagement. The framework encourages regulation and consistent enforcement by addressing three pathologies that have historically plagued environmental regulation: political lag, industry resistance, and regulatory inertia. These pathologies are why the environmental minimum pushes regulators towards establishing and consistently enforcing legal standards and applying established and emerging standards in their absence.

States must consistently update and enforce legal standards as well as act on harm subject to established and emerging evidence. This will be of particular benefit where environmental harm has not been previously regulated and ensure consistent enforcement within the boundaries drawn by good faith. There are conceivable acceptable reasons for falling short of the generally accepted standards, but in light of the strong evidence they represent, there is a rebuttable presumption that any such regulation was adopted in bad faith. Established standards operating in the absence of legal standards essentially implement this presumption for cases of regulatory inertia. With respect to emerging standards, there may be sound reasons for regulators to avoid regulatory action, or conversely act even on the basis of uncertain evidence. Barring cases of grave potential harm, states are afforded broad discretion.

In terms of enforcement, the environmental minimum is less deferential to states. It places a strong emphasis on compliance with those legal standards that have been adopted in domestic and international laws. Under the environmental minimum it would, for instance, not matter whether the Directive 2008/50/EC was intended to grant individuals an enforceable claim, as it would be conceptualised as a legal standard of review. Individuals are then empowered to seek regulatory intervention through rights-based challenges, provided they can establish a specific risk. There is some evidence to suggest that the mere existence of regulations can ensure compliance by larger corporations, as they are fearful of exposure through regulators, advocacy groups, and consumers.[80] However, this does not necessarily apply to smaller businesses in competitive markets and to systemically dominant corporations, or whenever the risk of enforcement and sanctions is generally seen as remote.[81] Enforcement practices

[80] Thornton and others, 'When Social Norms and Pressures Are Not Enough', 408.

[81] Ibid., 431.

must be robust and lead to a high plausibility of significant punishment for non-compliance to ensure widespread observance. Otherwise, private actors will prioritise economic interests over regulatory obligations.

In addressing industry resistance to regulation, the environmental minimum provides minimum standards for environmental harms constituting a specific risk, even in the absence of legal standards. Serious environmental harm thus cannot be ignored and kept off the agenda indefinitely. Should states, whether under pressure from established industries or out of other political considerations, not prescribe regulations, they are nonetheless held accountable. Should this prove unsatisfactory to lawmakers, states can engage in the regulatory process, which permits advocacy groups, as well as experts to engage and lobby regulators. Even if the outcome should ultimately fall short of optimal environmental protection standards, regulations adopted in good faith would nonetheless be a significant improvement over politically expedient and industry encouraged indifference. It can clear a path towards more ambitious regulation in the future, and focus public attention on the many overlooked environmental harms that are not adequately regulated.[82] For instance, one might think of the widespread use and unclear environmental impact of the antioxidant ethoxyquin in the fish industry,[83] or the use of environmentally harmful copper-based pesticides in organic farming and its nonetheless continued authorisation for use throughout the EU.[84]

[82] Hayward, *Constitutional Environmental Rights*, 36.

[83] *Safety and Efficacy of Ethoxyquin (6-ethoxy-1,2-dihydro-2,2,4-trimethylquinoline) for All Animal Species* (European Food Safety Authority Panel on Additives and Products or Substances used in Animal Feed, 2015), especially in the fish industry, *see* A. K. Lundebye and others, 'Levels of Synthetic Antioxidants (Ethoxyquin, Butylated Hydroxytoluene and Butylated Hydroxyanisole) in Fish Feed and Commercially Farmed Fish' (2010) 27 *Food Additives & Contaminants* 1652.

[84] *Conclusion on the Peer Review of the Pesticide Risk Assessment of Confirmatory Data Submitted for the Active Substance Copper (I), Copper (II) Variants Namely Copper Hydroxide, Copper Oxychloride, Tribasic Copper Sulfate, Copper (I) Oxide, Bordeaux Mixture* (European Food Safety Authority, 2013).

11

International Environmental Law

The earliest treaties that regulate environmental conduct between states date back to the late nineteenth and early twentieth centuries,[1] and typically focused on issues such as fisheries,[2] watercourses,[3] the protection of certain species of wild birds,[4] and whaling.[5] Environmental protection was mostly an incidental motive, as the treaties were primarily concerned with safeguarding territorial sovereignty and economic interests rather than recognising the intrinsic value of the environment.[6] Much of this changed after the 1972 Stockholm Conference on the Human Environment,[7] which is widely considered a turning point towards the so-called modern era in international environmental law.[8] The achievements of the conference are reflected in two central, albeit non-binding documents: the Stockholm Declaration, which lays out the principles of environmental protection, and the Action Plan consisting of 109 recommendations.[9]

1 Beyerlin and Marauhn, *International Environmental Law* 3.
2 Fisheries Convention between France and Great Britain of 11 November 1867; Agreement between Great Britain and the United States of America for a Modus Vivendi in Relation to the Fur Seal Fisheries in the Bering Seas of 15 June 1891.
3 Convention between Great Britain and the United States of America Relating to Boundary Waters and Boundary Questions between the USA and Canada of 11 January 1909.
4 Convention for the Protection of Birds Useful to Agriculture of 19 March 1902; the Convention for the Protection of Migratory Birds in the United States and Canada of 16 August 1916.
5 International Convention for the Regulation of Whaling of 24 September 1931.
6 Peter H. Sand, 'The Evolution of International Environmental Law' in Daniel Bodansky, Jutta Brunnée, and Ellen Hey (eds.), *The Oxford Handbook of International Environmental Law* (Oxford University Press 2007) 31; Sands and others, *Principles of International Environmental Law*, 22.
7 Established pursuant to General Assembly Resolution 2398 (XXIII) of 3 December 1968.
8 Sand, 'The Evolution of International Environmental Law', 33; Sands and others, *Principles of International Environmental Law*, 22, 30.
9 Beyerlin and Marauhn, *International Environmental Law*, 7.

While international environmental law matured towards greater acknowledgement of human rights concerns after 1972, it largely stuck to the limited transboundary and economic utility perspective on environmental harm. Nonetheless, against the backdrop of the political and policy debate over climate change, some scholars have mooted human rights as a vehicle for adding weight to the urgency of this prominent global transboundary environmental harm. The chapter begins by explaining that the environmental minimum is unlikely to make meaningful contributions to addressing climate change (Section 11.1). Temporal and geographic concerns present significant barriers, but ultimately the environmental minimum presupposes a robust commitment to human rights, environmental protection, and the rule of law that is simply not met by many of the most notorious global polluters.

The environmental minimum is overall more suited to specifying the scope, context, and relationship of competing obligations, notably among the SDGs (Section 11.2). Both economic utility and environmental protection have merged under the fashionable umbrella term of sustainable development following the Rio Declaration.[10] The concept as such and particularly the Sustainable Development Goals (SDGs) introduced by the UN General Assembly in 2015 struggle with their broad and indeterminate normative claims.[11]

Few would seriously disagree with the stated aspirations of the SDGs in principle. The real test of their usefulness lies in their practical implementation and the inevitable trade-offs and compromises. The uneasy merger of environmental protection and economic interests in these aspirational commitments produces conflicts that remain largely unresolved and threaten to scupper practical improvements on the ground. The 17 goals and 169 targets are so broad and normatively diverse that conceivably any state action can be presented as furthering at least some SDGs.

Absent a principled framework for specification, practical evaluation and resolving conflicts, the SDGs risk a similar normative redundancy as the weak forms of the precautionary principle we discussed in Chapter 3. The environmental minimum provides one potential avenue to clarifying and balancing the plethora of competing interests.

Even if the normative claims can be sufficiently specified and rendered practically meaningful, one would still have to overcome the mixed track

[10] *Rio Declaration on Environment and Development.*

[11] Transforming our world: the 2030 Agenda for Sustainable Development, 25 September 2015, UN Doc. A/Res/70/1.

record of state compliance with international environmental law.[12] In that vein, the environmental minimum can provide a viable framework to enforce otherwise non-binding or underenforced treaty obligations under international environmental law.

11.1 CLIMATE CHANGE

Climate change has attained a special status both within the broader debate in environmental ethics and in legal scholarship. To many scholars, lacklustre regulatory responses to climate change suggest an inability of the law in general and environmental regulations in particular to address an existential global challenge,[13] prompting varied suggestions for reform.[14] An important strand of reform suggestions focuses on the legal mechanisms of transboundary environmental harm. In a nutshell, a transboundary approach emphasises the role of inter-state environmental harm prevention as a mechanism to achieve global protection ambitions, particularly on climate change. This branch of international environmental law has a long pedigree, traditionally traced back to the *Trail Smelter Arbitration* case between the United States and Canada.[15] The dispute arose over the emissions of sulphur dioxide from a smelting plant and the resulting damages to agriculture and forestry in the US State of Washington. Canada was held under an obligation to compensate for past and future damages resulting from the smelting operation and to cease detrimental activities altogether if these continued to produce adverse transboundary effects.[16]

From a transboundary environmental law perspective, human rights protection regimes are seen as vehicles for overcoming existing enforcement deficits and broadening the normative reach of international environmental

[12] Beyerlin and Marauhn, *International Environmental Law*, 11; Sand, 'The Evolution of International Environmental Law', 40.

[13] Rolston, *A New Environmental Ethics:*, 216; Jamieson, *Reason in a Dark Time*, 11; Brennan and Yeuk-Sze, 'Environmental Ethics'.

[14] Elizabeth Fisher, Eloise Scotford, and Emily Barritt, 'The Legally Disruptive Nature of Climate Change' (2017) 80 *Modern Law Review* 173; Christine Bakker, 'Children's Rights Challenged by Climate Change: Is a Reconceptualization Required?' in Federico Lenzerini and Ana Filipa Vrdoljak (eds.), *International Law for Common Goods: Normative Perspectives on Human Rights, Culture and Nature* (2014) 361; Stephen Humphreys, 'Introduction: Human Rights and Climate Change' in Stephen Humphreys (ed.), *Human Rights and Climate Change* (Cambridge University Press 2010).

[15] *Trail Smelter Arbitration (United States* v. *Canada)* 3 UN Rep Int'l Arb Awards 1905 (1941).

[16] Ibid. at 1965.

law.[17] Specifically, supporters envision expanding the geographical scope and the jurisdictional reach of conventionally constrained domestic and regional human rights protection regimes to mobilise states against global environmental harm. For instance, the ECHR might require states to reduce their contributions to global greenhouse gas emissions due to the harmful impact of climate change on communities inside and outside of the Council of Europe.

In opposition to this view, the following section argues that existing human rights protection regimes ultimately cannot contribute decisively towards resolving transboundary environmental harms. This objection is best illustrated through the quintessential environmental harm with transboundary dimensions: anthropocentric climate change. Notwithstanding suggested modifications to human rights protection regimes, core difficulties centre on (1) issues of causality related to the degree of scientific certainty, (2) temporal constraints that hamper regulation, (3) disparate geographical impacts, and (4) a questionable level of global human rights commitments, especially amongst the most notorious polluters.

11.1.1 *Causality*

The anthropocentric causes of climate change are clearly established and widely accepted in contemporary scientific scholarship, as are their broad (regional) impacts.[18] A key difficulty for any human rights approach lies in establishing causality: specifically, attributing responsibility for any specific event (or broader climate trend) to any state action (or omission) that might be challenged by rightsholders. As an example, more prolonged heatwaves are projected in the coming century for parts of Southern and Eastern Europe as a result of climate change.[19] However, while the excessively timid climate change targets of one state may very well aggravate global warming and contribute to an increase in the worldwide number of natural disasters, there can be no certainty of a causal link in individual cases. Likewise, there can be

[17] Boyle, 'Human Rights and the Environment', 238; Viñuales, 'A Human Rights Approach to Extraterritorial Environmental Protection?', 177; Burns H. Weston and David Bollier, *Green Governance: Ecological Survival, Human Rights, and the Law of the Commons* (Cambridge University Press 2013) 95.

[18] John Cook and others, 'Quantifying the Consensus on Anthropogenic Global Warming in the Scientific Literature' 8 *Environmental Research Letters* 024024; *Climate Change 2014 – Synthesis Report* (United Nations Intergovernmental Panel on Climate Change, 2014); *Report of the Conference of the Parties on Its 21st Session Held in Paris, 30 November to 13 December 2015* (United Nations, FCCC/CP/2015/10/Add1, 2016).

[19] J. Alcamo and others, 'Europe' in Martin L. Parry and others (eds.), *Climate Change 2007: Impacts, Adaptation and Vulnerability*, vol 4 (Cambridge University Press 2007) 541.

no certainty that any specific regulatory intervention can prevent or at least mitigate these heatwaves. Even if a strict reduction of emissions were to occur throughout Europe, it may well have little overall impact if other continents do not implement similar measures. That leaves potential applicants in a double bind: they cannot demonstrate that any specific regulatory failure contributed to a climate change event, nor can they argue that a more ambitious regulatory target would have significantly reduced the likelihood of a specific climate change event.

In his influential work on international environmental law with Pierre-Marie Dupuy, and a subsequent chapter contribution, Jorge Viñuales suggests a workaround that would permit an applicant to establish the necessary causal connection between an environmental harm and human rights in a trans-boundary context.[20] The suggested 'spatial expansion' seeks to stretch human rights to cover more than individual states and go beyond the rights of individual rightsholders.[21] Specifically, Viñuales suggests a shift of focus in the context of climate change, from individual events to broad regional harm, and from harm caused to individuals to population-wide effects. However, there are a few concerns with this proposal. Ultimately, such an expansion might serve to further obscure, rather than to clarify, causality issues encountered with respect to climate change.

To appreciate why this might be the case, it is important to initially consider Viñuales' views on the causal relationship between human rights and environmental harm in general, as this ultimately informs his approach to spatial expansion on climate change. He conceptualises the relationship chiefly through two causality inquiries, which in his view are required to establish a human rights violation through environmental harm. The first inquiry is between the state action (or inaction) and the environmental harm in question and the second inquiry is between the environmental harm and the impairment of a human right.[22] This latter inquiry appears at odds with well-established doctrine on human rights as it interrogates a causal relationship outside the ambit of human rights law. As the section demonstrates, Viñuales' notion of spatial expansion raises important questions that can be considered without endorsing a problematic notion of two causality inquiries.

20 Dupuy and Viñuales, *International Environmental Law*, 331; Viñuales, 'A Human Rights Approach to Extraterritorial Environmental Protection?', 215.

21 An argument also broadly advanced by Weston and Bollier, *Green Governance*, chapter 3.

22 Viñuales, 'A Human Rights Approach to Extraterritorial Environmental Protection?', 215; first introduced in his earlier work, *see* Dupuy and Viñuales, *International Environmental Law*, 328.

11.1.1.1 Two Causality Inquiries under Human Rights?

It is useful at this stage to recall the doctrinal approach human rights lawyers conventionally take towards alleged violations. One begins by investigating whether there is a causal link between state action and the issue identified by the applicant.[23] Should there be such a causal link, one turns to the question of whether the state action infringes the scope of an established human right. Should there be an infringement, one asks in a final step, whether it can be justified through competing rights and interests. If the infringement cannot be justified, then a violation has been established. Against the backdrop of this conventional approach, the first causality inquiry suggested by Viñuales appears clear and useful: it examines whether a state action or inaction has led to an environmental harm, and hence whether there is any prima facie responsibility of the state for the harm. The second suggested causality inquiry is more problematic.

As the issue is one of considerable complexity, it can be useful to consider a slightly simplified analogy. Suppose that we are observing a football game and wish to determine whether an individual player has scored a regular goal from a penalty shot. At a basic level, we would need to ask ourselves two questions. First, whether the player kicked the ball, which is analogous to asking whether the state caused the environmental harm in question. Second, as simply kicking the ball is not sufficient to score, we further ask whether the ball landed in the goal. This latter question is no longer focused on causality, as we already know that the player succeeded in kicking the ball from the penalty spot. Hence, all relevant causal links are already established for our defined purpose. We now wish to ascertain whether the ball kicked by the player landed in the goal, that is to say, whether it crossed the goal line between the two posts and below the crossbar. In terms of human rights law, that question is settled by examining the scope of a given right. One makes a determination of whether the environmental harm caused by the state infringes the scope of protection offered by the human right in question, and whether any infringement is ultimately justified.

A second causality inquiry appears ill-equipped to address the question of scope. If one were committed to conducting such an inquiry, then one might ask whether the environmental harm caused by the state in turn *caused* a violation of a human right. In the terms of our football analogy, we would no longer propose to determine whether a player scored a goal. Instead, we are

[23] Or inaction, in the case of positive obligations.

asking something along the lines of whether, given that the player kicked the ball and that it landed behind the goal line, under the crossbar and between the two posts, a referee will acknowledge it as a regular goal. In legal terms, this suggests scrutiny of the judicial decision-making process. We might ask whether a judge would find a violation, given an environmental harm for which a state is causally responsible, and which infringes the scope of a human right without justification. Indeed, one would hope that any judge would find a violation under those circumstances, and examining why they may not is an important socio-legal question key to understanding the judicial process in any jurisdiction.[24] However, this inquiry stands apart, and is external to questions of conventional human rights doctrine and the goal of determining whether a violation of human rights law has occurred. Doctrinal work is concerned with systematising and establishing legal rules based on judgments, and socio-legal work is concerned with, amongst other things, the conduct of the people that apply these rules in practice and their broader societal impact.

Admittedly, environmental harm in the form of climate change adds another layer of complexity that may yet explain the modifications to conventional human rights doctrine that Viñuales suggests.[25] In addition to the problems with establishing an environmental harm and a threat to human rights, applicants are faced with the fact that it is unclear to what extent an individual state has contributed to climate change, even if an action (or inaction) could be seen as elevating the risk of the environmental harm coming to pass. In keeping with our football analogy, we might imagine that we are now observing a direct free kick, with three players lining up to take the kick, in an effort to confuse the opponent's defence. Our goal is still to determine whether a specific player scored, but we encounter problems in establishing who actually kicked the ball. This is of course a strong simplification of climate change, as a significant number of states are jointly contributing to the phenomenon to varying degrees – all kicking the ball simultaneously, as it were. However, because human rights inquiries always focus on a specific rights addressee, as the applicant is alleging a violation by a particular state, the question of responsibility for violations nonetheless always has a clear subject. The question as to the degree of responsibility of an

[24] See, for instance, the research in the United Kingdom by Hazel Genn and Cheryl Thomas, *Tribunal Decision-Making: An Empirical Study* (Nuffield Foundation – UCL Judicial Institute, 2013); Howard Gillman, 'What's Law Got to Do with It? Judicial Behavioralists Test the "Legal Model" of Judicial Decision Making' (2001) 26 *Law & Social Inquiry* 465.

[25] Viñuales, 'A Human Rights Approach to Extraterritorial Environmental Protection?', 215; Fisher and others, 'The Legally Disruptive Nature of Climate Change'.

individual state is not necessarily of concern, provided that some level of responsibility is established, or indeed where courts accept a material contribution to the risk as sufficient.[26] A violation of a human right is not justified simply because the state was not solely or even primarily responsible for the environmental harm.

This insight was crucial to the Dutch Appeal Court ruling in the *Urgenda* case we discussed in Chapter 3: the duty of care of the state was augmented through ECHR rights and specified by reference to commitments the Netherlands had made to reduce carbon emissions under international and EU law. The apparently more limited share of the Netherlands towards global emissions and the failure of other major polluters to act was rightly not accepted as a defence for lower domestic targets.[27] The court thereby implicitly rejected the need for further causality inquiries beyond the undisputed failure of the Dutch state to meet the emission targets deemed necessary under the Paris agreement. Indeed, from a functional doctrinal perspective this approach makes perfect sense: the question was and remains whether the player in question kicked (or at least contributed to kicking) the ball across the goal line.

Likewise, there is no meaningful legal difference in asking whether rising sea levels were caused by an individual state's action, versus whether a state has contributed to climate change, which in turn led to rising sea levels. We are still investigating whether the state caused (or at least contributed to) the environmental harm in question. Therefore, it is hard to see how the issue of climate change is fundamentally different from the general relationship between human rights and environmental harm, and likewise how it might benefit from an additional causality inquiry. Nonetheless, we can discuss the merits of a spatial expansion of human rights to overcome undoubtedly existing problems in addressing climate change, without adopting a problematic two-tiered approach to causality.

11.1.1.2 Spatial Expansion

To understand spatial expansion in terms of our football analogy, we must now imagine ourselves at a significant observational disadvantage. Assume that we are observing the football match not from the comfort of our sofa, but from a great altitude aboard an aircraft circling the stadium. With the naked eye, it is difficult to make out individual players and determine whether the relevant

26 Recall for instance the approach taken to liability in asbestos cases, see Chapter 3.
27 *The Netherlands* v. *Urgenda Foundation* 200178245/01, Court of Appeal, 9 October 2018.

rules were observed, much less whether the ball has landed in the goal at any given point in time. In order to compensate for this observational disadvantage, some scholars effectively suggest moving the goalposts closer to the touch lines and raising the crossbar to the stadium ceiling, as well as limiting ourselves to determining whether a goal was scored, dropping the requirement of identifying the individual player responsible. This broad regional view on environmental harm in the place of individual events, and population-wide effects instead of harm to individuals, suggests a broader (collectively held) right to an environment of a certain quality.[28]

In terms of human rights, this expansion conflicts with key characteristics of human rights law and doctrine. As we have seen, human rights reject broad statements of principle and primarily protect individual rightsholders, not loosely defined collectives. It remains doubtful that the prevalent doctrinal approach can be modified in the way some suggest without compromising the characteristics that we identified as central to human rights. Then again, this is exactly the point that sceptics make: if human rights as currently conceived are unable to address an issue of considerable complexity, urgency, and harm such as climate change, then they should be expanded to groups and regions, to correspond to the available level of scientific certainty.[29] The trouble with this view is that there is little evidence to suggest that the individual-focused nature of human rights is a key obstacle to addressing climate change.

There is nothing inconsistent about the idea that an individual can effectively challenge an undesirable state of affairs under individual rights, and thereby also benefit an unknown number of further individuals not directly involved in the legal proceedings.[30] As argued throughout Chapter 2, that is precisely the case with respect to environmental harm under human rights protection regimes. Claims under individual rights have important public functions and can act as protections for collective goods to the extent that individualised damage is caused to them.[31] This rationale is also central to many areas of domestic legal remedies for environmental harm, for instance in

28 Viñuales, 'A Human Rights Approach to Extraterritorial Environmental Protection?', 215; Dupuy and Viñuales, *International Environmental Law*, 331.

29 Dupuy and Viñuales, *International Environmental Law*, 324; Francioni, 'International Human Rights in an Environmental Horizon', 54–5; for a critique of this tendency, see Catherine Donnelly, 'Third Generation Rights' in Catherine Brölmann, R. Lefeber, and Marjoleine Zieck (eds.), *Peoples and Minorities in International Law* (Martinus Nijhoff 1993) 119.

30 Hayward, *Constitutional Environmental Rights*, 75–7; Raz, *The Morality of Freedom*, 256–63.

31 Raz, *The Morality of Freedom*, 258.

public interest litigation in Germany,[32] rights-based litigation in South Africa, and the concept of class action lawsuits.[33] At the international level, human rights litigation under regional systems such as the ECHR with its system of pilot judgments,[34] the protection of indigenous rights under the ACHR, and the protection of the environment under the ACHPR illustrate a similar approach.

That is of course no reason to object to the introduction of collective rights, or the incremental development of human rights doctrine in that direction. However, even with a suggested spatial expansion, interpreted primarily as an expansion of the scope rather than involving a problematic two-tier causality requirement, human rights are unlikely to contribute decisively to the debate on climate change. This conclusion rests on three practical concerns: temporal constraints, geography, and global rights commitments.

11.1.2 *Temporal Constraints*

The debate surrounding human rights approaches to climate change must reckon with the fact that significant and irreversible damage has already been done, and the responsibility lies with countless actors throughout history.[35] The Climate Agreement of Paris had the modest ambition to limit further increases of the global mean temperature to two degrees Celsius, and expressed ambitions towards a 1.5-degree goal.[36] There is no realistic prospect of reversing the already significant increases compared to pre-industrial levels. Even if the 2-degree goal is achieved, significant climate-based events, natural disasters, the rise of sea levels, and many other environmental harms are

[32] See § 64 of the Bundesnaturschutzgesetz – German Federal Nature Conservation Act, which modifies the general rule of § 42 of the Verwaltungsgerichtsordnung of 19 March 1991 (BGBl. I S. 686), last amended through Article 1 of the Law of 3 December 2020 (BGBl. I S. 2694).

[33] See, for instance, the more recent lawsuits in the United States surrounding the manipulation of diesel engine software through VW, 'Volkswagen "to pay $10bn" for US Emissions Scandal' *BBC News* (23 June 2016) www.bbc.co.uk/news/business-36611152, accessed 1 April 2021.

[34] See Rule 61 of the Rules of the Court; Lucius Caflisch, 'Rechtsfolgen von Normenkontrollen – Die Rechtsprechung des EGMR: Die Technik der Pilot-Fälle' [2006] Europäische Grundrechtezeitschrift 521; Stefanie Schmahl, 'Piloturteile des EGMR als Mittel der Verfahrensbeschleunigung' [2008] Europäische Grundrechtezeitschrift, 369; W. Sadurski, 'Partnering with Strasbourg: Constitutionalisation of the European Court of Human Rights, the Accession of Central and East European States to the Council of Europe, and the Idea of Pilot Judgments' 9 *Human Rights Law Review* 397.

[35] A point that essentially sealed the fate of the unsuccessful litigation against a private energy company in *Native Village of Kivalina* v. *Exxonmobil Corporation* 663 F Supp 2d 863 (ND Cal 2009) (Kivalina).

[36] *Paris Climate Agreement* [17].

inevitable. If applicants were to bring a case, some states could plausibly claim that the failures of the past era of unrestricted industrialisation, rather than any action of the present, led to the environmental harm. Moreover, states can point to the significant commitments undertaken to limit further emissions. To the extent that insufficient regulatory regimes of the past are (partially) to blame, these are historic mistakes that no conceivable contemporary regulation can rectify, and thus a human rights-based claim is difficult to justify.

11.1.3 *Geography*

The effects of climate change are most acutely felt in developing states, while the primary contributors to greenhouse gas emissions are found in industrialised nations. Industrialised states also have financial capabilities to adapt to any changes and challenges that climate change may present. This imbalance brings with it an additional problem when attempting to address climate change through human rights. Severe harm may well arise in parts of the world that have made no meaningful contribution to climate change. Consider a low-lying island nation that is threatened by rising sea levels attributable to global warming. These islands typically have no meaningful industry and contribute little per capita and in absolute terms to greenhouse gas emissions. Nonetheless, they are profoundly affected, in some cases existentially threatened. An individual bringing a case against such an island nation would be addressing a state that can reasonably claim to be doing everything within its power to combat climate change. The creation of an interest group of small islands nations at COP21 is a direct response to this dilemma and an attempt by those most affected to punch above their political weight in the negotiations dominated by global economic powerhouses.

Conversely, an individual living in a state that is a leading greenhouse gas emitter may well not experience significant harm. The state may take precautions, adopt mitigating measures, and through regulations and consistent enforcement prevent some climate change–related risks from materialising.[37] Thus, citizens of an industrialised state can challenge the insufficient measures and the significant historic contribution to global warming, but cannot necessarily point to an infringement of their rights. The two domestic human rights protection regimes often do not link up owing to their respective

[37] This point was – unsuccessfully – raised by the state to challenge the standing of the applicant in *Urgenda Foundation* v. *The Netherlands* C/09/456689 / HA ZA 13-1396, Den Haag Regional Court, 24 June 2015 [4.5]–[4.8].

geographic scope and both states are unlikely to be subject to the same enforceable regional protection regime. What remains is the international level, but as we observed in Chapter 8, the ICCPR and ICESCR are far from recognising environmental harm as an unequivocal threat to human rights, much less offering strong enforcement mechanisms.

11.1.4 *Global Rights Commitment*

In my view, a human rights–based approach to climate change is currently most likely to be impactful on the European continent: there are a number of domestic and regional human rights obligations that can conceivably accommodate individual claims. Even if developed states mitigate environmental harm to their citizens, cases such as *Urgenda* demonstrate that at the least they may be held accountable for the global harm caused through inadequate emission reduction targets. At the same time, this regional strength is also an inherent weakness, as it highlights the inability to address global problems, especially where commitments to human rights are weaker. Simply adopting a human rights–based approach and expecting tangible environmental improvements globally would be naïve. It wrongly presupposes that all states have enforceable human rights obligations, as well as the institutions, respect, and desire to uphold them in practice. The reality is that human rights are not universally recognised, nor fully respected, especially among some of the most persistent transboundary polluters. This may well be a less significant problem with respect to regional environmental harm, where causes and effects are relatively contained either in one state or in states subject to the same human rights protection regime. However, with transboundary harm of global dimensions this poses a significant barrier to the practical value of a human rights-based approach.

With respect to climate change specifically, it remains questionable whether a minority of states, conceivably motivated by human rights commitments, can hope to effectively tackle this global environmental emergency. Human rights do little to bring authoritarian states to the negotiating table, much less serve as an incentive for enforcing firm commitments. It is hardly surprising therefore that human rights have not been at the forefront in the context of UN climate change conferences. As one senior negotiator representing a major greenhouse gas emitting state remarked, the concerns of human rights advocates at these conferences are humoured, but ultimately seen as little more than tedious 'sideshows' to the main negotiations. It is not difficult to appreciate why this might be the prevailing view: human rights simply have little weight for states that are ideologically opposed to them, pay lip service to upholding them in practice, or who care little about being perceived internationally as not complying with them.

11.1.5 *Conclusion*

The example of climate change is symptomatic of a more general problem in the relationship between human rights and transboundary environmental harm. Long-term human interests in preserving the natural environment are often in conflict with short-term desires for prosperity and economic growth. States often ignore the wider environmental implications in favour of economic development, especially when the negative impacts are gradual and global, and the benefits are immediate and local. As laudable as the cause of tackling climate change through international cooperation is, human rights appear ill-suited to add much to this debate. Issues surrounding temporal, geographic, and rights commitments largely defeat the prospect of decisive contributions. Indeed, the primary barrier to improvements on transboundary environmental harm through human rights is not their individual-focused nature, their doctrinal approach to causality, nor the disconnect between available scientific evidence and scope. Rather, it is a lack of respect for human rights and an absence of institutions that are willing and able to enforce environmental regulations globally. Apart from failing to tackle environmental harm, this also leaves millions of victims of human rights abuses without an effective remedy. Until human rights law is universally enforceable and respected, states may well have a moral and a legal obligation to seek meaningful, collective action on climate change, but it cannot be based on human rights law. That does not, however, preclude human rights law from making important contributions to the enforcement of international environmental law more broadly.

11.2 ENFORCEMENT AND SUSTAINABLE DEVELOPMENT

It is often observed that ensuring compliance is a recurring concern in international environmental law. States can be unreliable partners in international cooperation, as even comparably simple procedural commitments, such as the filing of progress reports, remain unfulfilled.[38] Quite apart from the question of whether enforcement, the prevalent legal mode of ensuring

[38] Bodansky, *The Art and Craft of International Environmental Law*, 226; Martti Koskenniemi, 'Breach of Treaty or Non-Compliance? Reflections on the Enforcement of the Montreal Protocol' (1993) 3 *Yearbook of International Environmental Law* 123.

compliance, is consistently available at the international level, Daniel Bodansky laments that it has 'never been the strong suit of international environmental law.'[39] He argues that international environmental law is fundamentally a political process that balances competing interests and is therefore ill-suited to traditional avenues of legal enforcement. By contrast, he views human rights law as 'more legalistic in nature' as once 'an issue is conceived in terms of rights, it is removed from the political arena of competing interests and policies.'[40] As we have seen throughout the book, that is perhaps an overly strong statement. Far from removing issues from competing interests and policies, human rights law and the environmental minimum frame and facilitate the balancing of competing considerations. Nonetheless, we can take on board the general proposition that human rights law allowed access to enforcement mechanisms that are often unavailable in international environmental law. Scepticism towards legal enforcement mechanisms has led some authors to argue that market-based solutions are preferable,[41] and that non-compliance arising from an inability, as opposed to an unwillingness to comply, is best addressed through financial and technical assistance, as well as other non-adversarial methods,[42] in addition to a stronger reliance on NGOs[43]

While these concerns seem valid, they do not exclude recourse to legal enforcement through human rights law whenever it is available and appropriate. In that sense, the human rights–based approach of the environmental minimum should be considered an addition to other modes of enforcement: legal, political, market-based, or otherwise. Indeed, where international courts and tribunals have been engaged with the enforcement of international environmental law agreements, commentators have noted their influential role, both by clarifying obligations and providing a binding legal assessment of compliance.[44] The key is to operationalise existing enforcement mechanisms,

39 Bodansky, *The Art and Craft of International Environmental Law*, 226.

40 Ibid., 251.

41 Freeman and Kolstad, 'Prescriptive Environmental Regulations versus Market-Based Incentives', 3.

42 Dupuy and Viñuales, *International Environmental Law*, chapter nine; Chayes Abram and Antonia Handler Chayes, *The New Sovereignty: Compliance with International Regulatory Agreements* (Harvard University Press 1995); Edith Brown Weiss and Harold Karan Jacobson, *Engaging Countries: Strengthening Compliance with International Environmental Accords* (Massachusetts Institute of Technology Press 1998).

43 Bodansky, *The Art and Craft of International Environmental Law*, 234 and 237.

44 Harrison, 'Reflections on the Role of International Courts and Tribunals in the Settlement of Environmental Disputes and the Development of International Environmental Law', 513; Jorge E. Viñuales, 'The Contribution of the International Court of Justice to the Development

a task where the environmental minimum can act as a useful bridge. It offers a compliance framework for international environmental obligations, even where treaties do not expressly provide for enforcement mechanisms. Existing obligations become actionable through rights-based litigation in the form of legal standards, and states that fail to comply with international obligations would hence not only wrong other contracting states, but also act contrary to their human rights commitments. At least to the extent that individuals are affected by environmental harm that treaties regulate, the more robust enforcement mechanisms of human rights protection regimes can enforce obligations more consistently. This might be of particular use with respect to vaguely defined and mutually contradictory objectives that lack an inherent evaluation and enforcement mechanism: most prominently, this describes the Sustainable Development Goals.

As we have already observed in Chapter 3, the environmental minimum can be helpful in specifying obligations and mediating conflicts in international environmental law, notably those arising from the application of the precautionary principle. Many of the same qualities can also be brought to bear on the Sustainable Development Goals, which like the precautionary principle, suffer from vague normative claims and a general absence of mechanisms for conflict resolution.

The principle of sustainable development can be traced back to the 1972 Stockholm Declaration, particularly principles nine, ten, and eleven.[45] It was first expressly recognised in the 1992 Rio Declaration, which held that '[t]he right to development must be fulfilled so as to equitably meet developmental and environmental needs of present and future generations.'[46] Ever since, the status of the principle in International Law is the subject of on-going debate.[47] It featured in the case law of the ICJ, namely its decision in

of International Environmental Law: A Contemporary Assessment' (2008) 32 *Fordham International Law Journal* 232; Tim Stephens, *International Courts and Environmental Protection* (Cambridge University Press 2009) chapter 2.

45 Ulrich Beyerlin, 'Sustainable Development' in Rüdiger Wolfrum (ed.), *Max Planck Encyclopedia of Public International Law* (Oxford University Press 2013) [2].

46 *Rio Declaration on Environment and Development* [1.6]; Francesco Francioni, 'Principle 1: Human Beings and the Environment' in Jorge E. Viñuales (ed.), *The Rio Declaration on Environment and Development* (Oxford University Press 2015) 93; Virginie Barral and Pierre-Marie Dupuy, 'Principle 4: Sustainable Development through Integration' in Jorge E. Viñuales (ed.), *The Rio Declaration on Environment and Development* (Oxford University Press 2015) 157.

47 In favour of its status as a principle of International Law: Sands and others, *Principles of International Environmental Law*, 217; sceptical: Beyerlin, 'Sustainable Development' [17]–[18].

Gabčíkovo-Nagymaros,[48] and the *Pulp Mills* case.[49] Both judgments ultimately did little to clarify its status under International Law.

In *Gabčíkovo-Nagymaros* the ICJ accepted that the principle of sustainable development was increasingly important, particularly with respect to treaties that have clear environmental and economic implications.[50] Accordingly, Hungary and Slovenia were required to take the environmental impact on the Danube into account when evaluating the operation of the power plant. Crucially, the ICJ did not specifically pronounce itself on the status of the 'concept' of sustainable development in international law. The ICJ insisted that a solution must be sought within the established treaty, relying on the good faith provision it contained.[51] In the *Pulp Mills* case the ICJ likewise paid little attention to the principle of sustainable development: instead it held that the underlying essence of balancing economic and environmental concerns was contained in Article 27 of the Treaty between Argentina and Uruguay.[52]

Despite the unclear status of the principle, the UN General Assembly adopted 17 SDGs with targets covering a broad range of issues.[53] The main thrust of the SDGs is directed towards social and economic development by emphasising poverty eradication.[54] Nonetheless, the SDGs contain several goals that relate to environmental harm, such as ensuring good health and well-being (no.3), clean water and sanitation (no.6), affordable and clean energy (no.7), sustainable cities and communities (no.11), responsible consumption and production (no.12), climate action (no.13), and protecting aquatic (no.14) as well as land-based life (no.15). While the phrasing thus accommodates a broad range of issues and concerns, it equally muddies the waters on the specific claims and means to implement them. Some level of specification is achieved through the 169 targets and 304 proposed indicators of compliance. However, these still leave the obligations vague and offer little indication how inevitable conflicts between development goals ought to be resolved. Indeed, states are expected to specify the general goals and determine how these can be best implemented independently. This leaves three

48 *Gabčíkovo-Nagymaros Project (Hungary/Slovakia)* ICJ Reports 1997.

49 *Case Concerning Pulp Mills on the River Uruguay (Argentina* v. *Uruguay)* (Merits) [2010] ICJ Rep 167.

50 *Gabčíkovo-Nagymaros Project (Hungary/Slovakia)* ICJ Reports 1997 [140].

51 Ibid. [142].

52 *Case Concerning Pulp Mills on the River Uruguay (Argentina* v. *Uruguay)* (Merits) [2010] ICJ Rep 167 [177].

53 Transforming our world: the 2030 Agenda for Sustainable Development, 25 September 2015, UN Doc. A/Res/70/1.

54 Dupuy and Viñuales, *International Environmental Law*, 20.

primary sources of uncertainty unresolved that are of concern to the environmental minimum.[55]

- Defining the goal: many SDGs are phrased in broad terms that cover large, complex, and interconnected policy areas such as health and well-being (target 3.4). This renders it difficult to determine any specific policy initiatives and conceivably permits a host of contradicting interpretations. In turn, this raises the question to what extent SDGs offer any practical guidance beyond the expression of ideals and values. Crucially, given their broad scope, potentially any state action could be conceptualised as being in furtherance of some SDG, thus opening the door to cynical trade-offs and legitimising a host of interventions that achieve little tangible improvements for the environment.
- Defining progress: a central problem lies in the unclear definition of what constitutes progress on an SDG. Terms such as 'safe', 'effective', 'sustainable', and 'reliable' are commonly employed, but inevitably require further specification. Other targets are more specific, but are clearly not realistically achievable, as they require the complete absence of an undesirable or complete fulfilment of a desirable state of affairs, for instance target 8.5 which mandates 'full and productive employment (. . .) for all.'
- Conflicting goals: once the scope and requirements are specified, SDGs will inevitably conflict. For instance, the eradication of poverty (no.1) presupposes a certain level of economic development which is difficult to achieve without environmental harm to land- and sea-based organisms, as well as associated detriments to human health and well-being. For some economies, progress on the former might be mutually exclusive with progress on the latter. States have broad discretion to reconcile and prioritise conflicting goals as well as the extent to which deterioration on some goals can be justified through progress on other SDGs.

The environmental minimum can translate goals, targets, and indicators into regulatory and enforcement actions, and can offer a framework for prioritising and optimising conflicting demands. The environmental minimum achieves this by taking abstract goals of SDGs and providing concrete demands for environmental regulation through the specific risk principle and minimum standards: both build on the generally accepted standards of the relevant scientific community as formulated for instance by the WHO. On a

[55] Nina Weitz and others, *Sustainable Development Goals for Sweden: Insights on Setting a National Agenda* (Stockholm Environment Institute, Working Paper 2015-10, 2015) 10.

more limited scale, this has already been attempted. For instance, following the adoption of the SDGs, the Stockholm Environmental Institute systematically engaged with the SDGs relevant to the environment and specified these targets through reports and guidelines of international organisations such as the WHO and the OSCE.[56]

Crucially, the environmental minimum also provides a framework for reconciling conflicting claims through the framework of human rights. At the practical level, conflicts can be addressed through a balancing of interests and proportionality assessments, and regulatory action can be prioritised through a focus on the potential harm to humans, as opposed to the likelihood of any specific harm occurring.

Without such a framework, the danger is that states engage in mere 'SDG-washing' as conceivably any action can be presented as being in furtherance of some broadly phrased sustainable development goal. Furtherance of some SDGs is not valuable in and of itself, unless it is tied into a broader framework that allows for the reconciling and balancing of the competing policy demands. There is little environmental or sustainable development utility in denying the reality of trade-offs, nor in obscuring persistent disagreement through vague expressions of unobjectionable principles and values.

11.3 SUMMARY

The purpose of this chapter was to demonstrate some of the improvements that the environmental minimum may bring about for the purposes of international environmental law. In any case, blanket scepticism towards the current legal approaches generally and environmental regulation specifically seems misplaced. Real improvements are possible, even without far-reaching and intrusive changes to well-established legal categories and principles. In this respect, the Anthropocene presents itself more as a recognition of the level of human interference in global environmental systems and a call to adopt ambitious global regulations, than a credible basis for a recalibration of entire domestic and international legal orders. This is especially apparent in attempts to draw on human rights protection regimes to tackle global environmental challenges: the prime example is man-made climate change. Crucially, there is no indication that there is a global commitment to the protection of human rights that can persuade some of the most persistent polluters to commit themselves to strict emission targets. Human rights do little to bring

[56] Ibid.

authoritarian states to the negotiating table, much less serve as an incentive for enforcing firm commitments.

The environmental minimum can, however, specify obligations under the SDGs as well as mediate inevitable conflicts and bolster enforcement mechanisms. The environmental minimum can channel some of the vague, idealistic, and contradictory targets under the SDGs and provide a framework for balancing and managing normative claims. This evidently does not resolve the broader issues of vague expressions of principles and values found in the SDGs beyond those related to the environment. However, it can go some way towards specifying and balancing commitments in a way that is practically useful for environmental regulation and that is capable of holding states to account. Relying on the established guidelines of international organisations like the WHO is a sensible means and provides a readily available metric for measuring the performance of states on environmental harms and SDGs.

12

Conclusion and Outlook

This book has set out the case in favour of the environmental minimum as a framework that protects humans from the threats posed by environmental harm. The final chapter offers a concise summary of the core arguments of the book, specifically explaining that the environmental minimum constitutes a normatively attractive and practically viable framework: one that translates general and abstract commitments into specific and practical measures to protect the environment. This chapter then explores the overarching incremental theme of the environmental minimum, highlighting its preference for continuity with established legal categories and principles. It closes with an outlook that situates the framework in the broader legal and socio-political landscape.

Following the introduction, Part I of the book introduced and defended the environmental minimum as a framework for the protections (and correlative duties) that environmental human rights require.

Chapter 2 began by arguing that a commitment to human rights necessarily entails basic environmental protection duties as a matter of political morality. This is because egregious forms of environmental harm critically undermine the fundamental values that underpin human rights, chiefly human dignity and autonomy. Human rights must therefore contain a sub-category of protections which we can conceptualise as environmental human rights. The human interests that environmental human rights protect are the environmental conditions necessary for the preservation and flourishing of human life. Those who challenge these rights as vague overlook the significant room for agreement in the pursuit of diverging universal notions of a 'sound' environment. Meanwhile, concerns over potential conflicts with other rights are overstated, because balancing of competing interests is a pervasive and well-established feature of human rights law. Contemporary environmental regulations are already frequently challenged on the basis of competing rights, for instance to property.

Chapter 3 then introduced and defended the environmental minimum as a framework for the protections that environmental human rights require. The chapter distinguished between human rights, understood as an expression of idealised values and norms derived from political morality, and human rights *law*, contained in domestic and international protection regimes that entail practical limitations and trade-offs. The environmental minimum does not purport to facilitate the comprehensive protection of the environment in human rights law that might be justified as a matter of political morality. The chapter developed the guiding principles of the environmental minimum, modelled on the characteristics of human rights identified by James Nickel and as specified through the lens of environmental protection following Joseph Raz's interest theory of rights. The three principles of the environmental minimum that allow courts to make sense of environmental harm are as follows: (1) the specific risk principle, which acts as the trigger for the environmental minimum, requiring a specific risk to a recognised right arising from environmental harm; (2) the minimum standards, which specify the baseline standard of review that courts ought to employ, based on domestic and international legal norms, as well as established and emerging standards that are generally accepted in the relevant scientific community; and finally (3) the high-priority characteristic, which guides the balancing of environmental protection with competing interests based on human rights.

Part II then focused on the practical case in favour of the environmental minimum and applied the framework to address two key shortcomings in the relationship between environmental protection and human rights as conventionally presented: especially a lack of acknowledgement of the substantial engagement through courts with environmental harm, and the absence of a framework that can translate the plethora of general and abstract commitments into specific and practical regulations under the supervision of courts. This is most evident under the ECHR, which has dealt with a comparatively high volume of cases.

Chapter 4 introduced the doctrinal data analysis developed for this book to explore the compatibility of the ECHR with the specific risk principle. The principle is the trigger of the environmental minimum framework, determining whether the minimum standards of review apply to a given environmental harm. The principle requires that environmental harm is successfully linked to a threat to a specific human right recognised under the invoked protection regime. Compatibility with the ECHR can be determined through the extent the doctrine of the Court subscribes to two key claims: (1) in principle, any environmental harm can give rise to infringements of Convention rights; and (2) environmental harm has been successfully linked to the Convention rights and led to findings of violations. The chapter concludes that, notwithstanding some inconsistencies in earlier cases, the prevailing ECHR doctrine is compatible with the environmental minimum.

Chapter 5 examined the crucial question whether the ECtHR recognises that a reasonable hypothesis of harm is sufficient to establish the link between environmental harm and a specific risk to a Convention right. The chapter first set out the general principles developed with respect to the rules of evidence and victim status. The chapter recognised the difficulties that arise for applicants who seek to obtain relevant evidence and records from reluctant and secretive states. The ECHR does not require applicants to demonstrate an irrefutable scientific link between the environmental harm and an impact on their rights: in many recent decisions the ECtHR is content with evidence of a general population level risk arising from the environmental harm experienced by the applicant.

Chapter 6 explored the ECHR approach to the minimum standards of review. These standards apply once a specific risk has been established and are derived either from domestic and international legal norms, or in lieu of these, established and emerging standards. The ECHR generally meets these requirements, but the chapter acknowledged some enduring concerns, notably the failure of the Court to invariably find violations where states fail to enforce domestic law.

Finally, Chapter 7 investigated the claim that environmental interests must be balanced with competing considerations, and that neither can be presumed to have priority in the abstract. In principle, the ECtHR accepts that environmental regulations can give rise to infringements of Convention rights and acknowledges that regulations must be proportional to the established threat of environmental harm. The chapter then discussed an enduring area of concern, namely the acceptance of vague economic justifications for environmental harm and the problems surrounding the margin of appreciation doctrine.

Part III expanded our perspective beyond the ECHR by exploring further international protection regimes and two representative domestic case studies: South Africa and Germany.

Chapter 8 presented an overview of the case law of other international and regional human rights protection regimes examined for this book. The number of cases is generally limited; therefore, it is difficult to conclude with much certainty whether the environmental minimum is compatible with the prevailing doctrinal position under the UN Covenants, the African Charter on

Human and Peoples' Rights and the American Convention on Human Rights. Nonetheless, two recent decisions from the UN Human Rights Committee as well as an encouraging advisory opinion from the American Court of Human Rights may signal a move towards greater environmental protection. Overall, if the framework were consistently applied to conceptualise environmental harm in these protection regimes, it would bring about significant improvements towards a clearer case law aligned with the normative claims of environmental human rights.

Chapter 9 discussed the constitutional protections of the environment in South Africa and Germany, which represent two broad trends that go beyond the greener interpretation of existing rights through courts. South Africa codified an enforceable, individual right to an environment of a certain quality while Germany introduced a general constitutional mission statement. As the case studies demonstrate, both approaches have been limited in their practical impact: key conceptual and doctrinal questions remain unresolved and have failed to significantly impact environmental regulations. In South Africa, the environmental minimum could contribute towards the development of a clearer doctrine on section 24 and clarify its relationship to dominant environmental framework legislation. With respect to Germany, the chief contribution lies in limiting the deference granted to regulators and stimulating the development of more exacting positive obligations under fundamental rights.

Finally, Part IV of this book examined the potential contributions to environmental regulations under domestic and international law.

Chapter 10 shifted our attention to domestic and international efforts to regulate environmental harm and suggested that three pathologies have historically hampered regulatory success: political lag, which describes the gap between the best available scientific evidence and regulatory efforts that address environmental harm; industry resistance, which arises from the fruitful ground of political lag, permitting vested interests to entrench harmful, but profitable business models and practices; and finally, regulatory inertia, which means that regulators are less likely to burden vested economic interests with effective regulations or enforce them consistently. Historic atmospheric ozone and asbestos regulation provide contrasting examples of regulatory success, and the pervasive contemporary failures in addressing air pollution demonstrate that the pathologies remain salient concerns. Conventional regulatory approaches can be an effective tool to address environmental harm, provided that there is a close relationship between scientific evidence and regulatory action.

Chapter 11 considered two key implications of the environmental minimum for international environmental law. For one, there is considerable difficulty in applying the human rights-based framework of the book to issues surrounding transboundary environmental harm, particularly the global emergency of man-made climate change. The environmental minimum and its human rights framework presuppose a robust commitment to human rights, environmental protection, and the rule of law that is questionable with respect to many of the most notorious global polluters. The environmental minimum is more suited to specifying the scope, context, and relationship of competing environmental obligations, notably among the SDGs, as well as providing a viable framework to enforce otherwise non-binding or underenforced treaty obligations.

From the outset, the environmental minimum was presented as a modest framework. Even though it is firmly rooted in the normative claims of environmental human rights, it is only one possible expression of these normative demands. The crucial advantage of the environmental minimum over other possible incarnations is its practical approach and continuity with established legal categories and principles.

In the absence of a consensus on human rights and environmental protection, an environmental minimum that recognises the shared interests of humanity is a crucial foundation for more ambitious action. Accordingly, the book took it upon itself to close the gap between domestic constitutional and international law commitments on the one hand, and the practical challenges facing environmental regulation and adjudication on the other. The main obstacle in domestic and international protection regimes is the absence of a normative framework that can mediate the inevitable conflicts between environmental protection and competing interests. The key normative claim is that enacting consistent environmental regulation based on generally accepted standards is the only way to comply with commitments to human rights, and the key obligation is their consistent enforcement.

The framework thus stands against the prevailing winds that often seek more radical solutions in response to climate change, and in understandable dismay over well-documented failures of conventional environmental regulation. Bold steps are undoubtedly required to mitigate and address the profound harm that anthropocentric climate change has brought, and it is not with much hyperbole that environmentalists refer to this challenge as an existential threat to humanity. The Anthropocene is the watchword for those who believe a radical overhaul of the economy, environmental decision making and legal regulation is required in the face of systemic failures to address dire environmental challenges.

The supporters of this movement will likely be disappointed by the environmental minimum, lamenting that it does not go further in framing and advocating the normative claims of environmental human rights, or indeed, recognise rights of nature itself. The animating moral insight at the heart of these concerns is certainly sound. The world is ultimately one grand interdependent ecosystem, making it impossible to regionally contain environmental degradation and pollution indefinitely. Eventually, any serious environmental harm impacts the global ecosystem of which humans from all corners of the world form an inseparable part. Current environmental harm is so pervasive and ecologically unsustainable that it jeopardises the future of humanity itself. The challenge is to achieve a delicate balance between ensuring the long-term ecological survival of humankind and the sustainable use of environmental resources that support modern life and societies.

For this reason, it is crucial to resist the desire for an only seemingly clean and simple break with established legal categories and principles. Human rights law is essential, especially where the demand for optimal protection of the natural environment must be balanced with a plethora of conflicting, and legitimate interests. All the while, we should remember that human rights are not a silver bullet for all environmental woes.

For one, human rights are ultimately grounded on harms and threats to humans: they are inescapably anthropocentric. The human development necessary to ensure the enjoyment of a whole host of human rights rests in no small part on economic exploitation of the environment, which inevitably entails considerable environmental harm. The environmental minimum acknowledges this reality and the difficult trade-offs that challenge even the greenest of regulatory intentions. Proposed solutions that seek to theoretically disentangle human rights from harm to humans, somehow recognising the environment as a rightsholder, or denying the need for difficult balancing of competing interests altogether, are simply not legally or practically credible.

Considering all the legal tools available to achieve environmental protection, human rights are the least suited to securing meaningful, globally coordinated action on climate change. The rights commitment of many of the most persistent polluters is questionable and human rights have been a distant concern at the climate negotiations. International solidarity and commitments to greenhouse gas emissions reduction will hopefully avoid the more catastrophic calamities of climate change, but their chief motivation has undeniably not been a consistent and principled commitment to human rights.

Admittedly, most improvements that the environmental minimum facilitates are regional and incremental. Changes are unlikely to occur overnight and are almost certain to fall short of grander ambitions. Incremental efforts to protect humans from environmental harm may appear to some as piecemeal responses to a global ecosystem in crisis. Cumulatively, these smaller changes can nonetheless achieve many improvements, without waiting for improbable, fundamental reforms to domestic and international legal orders. Especially with respect to climate change, time is in short supply: humanity simply cannot wait for the re-design of entire legal systems that some more radical proposals require.

This is especially true given that in many states the legal system as such, the available scientific evidence, or limited resources are not the primary causes of inertia on environmental protection. Domestic and international regulations for the protection of the ozone layer have demonstrated that decisive action on key environmental harms is achievable within familiar legal orders. Instead, it appears most environmental inertia is maintained through industry resistance, enabled at least in part through an absence of actionable constitutional principles, treaty commitments, and environmental human rights.

The environmental minimum promises to empower regulators to enact and enforce measures against industries that seek to maximise private profit and shift the burdens of their business model onto the general public. The history of asbestos regulation provides a cautionary tale, while at the same time demonstrating that it is possible to regulate persistently harmful industries out of existence. If we are to overcome the three pathologies that plague environmental regulation, then we will require a solid legal foothold.

The book has hopefully shown that the environmental minimum is exactly the kind of practical legal framework that can serve as a crucial starting point for grander environmental policy ambitions. The framework promises the political urgency that pressure groups struggle to nourish and the consistent enforcement of existing laws that has eluded many regulators: it is sufficiently demanding to facilitate practical improvements that benefit humans and the environment here and now, but not so radical as to require a reengineering of established legal categories and principles. Perhaps more than the large and global, an apt goal is for human rights to tackle the smaller and local environmental harm and thus play their modest part in seeing humanity thrive well beyond the challenges of the 21st century.

Appendix: Data Methodology

The Appendix outlines and explains in detail the methodological approach to gathering and organising the dataset that ultimately facilitated the doctrinal data analysis in this book. It gives specific details on data sources, timespans considered, and crucial keywords for all examined jurisdictions (European Convention on Human Rights, African Charter on Human and Peoples' Rights, American Convention on Human Rights, American Declaration of the Rights and Duties of Man, and the two UN Covenants) before detailing the data recording choices and MS Excel formulas used to analyse and draw conclusions from the dataset.

The complete dataset is available from the author on request (contact email: stefan.theil@law.ox.ac.uk) and will be made more widely available on the author's ResearchGate profile (www.researchgate.net/profile/Stefan_Theil) in the near future. The dataset may be freely used for research and teaching purposes, subject to appropriate attribution and referencing of the source material.

A.1 EUROPEAN CONVENTION ON HUMAN RIGHTS

Item	Methodology
Parent Lists	The initial datasets were obtained through the search engine provided by the European Court of Human Rights (www.hudoc .echr.coe.int). The search engine accesses the following records: • ECtHR: all court judgments and screening panel decisions from 1959 to October 1998, and all judgments and admissibility decisions from November 1998 to present. • EComHR: all public reports from 1963 onwards, some published admissibility decisions from 1955 to 1986, and all admissibility decisions from 1986 to present. The following search parameters and filters were employed in the Boolean search: • all document collections • all language versions • no date limitation • keywords (any of these sufficed for a hit, i.e. the 'or' option was selected): Article 2, Article 3, Article 6, Article 8, Article 13, Article 1 AP-I • text must include either 'climate change', 'global warming', 'pollution', 'environment', 'nuclear', or 'ecosystem', and their respective French equivalents ('changement climatique', 'réchauffement climatique', 'pollution', 'environnement', 'nucléaire', or 'écosystème') This returned approximately 3,091 hits, which were then manually examined for the existence of an environmental issue within the defined scope of the book. If the same case was considered by different Convention entities over time (Commission, Court, Grand Chamber) only the final ruling was considered. Cases that were somehow linked to the environment, but where no party made any environmental claims were excluded. See, for instance, the case of *Piermont v. France*, 15773/89, Court, 27.4.1995, where the applicant was expelled from French Polynesia for taking part in protests against nuclear weapon testing by France. Upon confirmation of an environmental issue, the cases were categorised and 23 data points were collected: • Category one if the applicant complained of an environmental harm OR sought information, procedural access, or enforcement in environmental proceedings • Category two if the applicant claimed a violation of Convention rights through a state action designed to protect the environment OR if the state sought to justify an infringement of Convention rights by invoking environmental protection

Item	Methodology
	It was then noted whether the individual provisions invoked were violated, justified, dismissed, settled, dismissed for non-exhaustion of domestic remedies, or struck out. The meaning of 'dismissed' was expanded to include the use of any of the following phrases by the Convention entities (ECtHR and EComHR): 'no violation', 'not necessary to examine', 'no need to examine', 'rejected', 'not applicable', 'gives rise to no separate issue.' The list includes all cases available through the search engine on 31 December 2019.
Dataset	The following columns make some general points about the data gathering and recording. The book is based on version 0.2 of the dataset.
Joined decisions	The automated analysis has an important limitation when dealing with joined decisions. Joined decisions arise when Convention entities join multiple applications into one decision, usually because they deal with similar factual circumstances and questions of law involving the same respondent state. This affects the figures and the underlying data analysis in three important ways. First, we require a justification for selecting amongst multiple possible filing dates when calculating the median time from initial application to final decision. With respect to joined decisions, the question arises which year should be relevant if the applications arise from different years: the article has opted for the earliest filing year of a joined decision as the decisive one. Second, joined decision necessitates a caveat on the number of violations given in the figures used throughout the book. The analysis is based on data extracted from final decisions and does not separately account for each application covered by a joined decision. This means that the number of violations may in fact be higher than stated, depending on the number of initially distinct applications joined and the specific findings and grounds of a violation found by the Court. To illustrate, consider the following example. Suppose that five individual applications are filed against the same respondent state regarding largely similar subject matters. These applications are initially admitted separately, but later joined by the Court. The Court then finds a violation of Article 8 ECHR in three, but not all five applications. The dataset would record this information as follows: five applications filed and one decision issued, both accurate, but only one violation found. The latter is incorrect as there were in fact three violations and thus the number of violations is understated by two.

(*continued*)

(*continued*)

Item	Methodology
	Accounting correctly for distinct outcomes on violations in joined decisions goes beyond the capabilities of the automated data analysis currently employed. It remains possible to manually account for distinct findings, but integration into the automated analysis has proven difficult and as yet remains an unresolved issue. That being said, instances of joined applications in the dataset are comparatively rare affecting 21 out of 371 decisions. Moreover, the Court only found a violation in 12 of these decisions; therefore, the discrepancies in the dataset are limited. Finally, an issue arises if joined decisions invoke the same Convention articles and the Court reaches distinct conclusions. Under those circumstances, the dataset recorded both outcomes, thereby inflating the number of decisions on Convention articles.
Cases with no mention of filing date	Where a decision does not give the filing date of the application, this was substituted with an estimate. The filing date was estimated as falling on the 1 January of the year given by the final two digits of the application number. For instance, the filing date for the application number 7407/76 (X and Y v. Germany) is estimated as 1 January 1976 using this method. This approach is naturally imprecise and may in extreme cases result in a margin of error of 364 days, if the actual filing date was, for instance, 31 December of the year indicated by the application number. However, given that the number of affected cases without a mention of the filing date is low (13 out of 371 applications, or approximately 3.5 per cent), these cases are unlikely to significantly alter the underlying median time from initial application to final decision. The cases where an estimate filing date was used are: 7407/76 (X and Y v. Germany); 9234/81 (Weltbund zum Schutze des Lebens eV v. Germany); 9515/81 (X v. United Kingdom); 9278/81 (G and E v. Norway); 13728/88 (S v. France); 14563/89 (M v. Italy); 28204/95 (Tauria v. France); 46346/99 (Noak and others v. Germany); 35392/97 (Pagliccia and others v. Italy); 9465/81 (M v. Austria); 11185/84 (Muriel Herrick v. United Kingdom); 10395/83 (N v. Austria); 7151/75 (Sporrong and Lönnroth v. Sweden).
Excluded prison cases	The following cases in the initial list of hits relating to prison conditions were removed from the dataset as they were judged to not fall within the environmental definition adopted for this book: 17525/90 (Delazarus v. United Kingdom); 6954/02 (Maltabar and Maltabar v. Russia); 5816/10 (Sotner v. Slovenia); 11303/12 (Novikov v. Russia); 63993/00 (Romanov v. Russia); 65550/01 (Koval v. Ukraine); 22088/04 (Bragadireanu v. Romania); 66820/01 (Dorokhov v. Russia); 3522/04 (Salmanov v. Russia); 1555/04

Item	Methodology
	(Zakharkin v. Russia); 24202/05 (Veliyev v. Russia); 14248/05 (Trepashkin v. Russia); 20641/04 (Chudun v. Russia); 918/02 (Solovyevy v. Russia); 19936/04 (Lopata v. Russia); 23185/03 (Petrov v. Russia); 16133/08 (Insanov v. Azerbaijan); 52624/10 (Hagyó v. Hungary); 66630/10 (Mihalia v. Romania); 63378/00 (Mayzit v. Russia)
Results Table	The Results Table automates the extraction of data from the Parent Lists through the use of Excel formulas focusing on individual, or a culmination of data points. The formulas employed in the Results Table are described below, along with representative examples.

Overview	This determines the number of cases in category one though this formula, which simply counts the number of entries:
	`=COUNTIF('Category 1 - Parent List'!S:S;"1")`
	This formula counts the number of cases that were decided by the Grand Chamber of the European Court of Human Rights:
	`=SUM(COUNTIFS('Category 1 - Parent List'!S:S;"1";'Category 1 - Parent List'!F:F;"*GC*"))`
	The following formula determines the number of cases where the applicants prevailed, either through the finding of a violation or a settlement with the respondent state:
	`=SUM(COUNTIFS('Category 1 - Parent List'!R:R;"*yes*"))`

Environmental analysis	This formula counts the number of cases in category one that were tagged as falling into a specified environmental category. A case may fall into several categories, depending on the environmental harm encountered. The example below counts the number of cases where the environmental harm related to noise: `=SUM(COUNTIFS('Category 1 - Parent List'!O:O;"*[noise*"))`
Outcomes by Article invoked	This formula determines the outcomes of individual articles invoked by the applicants in categories one and two. Depending on the outcome, the article was recorded in a different data column corresponding to the result: violated, justified, dismissed,

(continued)

(*continued*)

Item	Methodology
	settled, non-exhaustion, struck out. The example below counts the number of times Article 2 ECHR was found violated: =SUM(COUNTIF('Category 1 - Parent List'!H:H;"*2;*"))
Country Analysis	This formula counts the number of cases involving a specific country and the outcomes of these cases in categories one and two. The example below counts the total number of cases involving Germany: =SUM(COUNTIFS('Category 1 - Parent List'!E:E;"*Germany*"))
Treatment of Article 1 AP-I	These formulas seek to determine the relevance of Article 1 AP-I to the outcome of a case, by recording instances of divergences of outcome between Article 1 AP-I and Article 8 ECHR in category one cases. In order to determine the number of cases where the ECtHR arrived at the same outcome for Article 1 AP-I and Article 8, the formula counts the number of time that both Articles ended up in the same column, indicating an identical result: =SUM(COUNTIFS('Category 1'!H:M;"*1 AP-I;*";'Category 1'!H:M;"*8;*")) The formula to determine different outcomes denotes all possible combinations of different outcomes between Article 1 AP-I and Article 8 ECHR. For instance, violation with dismissed, justified with non-exhaustion, dismissed, justified, and so on: =SUM(COUNTIFS('Category 1 - Parent List'!H:H;{"*1 AP- I;*";"*8;*"}; 'Category 1 - Parent List'!J:J;{"*1 AP- I;*"\"*8;*"})+COUNTIFS('Category 1 - Parent List'!H:H;{"*1 AP-I;*";"*8;*"}; 'Category 1 - Parent List'!I:I;{"*1 AP-I;*"\"*8;*"})+COUNTIFS('Category 1 - Parent List'!H:H;{"*1 AP-I;*";"*8;*"}; 'Category 1 - Parent List'!L:L;{"*1 AP-I;*"\"*8;*"})+COUNTIFS('Category 1 - Parent List'!J:J;{"*1 AP-I;*";"*8;*"}; 'Category 1 - Parent List'!I:I;{"*1 AP-I;*"\"*8;*"})+COUNTIFS('Category 1 - Parent List'!J:J;{"*1 AP-I;*";"*8;*"}; 'Category 1 - Parent List'!

Item	Methodology

	`L:L;{ "*1 AP-I;*"\"*8;*"} )+COUNTIFS ('Category 1 - Parent List' !I:I;{ "*1 AP-I;*";"*8;*"} ; 'Category 1 - Parent List' !L:L;{ "*1 AP-I;*";"*8; *"} ))`
	
	In order to determine whether Article 1 AP-I was responsible for the success of the applicant, this formula counts the number of instances where Article 1 AP-I was violated and Article 8 was either dismissed, failed to exhaust domestic remedies, or found justified:
	
	`=SUM(COUNTIFS ('Category 1 - Parent List' !H:H;"*1 AP-I;*"; 'Category 1 - Parent List' !J:J;"*8;*")+COUNTIFS ('Category 1 - Parent List' !H:H;"*1 AP-I;*"; 'Category 1 - Parent List' !L:L;"*8;*")+COUNTIFS ('Category 1 - Parent List' !H:H;"*1 AP-I;*"; 'Category 1 - Parent List' !I:I;"*8;*"))`

Time from initial application to final decision	This determines the median time in months from initial application to final decision. The formula calculates the arithmetic median of number of months between initial filing of an application and a final decision, conditioned through the use of IF function to certain timeframes. Should the decision not give an initial filing date for an application, the filing date is inferred from the application number, with a margin of error of up to 364 days. The IF condition ensures that only applications are factored into the median calculation that were decided in a certain data range, while the latter part of the formula captures the number of months between the application filing day and the day of the final decision in full months, as determined through the following formula: `=DATEDIF(D189;B189;"m")` In the example below, an entry is factored into the median calculation if the final decision was made before the year 2000 `=MEDIAN(IF(YEAR('Category 1 - Parent List' !B2:B214)<2000;' Category 1 - Parent List' !U2:U214))`
Joined Applications	The first formula gives the number of decision that resolve more than one application by making use of the case naming conventions employed by the registry of the court: `=COUNTIF('Category 1 - Parent List' !E:E;"*others*")`

(*continued*)

(*continued*)

Item	Methodology
	A second formula calculates the number of applications that were joined by Convention entities, relying on the convention of the registry of the court preserving their respective application numbers (the number was counted and recorded manually): `=COUNTIF('Category 1 - Parent List' !T:T;">0")`

Number of decisions and violations found in cases in a given year	This formula determines the number of decisions in a given year by the date of the final judgment in categories one and two. The example below counts the number of cases where the final judgment was issued between 1 January and 31 December 1990 in category one:
	`=SUM(COUNTIFS('Category 1 - Parent List' !B:B;"<31/12/1990";' Category 1 - Parent List' !B:B;">01/01/1990"))`
	
	A further formula extracts the number of violations found in a given year in categories one and two. The example below counts the number of cases where violations were found between 1 January and 31 December 1990:
	
	`=SUM(COUNTIFS('Category 1 - Parent List' !R:R;"yes";' Category 1 - Parent List' !B:B;"<31/12/1960";' Category 1 - Parent List' !B:B;">01/01/1960"))`

Nature of environmental regulation	This formula counts the number of cases in category two that were tagged as falling into a specific regulatory category. A case may fall into several categories, depending on the environmental regulation challenged. The example below counts the number of cases where the environmental regulation related to planning permission. `=SUM(COUNTIFS('Category 2' !P:P;"*[ planning permission* "))`
Figures	The tables and graphs all present and visualise the results determined on the Results Table. They are described in more detail below, in the order in which they appear in the book.
Coefficient of determination (R^2 value)	The coefficient of determination is the proportion of the variance in the dependent variable (usually the number of applications/ cases or violations) that is predictable from the independent variable (usually time): • an R^2 value of 0 indicates that the dependent variable cannot be predicted from the independent variable at all;

Item	Methodology
	• an R^2 value of 1 indicates that the dependent variable can be predicted entirely on the basis of the independent variable. Values between 0 and 1 denote degrees of predictability, for instance an R^2 value of 0.2 means that 20 per cent of the variance in the dependent variable is predictable from the independent variable
Trend line	The trend line visualises the predicted development of the dependent value based on the independent value (examined years).
Value n	The value given for 'n' indicates the sample size represented in the figure.
Figure 4.1	Number of environmental applications filed (categories one and two) per thousand overall applications filed from 1985–2014, n = 384. The number of total applications filed is based on the annual reports of the ECtHR, which are available from 1985 onwards. As two-thirds of applications require at least between four years (and up to eight years) before a final decision is reached, the scope of the represented dataset was limited to the year 2014.
Figure 4.2	Overall number of applications filed per year in thousands 1985–2018 based on official figures in ECHR annual report, n = 889,573. The figures are based on those given in the annual reports of the ECtHR. A linear graph gives the trend line along with the coefficient of determination (R^2 value).
Figure 4.3	Environmental issues encountered in category one cases and violations found, n(total) = 280; n(violations) = 98. The figure represents the environmental harm that forms the subject of the application in category one cases. Cases can fall into more than one category.
Figure 4.4	Number of environmental cases filed, decisions issued, and violations found in final judgments per year in category one from 1960–2014, n(applications filed) = 239; n(decisions issued) = 187; n (violations found) = 62. The years between 1960 and 1974 are omitted to shorten the represented timespan, as no environmental cases arose in those years. The values given for 'n' indicate the sample size represented in the figure. The discrepancy between number of decisions issued and applicants filed arises because the court routinely joins applications and rules on them with one decision: therefore, one decision can resolve multiple and originally distinct applications.

(*continued*)

(*continued*)

Item	Methodology
	The dataset is limited to the year 2014 for the same reasons stated above in the description for Figure 4.1.
Figure 4.5	Number of times an Article was invoked in category one and the number of corresponding violations and settlements from 1960–2019, n(total invocations) = 512; n(violations and friendly settlements) = 94.
Figure 7.1	Number of times an Article was invoked in category two and the number of corresponding violations and friendly settlements, n (total invocations) = 298; n(violations and friendly settlements) = 72.
Figure 7.2	Number of environmental cases filed and violations found in final judgments per year in category two from 1960–2014, n(number of applications filed) = 166; n(decisions issued) = 140; n(violations found) = 47. The figure presents the total number of cases and the number of violations found in category two cases decided by the Convention entities per year from 1960 to 2014. Again, the dates for the years from 2014 onwards were omitted for the reasons outlined in the description for Figure 4.1. Two graphs give the linear trend for both the total number of cases and the number of cases where violations were found, respectively, along with the coefficient of determination (R^2 value).
Figure 7.3	Nature of environmental regulations challenged in category two cases, n(total) = 257; n(violations) = 92. Cases may fall into more than one category.

A.2 AFRICAN CHARTER ON HUMAN AND PEOPLES' RIGHTS

Item	Methodology
Parent List	Initial datasets were obtained through the search engine provided by the Institute for Human Rights and Development (IHRDA) (http://caselaw.ihrda.org). The following search parameters and filters were employed in that search: • all document collections • all language versions • no date limitation • keywords (any of these sufficed for a hit, i.e. the '*or*' option was selected): 'Economic, social and cultural rights'; 'Indigenous peoples'; 'Natural resources'; 'Right to a healthy environment'; 'Right to food'; 'Right to health'; 'Right to life'; 'None'. • Mechanism: 'African Commission on Human and Peoples' Rights'; 'African Court on Human and Peoples' Rights' This returned approximately 61 hits. If the same case was considered by different Charter entities over time (AfComHR, AfCtHR) only the final ruling was considered. The decisions were manually examined to confirm the existence of an environmental issue within the meaning outlined in the book. Cases that were somehow linked to the environment, but where no party made any environmental claims were excluded. Upon confirmation, the cases were summarised and categorised into category one and category two according to the definitions set out under the ECHR. Data points were recorded in the same manner. The list includes all cases available on 31 December 2019.

A.3 AMERICAN CONVENTION ON HUMAN RIGHTS/AMERICAN DECLARATION OF THE RIGHTS AND DUTIES OF MAN

Item	Methodology
Parent List	Initial datasets were obtained through the search engine Buscador Jurídico de Derechos Humanos (www.bjdh.org.mx/interamericano/busqueda). The following search parameters and filters were employed in that search: • all document collections • all language versions • no date limitation • keyword: 'medio ambiente' or 'environment' This returned approximately 34 hits. If the same case was considered by different Convention entities over time (IAComHR, IACtHR) only the final ruling was counted. The decisions were then manually examined to confirm the existence of an environmental issue within the meaning of the book. Upon confirmation, the cases were summarised and categorised into category one and category two according to the definition set out under the ECHR. Data points were recorded in the same manner. The list includes all cases available through the search engine on 31 December 2019.

A.4 UN COVENANTS

Item	Methodology
Parent List	Initial datasets were obtained through the search engine OHCHR Jurisprudence (http://juris.ohchr.org/). The following search parameters and filters were employed in that search: • no date limitation • Body: CCPR; CESCR • Articles: 1, 6, 14, 17, 26, 27 of CCPR; no filtering by Article for the CESCR • keyword: 'climate change', 'global warming', 'pollution', 'environment', 'nuclear', or 'ecosystem' Due to technical limitations, the searches for the CCPR and CESCR case law had to be conducted separately. The search for CCPR cases returned approximately 125 and the search for CESCR cases returned approximately 4 hits. The decisions were then manually examined to confirm the existence of an environmental issue. Upon confirmation, the cases were summarised and categorised into category one and category two according to the definitions set out under the ECHR. Data points were recorded in the same manner. The list includes all cases available through the search engine on 31 December 2019.

Bibliography

REPORTS

25th Session of the Human Rights Council (3–28 March 2014) Resolution – Human Rights and the Environment (UN A/HRC/25/L31)

'Ambient (Outdoor) Air Quality and Health', *World Health Organization* (2014) www.who.int/mediacentre/factsheets/fs313/en/, accessed 1 April 2021

Annual Report 2018 (Inter-American Commission on Human Rights, 2018)

Asbestos – Final Report of the Advisory Committee, vol 1 (HMSO 1979)

Atmospheric Ozone, 1985: Assessment of Our Understanding of the Processes Controlling Its Present Distribution and Change (World Meteorological Organization, 1985)

'Babies' Brains Damaged by Pollution, UNICEF Says' *BBC News* (6 December 2017) www.bbc.co.uk/news/health-42250558, accessed 1 April 2021

Bericht der Gemeinsamen Verfassungskommission (Bundestag Drucksache 12/6000, 1993)

Beschlussempfehlung und Bericht – 5. Untersuchungsausschuss der 18. Wahlperiode (Deutscher Bundestag, 2017)

'Brussels Ordered to Get Real on Its Pollution Problem', *Politico* (27 June 2019) www.politico.eu/article/brussels-ordered-to-get-real-on-its-pollution-problem/, accessed 1 April 2021

Burden of Disease from Ambient Air Pollution for 2012 (World Health Organization, 2014)

'CDC Concludes Zika Causes Microcephaly and Other Birth Defects', *Center for Disease Control and Prevention* (13 April 2016) www.cdc.gov/media/releases/2016/s0413-zika-microcephaly.html, accessed 1 April 2021

Climate Change 2014 – Synthesis Report (United Nations Intergovernmental Panel on Climate Change, 2014)

Collected Edition of the 'Travaux préparatoires' of the European Convention on Human Rights – Consultative Assembly, Committee of Experts, Committee of Ministers, Legal Committee, Advisers to the Ministers, 1 October 1951–20 March 1952 (Martinus Nijhoff, vol 8, 1985)

'Commission Calls on Belgium to Comply with the EU Legislation on Air Quality' *European Commission* (8 November 2018) https://ec.europa.eu/commission/presscorner/detail/en/MEMO_18_6247, accessed 1 April 2021

'Commission Warns Germany, France, Spain, Italy and the United Kingdom of Continued Air Pollution Breaches' *European Commission* (Brussels, 15 February 2017) http://europa.eu/rapid/press-release_IP-17-238_en.htm, accessed 1 April 2021

Conclusion on the Peer Review of the Pesticide Risk Assessment of Confirmatory Data Submitted for the Active Substance Copper (I), Copper (II) Variants Namely Copper Hydroxide, Copper Oxychloride, Tribasic Copper Sulfate, Copper (I) Oxide, Bordeaux Mixture (European Food Safety Authority, 2013)

'Consumer's Guide to Asbestos', *Asbestos Nation* (2020) www.asbestosnation.org/facts/consumers-guide-to-asbestos/, accessed 1 April 2021

Cyanide in Drinking-Water – Background Document for Development of WHO Guidelines for Drinking-Water Quality (World Health Organization, 2003)

Daring Cities Make Bold Air Quality Commitment to Remove All Diesel Vehicles by 2025 – Mayors of Paris, Mexico City, Madrid and Athens Sign Air Quality Declaration at C40 Mayors Summit, 30 November–2 December 2016 (C40 Mayors Summit, Mexico City, 2016)

Economic Cost of the Health Impact of Air Pollution in Europe (World Health Organization, Regional Office for Europe, 2015)

Effects of Nuclear War on Health and Health Services (World Health Organization, 1987)

Environmental Law and Local Government in South Africa (du Plessis A ed., Juta 2016)

Evaluation of Certain Food Additives (World Health Organization – Food and Agriculture Organization of the United Nations, 2014)

'Fahrverbot für Dieselautos in Stuttgart' *Zeit* (21 February 2017) www.zeit.de/mobilitaet/2017-02/feinstaub-stuttgart-dieselfahrzeuge-schadstoffbelastung-fahrverbot, accessed 1 April 2021

General Comment No. 36 (2018) on Article 6 of the International Covenant on Civil and Political Rights, on the Right to Life, Adopted by the Human Rights Committee on 30 October 2018 (CCPR/C/GC/36, 2000)

Global Environment Outlook – Geo-6 Assessment for the Pan-European Region (United Nations Environment Programme, 2016)

Guidelines for Drinking-Water Quality (World Health Organization, 2011)

'How Brazil's "Valley of Death" drastically cut pollution' *BBC News* (9 March 2017) www.bbc.co.uk/news/science-environment-39186018, accessed 1 April 2021

Joint FAO/WHO Meeting on Pesticide Residues, Geneva 9–13 May 2016 – Summary Report (World Health Organization – Food and Agriculture Organization of the United Nations, 2016)

'Legislative resolution on the proposal for a regulation of the European Parliament and of the Council setting emission performance standards for new passenger cars as part of the Community's integrated approach to reduce CO_2 emissions from light-duty vehicles', *European Parliament COM(2007)0856 – C6–0022/2008 – 2007/0297 (COD)* (17 December 2008)

Luftreinhalteplan für die Stadt München (Bayerisches Staatsministerium für Umwelt und Verbraucherschutz, 2019)

National Asbestos Profile for Germany (Federal Institute for Occupational Safety and Health, 2014)

National Planning Policy Framework (Department for Communities and Local Government, 2012)

'Netherlands Ordered to Cut Greenhouse Gas Emissions' *BBC News* (24 June 2016) www.bbc.co.uk/news/world-europe-33253772, accessed 1 April 2021

Neue Meta-Analyse zu glyphosathaltigen Pflanzenschutzmitteln ändert die Bewertung des Wirkstoffs nicht – Stellungnahme Nr 008/2019 (Bundesinstitut für Risikobewertung, 2019).

'New Brussels Air Pollution Figures Mask Health Risk Reality' *ClientEarth* (3 October 2019) www.clientearth.org/press/new-brussels-air-pollution-figures-mask-health-risk-reality/, accessed 1 April 2021

'Paris Imposes Car Restrictions to Fight Pollution' *BBC News* (23 March 2015) www.bbc.co.uk/news/world-europe-32016730, accessed 1 April 2021

Pesticide Risk Assessment of the Active Substance Glyphosate (European Food Safety Authority, 2015)

'President Trump Announces U.S. Withdrawal from the Paris Climate Accord' *The White House* (Washington, DC, 1 June 2017) www.whitehouse.gov/blog/2017/06/01/president-donald-j-trump-announces-us-withdrawal-paris-climate-accord, accessed 1 April 2021

'Proposal for a Regulation of the European Parliament and of the Council setting emission performance standards for new passenger cars as part of the Community's integrated approach to reduce CO_2 emissions from light-duty vehicles', *European Commission Communication COM/2007/0856* (19 December 2007)

Report of the Conference of the Parties on Its 21st Session Held in Paris, 30 November to 13 December 2015 (United Nations, FCCC/CP/2015/10/Add1, 2016)

Report of the World Commission on Environment and Development – Our Common Future (1987)

Review of Evidence on Health Aspects of Air Pollution – REVIHAAP Project (World Health Organization, 2013)

Rio Declaration on Environment and Development, UNCED Doc A/CONF/151/5/Rev. 1 (United Nations Conference on Environment and Development, 1992)

Safety and Efficacy of Ethoxyquin (6-ethoxy-1,2-dihydro-2,2,4-trimethylquinoline) for All Animal Species (European Food Safety Authority Panel on Additives and Products or Substances used in Animal Feed, 2015)

Updated Projections for Nitrogen Dioxide (NO_2) Compliance (Department for Environment Food & Rural Affairs, 2014)

'Volkswagen "to Pay $10bn" for US Emissions Scandal' *BBC News* (23 June 2016) http://www.bbc.co.uk/news/business-36611152, accessed 1 April 2021

'Volkswagen: The Scandal Explained' *BBC News* (10 December 2015) http://www.bbc.co.uk/news/business-34324772, accessed 1 April 2021

Water Fluoridation – Health Monitoring Report for England 2014 (Public Health England, 2014)

Water Fluoridation Basics (Centers for Disease Control and Prevention, 2016)

'Where Can You Find Asbestos?' (*Health and Safety Executive*, 2017) www.hse.gov.uk/asbestos/building.htm, accessed 1 April 2021

WHO Air Quality Guidelines for Particulate Matter, Ozone, Nitrogen Dioxide and Sulfur Dioxide (World Health Organization, 2005)

'WHO Chief: Air Pollution Link to 600,000 Deaths in Children' *BBC News* (6 March 2017) www.bbc.co.uk/news/health-39175410, accessed 1 April 2021

'WHO Global Urban Ambient Air Pollution Database' (*World Health Organization*, 2016) www.who.int/phe/health_topics/outdoorair/databases/cities/en/, accessed 1 April 2021

SCHOLARSHIP

Abram C and Chayes AH, *The New Sovereignty: Compliance with International Regulatory Agreements* (Harvard University Press 1995)

Adler JH, 'More Sorry Than Safe: Assessing the Precautionary Principle and the Proposed International Biosafety Protocol' (2000) 35 *Texas International Law Journal* 173

Aguilar CD and others, *The Effects of Climate Change on the Full Enjoyment of Human Rights* (United Nations Human Rights Council, 2015)

Agusto A and Trindade C, 'Principle 15: Precaution' in Viñuales JE (ed.), *The Rio Declaration on Environment and Development* (Oxford University Press 2015)

Ahrens H-J, '§ 402' in Wieczorek B and Schütze RA (eds.), *Zivilprozessordnung und Nebengesetze: Großkommentar*, vol 6 (4th ed., De Gruyter 2015)

Aiken W, 'Human Rights in an Ecological Era' (1992) 1 *Environmental Values* 191

Alam S and others (eds.), *Routledge Handbook of International Environmental Law* (Routledge 2013)

Alcamo J and others, 'Europe' in Parry ML and others (eds.), *Climate Change 2007: Impacts, Adaptation and Vulnerability*, vol 4 (Cambridge University Press 2007)

Alexy R, 'Ermessensfehler' (1986) 41 *JuristenZeitung* 701

Alfredsson G and Ovsiouk A, 'Human Rights and the Environment' (1991) 60 *Nordic Journal of International Law* 19

Allan TRS, *Constitutional Justice: A Liberal Theory of the Rule of Law* (Oxford University Press 2003)

Law, Liberty, and Justice: The Legal Foundations of British Constitutionalism (Oxford University Press 1995)

Alston P, 'Conjuring up New Human Rights: A Proposal for Quality Control' (1984) 78 *American Journal of International Law* 607

Alston P, Goodman R, and Steiner HJ, *International Human Rights: Text and Materials* (Oxford University Press 2013)

Amos M, *Human Rights Law* (2nd ed., Hart Publishing 2014)

Arblaster L and others, 'Occupational and Environmental Links to Mesothelioma Deaths Occurring in Leeds during 1971–1987' (1995) 17 *Journal of Public Health Medicine* 297

Auribault M, *Note sur L'hygiène et la Sécurité des Ouvriers Dans les Filatures et Tissages D'amiante* (Bulletin de l'inspection du travail, 1906)

Bakker C, 'Children's Rights Challenged by Climate Change: Is a Reconceptualization Required?' in Lenzerini F and Vrdoljak AF (eds.), *International Law for Common Goods: Normative Perspectives on Human Rights, Culture and Nature* (Hart 2014)

Baris Y, 'The Clinical and Radiological Aspects of 185 Cases of Malignant Pleural Mesothelioma' [1979] IARC Scientific Publications 937

Barral V and Dupuy P-M, 'Principle 4: Sustainable Development through Integration' in Viñuales JE (ed.), *The Rio Declaration on Environment and Development* (Oxford University Press 2015)

Bartrip P, 'Too Little, Too Late? The Home Office and the Asbestos Industry Regulations, 1931' (1998) 42 *Medical History* 421

Batekas I and Oette L, *International Human Rights Law and Practice* (Cambridge University Press 2013)

Baur X, 'Asbestos-Related Disorders in Germany: Background, Politics, Incidence, Diagnostics and Compensation' (2018) 15 *International Journal of Environmental Research and Public Health* 143

Beckermann W, 'Sustainable Development and Our Obligations to Future Generations' in Dobson A (ed.), *Fairness and Futurity: Essays on Environmental Sustainability and Social Justice* (Oxford University Press 1999)

Beitz CR, *The Idea of Human Rights* (Oxford University Press 2011)

Bell S, McGillivray D, and Pedersen OW, *Environmental Law* (8th ed., Oxford University Press 2013)

Benvenisti E, 'The Margin of Appreciation, Subsidiarity, and Global Challenges to Democracy' [2016] Global Trust Working Paper

Berglund B, Lindvall T, and Schwela DH, *Guidelines for Community Noise* (World Health Organization, 1999)

Bernsdorff N, 'Artikel 20a' in Clemens T and Umbach DC (eds.), *Grundgesetz Kommentar*, vol 1 (C. F. Müller 2002)

Beyerlin U, 'Sustainable Development' in Wolfrum R (ed.), *Max Planck Encyclopedia of Public International Law* (Oxford University Press 2013)

'Umweltschutz und Menschenrechte' (2005) 65 *Zeitschrift für ausländisches öffentliches Recht und Völkerrecht* 525

Beyerlin U and Marauhn T, *International Environmental Law* (Hart Publishing – C. H. Beck – Nomos 2011)

Birnie PW, Boyle AE, and Redgwell C, *International Law and the Environment* (3rd ed., Oxford University Press 2009)

Bjørge E, *The Evolutionary Interpretation of Treaties* (Oxford University Press 2014)

Bodansky D, *The Art and Craft of International Environmental Law* (Harvard University Press 2010)

'Deconstructing the Precautionary Principle' in Caron DD and Scheiber HN (eds.), *Bringing New Law to Ocean Waters* (Brill 2004)

Bookchin M, *The Ecology of Freedom – The Emergence and Dissolution of Hierarchy* (Cheshire Books 1982)

Bosselmann K, *When Two Worlds Collide: Society and Ecology* (Rsvp Pub 1995)

Bosselmann K and Richardson B, 'Introduction: New Challenges for Environmental Law and Policy' in Bosselmann K and Richardson B (eds.), *Environmental Justice and Market Mechanisms* (Kluwer Law International 1999)

Boyd DR, *The Environmental Rights Revolution – A Global Study of Constitutions, Human Rights and the Environment* (University of British Columbia Press 2012)

Right to a Healthy Environment: Good Practices (UN A/HRC/43/53, 2020)

Boyle A, 'Human Rights and the Environment – Where Next?' in Boer B (ed.), *Environmental Law Dimensions of Human Rights* (Oxford University Press 2015)

'Environment and Human Rights' in Wolfrum R (ed.), *Max Planck Encyclopedia of Public International Law* (Oxford University Press 2015)

'The Role of International Human Rights Law in the Protection of the Environment' in Anderson MR and Boyle AE (eds.), *Human Rights Approaches to Environmental Protection* (Clarendon 1996)

Brandl E and Bungert H, 'Constitutional Entrenchment of Environmental Protection: A Comparative Analysis of Experiences Abroad' (1992) 16 *Harvard Environmental Law Review* 1

Brennan A and Yeuk-Sze L, 'Environmental Ethics' in Zalta EN (ed.), *The Stanford Encyclopedia of Philosophy* (Standford University Press 2015)

Caflisch L, 'Rechtsfolgen von Normenkontrollen – Die Rechtsprechung des EGMR: Die Technik der Pilot-Fälle' [2006] Europäische Grundrechtezeitschrift 521

Carrington D, 'London Breaches Annual Air Pollution Limit for 2017 in Just Five Days' *The Guardian* (6 January 2017) www.theguardian.com/environment/2017/jan/06/london-breaches-toxic-air-pollution-limit-for-2017-in-just-five-days#comment-90624590, accessed 1 April 2021

Carter SM and Chapman S, 'Smoking, Disease, and Obdurate Denial: The Australian Tobacco Industry in the 1980s' (2003) 12 *Tobacco Control* 23

Chayes A, 'The Role of the Judge in Public Law Litigation' (1976) 89 *Harvard Law Review* 1281

Chazan D, 'How Paris Is Stepping Up Its Drive against the Car' *BBC News* (2 May 2016) www.bbc.co.uk/news/magazine-36169815, accessed 1 April 2021

Christiansen EC, 'Empowerment, Fairness, Integration: South African Answers to the Question of Constitutional Environmental Rights' (2013) 32 *Stanford Environmental Law Journal* 215

Churchill RR, 'Environmental Rights in Existing Human Rights Treaties' in Anderson MR and Boyle AE (eds.), *Human Rights Approaches to Environmental Protection* (Clarendon 1996)

Cockburn H, 'Turkey Coup: Erdogan Bans All Academics from Leaving Country as Government Crackdown Intensifies' *Independent* (20 July 2016) www.independent.co.uk/news/world/europe/turkey-coup-erdogan-academics-ban-leaving-country-government-crackdown-latest-a7146591.html, accessed 1 April 2021

Cole Matthew A, 'Corruption, Income and the Environment: An Empirical Analysis' (2007) 62 *Ecological Economics* 637

Cook J and others, 'Quantifying the Consensus on Anthropogenic Global Warming in the Scientific Literature' (2013) 8 *Environmental Research Letters* 024024

Cooke WE, 'Fibrosis of the Lungs due to the Inhalation of Asbestos Dust' (1924) 2 *British Medical Journal* 147

Craig P, 'Formal and Substantive Conceptions of the Rule of Law: An Analytical Framework' [1997] *Public Law* 467

Cranor CF, *Toxic Torts: Science, Law, and the Possibility of Justice* (2nd ed., Cambridge University Press 2016)

Crawford J, *Brownlie's Principles of Public International Law* (9th ed., Oxford University Press 2019)

Crutzen PJ and Stoermer EF, 'The Anthropocene' (2000) 41 *Global Change Newsletter* 17

Dalton A, *Asbestos Killer Dust* (British Society for Social Responsibility in Science 1979)

Daly E and May J, 'Bridging Constitutional Dignity and Environmental Rights Jurisprudence' (2016) 7 *Journal of Human Rights and the Environment* 16

Darnton A, *Mesothelioma in Great Britain – Mesothelioma Mortality in Great Britain 1968 –2015* (Health and Safety Executive 2016) de Graaf KJ and Jans JH, 'The Urgenda Decision: Netherlands Liable for Role in Causing Dangerous Global Climate Change' (2015) 27 *Journal of Environmental Law* 517

De Schutter O, *International Human Rights Law – Cases, Materials, Commentary* (2nd ed., Cambridge University Press 2014)

de Shalit A, *Why Does Posterity Matter* (Routledge 1994)

Deane L, *Report on the Health of Workers in Asbestos and Other Dusty Trades* (HM Chief Inspector of Factories and Workshops, 1899, 1899)

Dèjeant-Pons M, 'Le Droit de L'homme à L'environnement et la Convention Européenne de Sauvergarde des Droits L'homme et des Libertés Fondamentales' in Cohen-Jonathan G, Flauss J-F, and Lambert P (eds.), *Liber Amicorum Marc-Andre Eissen* (Bruylant 1995)

DeMerieux M, 'Deriving Environmental Rights from the European Convention for the Protection of Human Rights and Fundamental Freedoms' (2001) 21 *Oxford Journal of Legal Studies* 521

Desgagné R, 'Integrating Environmental Values into the European Convention on Human Rights' (1995) 89 *American Journal of International Law* 263

Docherty B and Giannini T, 'Confronting a Rising Tide: A Proposal for a Convention on Climate Change Refugees' (2009) 33 *Harvard Environmental Law Review* 349–403

Donnelly C, 'Third Generation Rights' in Brölmann C, Lefeber R, and Zieck M (eds.), *Peoples and Minorities in International Lw* (Martinus Nijhoff 1993)

Donnelly CA and others, 'Positive and Negative Effects of Widespread Badger Culling on Tuberculosis in Cattle' (2006) 439 *Nature* 843

Donnelly J, 'Human Rights as Natural Rights' (1982) 4 *Human Rights Quarterly* 391

Donoghue A, 'Occupational Health Hazards in Mining: An Overview' (2004) 54 *Occupational Medicine* 283

Douglas-Scott S, 'Environmental Rights in the European Union – Participatory Democracy or Democratic Deficit?' in Anderson MR and Boyle AE (eds.), *Human Rights Approaches to Environmental Protection* (Clarendon 1996)

Du Plessis A, 'South Africa's Constitutional Environmental Right (Generously) Interpreted: What Is in It for Poverty?' (2011) 27 *South African Journal on Human Rights* 279

Du Plessis L, 'Subsidiarity: What's in the Name for Constitutional Interpretation and Adjudication' (2006) 17 *Stellenbosch Law Review* 207

Dunlap RE and Jacques PJ, 'Climate Change Denial Books and Conservative Think Tanks Exploring the Connection' (2013) 57 *American Behavioral Scientist* 699

Dunlap RE and McRight AM, 'Organized Climate Change Denial' in Dryzek JS, Norgaard RB, and Schlosberg D (eds.), *The Oxford Handbook of Climate Change and Society* (Oxford University Press 2011)

Dupuy P-M and Viñuales JE, *International Environmental Law* (Cambridge University Press 2015)

Dworkin R, *Freedom's Law: The Moral Reading of the American Constitution* (Oxford University Press 1996)

Justice for Hedgehogs (Harvard University Press 2011)

Law's Empire (Hart Publishing 1998)

A Matter of Principle (Harvard University Press 1985)

Ebbesson J, 'Public Participation' in Bodansky D, Brunnée J, and Hey E (eds.), *The Oxford Handbook of International Environmental Law* (Oxford University Press 2007)
Ekardt F, 'Umweltverfassung und Schutzpflichten' (2013) 17 *Neue Zeitschrift für Verwaltungsrecht* 1105
Ekeli KS, 'Green Constitutionalism: The Constitutional Protection of Future Generations' (2007) 20 *Ratio Juris* 378
Elliott M and Thomas R, *Public Law* (3rd ed., Oxford University Press 2017)
Ely JH, *Democracy and Distrust: A Theory of Judicial Review* (Harvard University Press 1980)
Epiney A, 'Artikel 20a' in von Mangoldt H, Klein F, and Stark C (eds.), *Kommentar zum Grundgesetz*, vol 2 (6th ed., Franz Vahlen 2010)
Farber DA, 'Basic Compensation for Victims of Climate Change' (2007) 155 *University of Pennsylvania Law Review* 1605
Faßbender K, 'Der Dieselskandal und der Gesundheitsschutz' [2017] *Neue Juristische Wochenschrift* 1995
Fawell J and others, Fluoride in Drinking-Water, World Health Organization (2006) www.who.int/water_sanitation_health/publications/fluoride_drinking_water_full.pdf, accessed 1 April 2021
Feldman D, *Civil Liberties and Human Rights in England and Wales* (2nd ed., Oxford University Press 2002)
Feris L, 'Constitutional Environmental Rights: An Under-utilised Resource' (2008) 24 *South African Journal on Human Rights* 29
Fisher DE, 'Jurisprudential Challenges to the Protection of the Natural Environment' in Maloney MM and Burdon P (eds.), *Wild Law: In Practice* (Routledge 2014)
Fisher E, 'Precaution, Precaution Everywhere: Developing a Common Understanding of the Precautionary Principle in the European Community' (2002) 9 *Maastricht Journal of European and Comparative Law* 7
Risk Regulation and Administrative Constitutionalism (Hart Publishing 2007)
Fisher E, Scotford E, and Barritt E, 'The Legally Disruptive Nature of Climate Change' (2017) 80 *Modern Law Review* 173
Fiss OM, 'The Supreme Court, 1978 Term' [1979] *Harvard Law Review* 1
Fitz-Henry E, 'Decolonizing Personhood' in Maloney MM and Burdon P (eds.), *Wild Law: In Practice* (Routledge 2014)
Flanagan DM, 2015 *Minerals Yearbook – Asbestos* (United States Geological Survey 2016)
Føllesdal A, 'Appreciating the Margin of Appreciation' in Etinson A (ed.), *Human Rights: Moral or Political?* (Oxford University Press 2018).
Forsyth CF, *Judicial Review and the Constitution* (Hart Publishing 2000)
Foster CE, *Science and the Precautionary Principle in International Courts and Tribunals: Expert Evidence, Burden of Proof and Finality* (Cambridge University Press 2011)
Francioni F, 'International Human Rights in an Environmental Horizon' (2010) 21 *European Journal of International Law* 41
'Principle 1: Human Beings and the Environment' in Viñuales JE (ed.), *The Rio Declaration on Environment and Development* (Oxford University Press 2015)

Frank AL and Joshi TK, 'The Global Spread of Asbestos' (2014) 80 *Annals of Global Health* 257

Freeman J and Kolstad CD, 'Prescriptive Environmental Regulations versus Market-Based Incentives' in Freeman J and Kolstad CD (eds.), *Moving to Markets in Environmental Regulation: Lessons from Twenty Years of Experience* (Oxford University Press 2006)

Freestone D and Hey E, 'Origins and Development of the Precautionary Principle' in Freestone D and Hey E (eds.), *The Precautionary Principle and International Law: The Challenge of Implementation* (Kluwer Law International 1996)

Friedman L, 'Testing the Limits: Judicial Enforcement of Positive State Constitutional Rights' (2015) 53 *Duquesne Law Review* 437

Fritschi L and others, 'Carcinogenicity of Tetrachlorvinphos, Parathion, Malathion, Diazinon, and Glyphosate' (2015) 16 *The Lancet* 490

Führ M, 'Der Dieselskandal und das Recht – Ein Lehrstück zum technischen Sicherheitsrecht' (2017) 36 *Neue Zeitschrift für Verwaltungsrecht* 265

Gutachterliche Stellungnahme (Öffentliche Sachverständigenanhörung des 5 Untersuchungsausschusses der 18 Wahlperiode, 2016)

Fuller LL and Winston KI, 'The Forms and Limits of Adjudication' (1978) 92 *Harvard Law Review* 353

Garcia San José D, *Environmental Protection and the European Convention on Human Rights* (Council of Europe Publishing 2005)

Gearty C, 'Do Human Rights Help or Hinder Environmental Protection?' (2010) 1 *Journal of Human Rights and the Environment* 7

Gee D and Greenberg M, 'Asbestos: From 'Magic' to Malevolent Mineral' in Harremoes P and others (eds.), *The Precautionary Principle in the 20th Century: Late Lessons from Early Warnings* (Routledge 2002)

Gellers JC, *The Global Emergence of Constitutional Environmental Rights* (Routledge 2017)

Genn H and Thomas C, *Tribunal Decision-Making: An Empirical Study* (Nuffield Foundation – UCL Judicial Institute, 2013)

Gerber P and Chen B, 'Recognition of the Human Right to Water – Has the Tide Turned?' (2011) 36 *Alternative Law Journal* 21

Giagnocavo C and Goldstein H, 'Law Reform or World Re-form: The Problem of Environmental Rights' (1989) 35 *McGill Law Journal* 345

Gillman H, 'What's Law Got to Do with It? Judicial Behavioralists Test the "Legal Model" of Judicial Decision Making' (2001) 26 *Law & Social Inquiry* 465

Giorgetta S, 'The Right to a Healthy Environment, Human Rights and Sustainable Development' (2002) 2 *International Environmental Agreements* 171

Glazewski J, *Environmental Law in South Africa* (Butterworths 2000)

Goodkin SL, 'The Evolution of Animal Rights' (1986–1987) 18 *Columbia Human Rights Law Review* 259

Gowers AM, Miller BG, and Stedman JR, *Estimating Local Mortality Burdens associated with Particulate Air Pollution* (Public Health England 2014)

Grant E, 'Re-imagining Adjudication: Human Rights Courts and the Environment' in Grear A and Grant E (eds.), *Thought, Law, Rights and Action in the Age of Environmental Crisis* (Edward Elgar 2015)

Grear A, 'Towards "Climate Justice"? A Critical Reflection on Legal Subjectivity and Climate Injustice: Warning Signals, Patterned Hierarchies, Directions for Future Law and Policy' (2014) 5 *Journal of Human Rights and the Environment* 103

'Towards New Legal Futures? In Search of Renewing Foundations' in Grear A and Grant E (eds.), *Thought, Law, Rights and Action in the Age of Environmental Crisis* (Edward Elgar 2015)

'The Vulnerable Living Order: Human Rights and the Environment in a Critical and Philosophical Perspective' (2011) 2 *Journal of Human Rights and the Environment* 23

Greenberg M, 'Knowledge of the Health Hazard of Asbestos Prior to the Merewether and Price Report of 1930' (1994) 7 *The Society for the Social History of Medicine* 493

Griffin J, *On Human Rights* (Oxford University Press 2008)

Griffin S. O. and others, 'Effectiveness of Fluoride in Preventing Caries in Adults' (2007) 86 *Journal of Dental Research* 410

Haakonssen K, 'Republicanism' in Goodin RE, Pettit P, and Pogge T (eds.), *A Companion to Contemporary Political Philosophy* (2nd ed., Wiley-Blackwell 2012)

Habel E, 'Menschenwürde und natürliche Lebensgrundlagen' (1995) 17 *Natur und Recht* 165

Habermas J, *Faktizität und Geltung: Beiträge zur Diskurstheorie des Rechts und des Demokratischen Rechtsstaats* (2nd ed., Suhrkamp 1992)

'Remarks on Legitimation through Human Rights' (1998) 24 *Philosophy & Social Criticism* 157

Hamner S and others, 'The Role of Water Use Patterns and Sewage Pollution in Incidence of Water-Borne/Enteric Diseases along the Ganges River in Varanasi, India' (2006) 16 *International Journal of Environmental Health Research* 113

Handl G, 'Human Rights and Protection of the Environment, a Mildly "Revisionist" View' in Canrado Trindade AA (ed.), *Human Rights, Sustainable Development and the Environment* (Instituto Interamericano de Derechos Humanos 1992)

'Human Rights and Protection of the Environment' in Eide A, Krause C, and Rosas A (eds.), *Economic, Social and Cultural Rights – A Textbook* (2nd ed., Martinus Nijhoff 2001)

Harris DJ and others, *Harris, O'Boyle & Warbrick: Law of the European Convention on Human Rights* (3rd ed., Oxford University Press 2014)

Harrison J, 'Reflections on the Role of International Courts and Tribunals in the Settlement of Environmental Disputes and the Development of International Environmental Law' (2013) 25 *Journal of Environmental Law* 501

Hart D and Wheeler M, 'Night Flights and Strasbourg's Retreat from Environmental Human Rights' (2004) 16 *Journal of Environmental Law* 100

Hayward T, *Constitutional Environmental Rights* (Oxford University Press 2005)

Political Theory and Ecological Values (Polity 1998)

Hendryx M and Ahern MM, 'Relations between Health Indicators and Residential Proximity to Coal Mining in West Virginia' (2008) 98 *American Journal of Public Health* 669

Hilson C, 'The Margin of Appreciation, Domestic Irregularity and Domestic Court Rulings in ECHR Environmental Jurisprudence: Global Legal Pluralism in Action' (2013) 2 *Global Constitutionalism* 262

'Risk and the European Convention on Human Rights: Towards a New Approach' [2009] Working Paper

Hiskes RP, *The Human Right to a Green Future – Environmental Rights and Intergenerational Justice* (Cambridge University Press 2009)

Hobe S, 'Menschenrechte auf Umweltschutz? – Bestand und Begründungsmöglichkeiten' [1994] *Zeitschrift für Umweltrecht* 15

Hoffman FL, *Mortality from Respiratory Diseases in Dusty Trades (Inorganic Dusts)* (US Department of Labor, Bureau of Labor Statistics 1918)

Hoffmann-Riem W, 'Kontrolldichte und Kontrollfragen bei nationalen und europäischen Schutz von Freiheitsrechten in mehrpoligen Rechtsverhältnissen' [2006] *Europäische Grundrechtezeitschrift* 492

Howarth W, 'Environmental Human Rights and Parliamentary Democracy' (2002) 14 *Journal of Environmental Law* 377

Hsiao EC, 'Whanganui River Agreement' (2012) 42 *Environmental Policy and Law* 371

Humphreys S, 'Introduction: Human Rights and Climate Change' in Humphreys S (ed.), *Human Rights and Climate Change* (Cambridge University Press 2010)

Huster S and Rux J, 'Artikel 20a' in Epping V and Hillgruber C (eds.), *Grundgesetz Kommentar* (2nd ed., C. H. Beck 2013)

Hyam J, 'Hatton v United Kingdom in the Grand Chamber: One Step Forward, Two Steps Back?' [2003] *European Human Rights Law Review* 631

Jamieson D, *Morality's Progress: Essays on Humans, Other Animals, and the Rest of Nature* (Oxford University Press 2002)

Reason in a Dark Time – Why the Struggle Against Climate Change Has Failed – and What It Means for Our Future (Oxford University Press 2014)

Jarass HD and Pieroth B, *Grundgesetz für die Bundesrepublik Deutschland – Kommentar* (14th ed., C. H. Beck 2016)

Jasanoff S, 'A World of Experts: Science and Global Environmental Constitutionalism' (2013) 40 *Boston College Environmental Affairs Law Review* 439

Jeremy DJ, 'Corporate Responses to the Emergent Recognition of a Health Hazard in the UK Asbestos Industry: The Case of Turner & Newall, 1920–1960' (1995) 24 *Business and Economic History* 254

Johnson DL and others, 'Meanings of Environmental Terms' (1997) 26 *Journal of Environmental Quality* 581

Kaiser K, 'Artikel 1 ZP I' in Karpenstein U and Meyer FC (eds.), *Europäische Menschenrechtskonvention – Kommentar* (2nd ed., C. H. Beck 2015)

Karagulian F and others, 'Contributions to Cities' Ambient Particulate Matter (PM): A Systematic Review of Local Source Contributions at Global Level' (2015) 120 *Atmospheric Environment* 475

Karp DJ, *Responsibility for Human Rights: Transnational Corporations in Imperfect States* (Cambridge University Press 2014)

Katyal NK and Schmidt TP, 'Active Avoidance: The Modern Supreme Court and Legal Change' (2014) 128 *Harvard Law Review* 2109

Kelsen H, *Reine Rechtslehre: mit einem Anhang: das Problem der Gerechtigkeit* (2nd ed., Deuticke 1960)

Kidd M, 'Environment' in Currie I and de Waal J (eds.), *The Bill of Rights Handbook* (6th ed., Juta 2016)

King J, *Judging Social Rights* (Cambridge University Press 2012)

'Parliament's Role Following Declarations of Incompatibility under the Human Rights Act' in Hunt M, Hooper H, and Yowell P (eds.), *Parliaments and Human Rights: Redressing the Democratic Deficit* (Bloomsbury Publishing 2015)

'The Pervasiveness of Polycentricity' [2008] *Public Law* 101

Kleiber M, *Der Grundrechtliche Schutz Künftiger Generationen* (Mohr Siebeck 2014)

Klein O, 'Straßburger Wolken am Karlsruher Himmel – Zum geänderten Verhältnis zwischen Bundesverfassungsgericht und Europäischen Gerichtshof für Menschenrechte seit 1998' [2010] *Neue Zeitschrift für Verwaltungsrecht* 221

Klinger R, 'Dieselgate öffentlich-rechtlich – EG-Typengenehmigung, Übereinstimmungsbescheinigung und Konsequenzen für deutsche Pkw-Halter' (2017) 22 *Zeitschrift für Umweltrecht* 131

Kloepfer M, 'Droht der Autoritäre Ökologische Staat?' (1993) 48 *Universitas* 236

Klug H, *The Constitution of South Africa – A Contextual Analysis* (Hart Publishing 2010)

Knox JH, *Final Report of the Special Rapporteur on the Issue of Human Rights Obligations Relating to the Enjoyment of a Safe, Clean, Healthy and Sustainable Environment* (UN A/HRC/37/59, 2018)

Report of the Special Rapporteur on the Issue of Human Rights Obligations Relating to the Enjoyment of a Safe, Clean, Healthy and Sustainable Environment (UN A/HRC/34/49, 2017)

Koskenniemi M, 'Breach of Treaty or Non-Compliance? Reflections on the Enforcement of the Montreal Protocol' (1993) 3 *Yearbook of International Environmental Law* 123

Kotelchuck D, 'Asbestos: The Funeral Dress of Kings' – and Others' in Rosner D and Markowitz G (eds.), *Dying for Work: Workers' Safety and Health in Twentieth-Century America* (Indiana University Press 1987)

Kotzé LJ, *Global Environmental Constitutionalism in the Anthropocene* (Hart Publishing 2016)

'The Judiciary, the Environmental Right and the Quest for Sustainability in South Africa: A Critical Reflection' (2007) 16 *Review of European Community & International Environmental Law* 298

Kotzé LJ and du Plessis A, 'Some Brief Observations on Fifteen Years of Environmental Rights Jurisprudence in South Africa' (2010) 3 *Journal of Court Innovation* 157

Langford M and Russell AFS, *The Human Right to Water: Theory, Practice and Prospects* (Cambridge University Press 2017)

Laslett P and Fishkin JS, *Justice between Age Groups and Generations* (Yale University Press 1992)

Lauterpacht H, *International Law and Human Rights* (Stevens 1950)

Lee J, 'The Underlying Legal Theory to Support a Well-Defined Human Right to a Healthy Environment as a Principle of Customary International Law' (2000) 25 *Columbia Journal of Environmental Law* 283

Letsas G, 'The ECHR as a Living Instrument: Its Meaning and Legitimacy' in Føllesdal A, Peters B, and Ulfstein G (eds.), *Constituting Europe: The European Court of Human Rights in a National, European and Global Context* (Cambridge University Press 2013)

'The Margin of Appreciation Revisited: A Response to Føllesdal' in Etinson A (ed.), *Human Rights: Moral or Political?* (Oxford University Press 2018)

A Theory of Interpretation of the European Convention on Human Rights (Oxford University Press 2007)

'Two Concepts of the Margin of Appreciation' (2006) 4 *Oxford Journal of Legal Studies* 705

Lewis J, 'Lead Poisoning: A Historical Perspective' *EPA Journal* (1985), https://archive.epa.gov/epa/aboutepa/lead-poisoning-historical-perspective.html, accessed 1 April 2021

Light A, 'Callicott and Naess on Pluralism' (1996) 39 *Inquiry* 273

'Compatibilism in Political Ecology' in Katz E and Light A (eds.), *Environmental Pragmatism* (Routledge 1996)

Light A and Katz E, 'Introduction: Environmental Pragmatism and Environmental Ethics as Contested Terrain' in Katz E and Light A (eds.), *Environmental Pragmatism* (Routledge 1996)

Lilienfeld DE, 'The Silence: The Asbestos Industry and Early Occupational Cancer Research – A Case Study' (1991) 81 *American Journal of Public Health* 791

Lind EA and Tyler TR, *The Social Psychology of Procedural Justice* (Plenum Press 1988)

Lixinski L, 'Treaty Interpretation by the Inter-American Court of Human Rights: Expansionism at the Service of the Unity of International Law' (2010) 21 *European Journal of International Law* 585

Lovett F, *A Republic of Law* (Cambridge University Press 2016)

Ludwigs M, 'Kooperativer Grundrechtsschutz zwischen EuGH, BVerfG und EGMR' [2014] *Europäische Grundrechtezeitschrift* 273

Lundebye AK and others, 'Levels of Synthetic Antioxidants (Ethoxyquin, Butylated Hydroxytoluene and Butylated Hydroxyanisole) in Fish Feed and Commercially Farmed Fish' (2010) 27 *Food Additives & Contaminants* 1652

MacDonald KE, 'A Right to a Healthful Environment – Humans and Habitats' [2008] *European Energy and Environmental Law Review* 213

Macrory R, 'Environmental Citizenship and the Law: Repairing the European Road' (1996) 8 *Journal of Environmental Law* 219

Maher BA and others, 'Magnetite Pollution Nanoparticles in the Human Brain' (2016) 113 *Proceedings of the National Academy of Sciences* 10797

Mandel GN and Gathii JT, 'Cost-Benefit Analysis versus the Precautionary Principle: Beyond Cass Sunstein's Laws of Fear' [2006] *University of Illinois Law Review* 1037

Markowitz G, 'Cater to the Children: The Role of the Lead Industry in a Public Health Tragedy, 1900–1955' (2000) 90 *American Journal of Public Health* 36

Martin R, *A System of Rights* (Oxford University Press 1997)

May JR and Daly E, *Global Environmental Constitutionalism* (Cambridge University Press 2015)

Mazor J, 'Liberal Justice, Future People, and Natural Resource Conservation' (2010) 38 *Philosophy & Public Affairs* 380

McCrudden C and King J, 'The Dark Side of Nudging: The Ethics, Political Economy, and Law of Libertarian Paternalism' [2015] Queen's University Belfast Law Research Paper No 16

McIntyre O and Mosedale T, 'The Precautionary Principle as a Norm of Customary International Law' (1997) 9 *Journal of Environmental Law* 221

Merewether ERA and Price CW, *Report on Effects of Asbestos Dust on the Lungs and Dust Suppression in the Asbestos Industry. Part II. Processes Giving Rise to Dust and Methods for Its Suppression* (HMSO 1930)

Merrills JG, 'Environmental Rights' in Bodansky D, Brunnée J, and Hey E (eds.), *The Oxford Handbook of International Environmental Law* (Oxford University Press 2007)

Michaels D, *Doubt Is Their Product* (Oxford University Press 2008)

Miller C, 'Case Law Analysis: The European Convention on Human Rights: Another Weapon in the Environmentalist's Armoury' (1999) 11 *Journal of Environmental Law* 157

'Environmental Rights in a Welfare State? A Comment on DeMerieux' (2003) 23 *Oxford Journal of Legal Studies* 111

Environmental Rights: Critical Perspectives (Routledge 1998)

Misiedjan D, *Towards a Sustainable Human Right to Water: Supporting Vulnerable People and Protecting Water Resources* (Intersentia 2019)

Molina MJ and Rowland FS, 'Stratospheric Sink for Chlorofluoromethanes: Chlorine Atom-Catalysed Destruction of Ozone' (1974) 249 *Nature* 810

Montini M, 'The Double Failure of Environmental Regulation and Deregulation and the Need for Ecological Law' [2016] C-EENRG Working Papers 1

Montini M and Volpe F, 'Rethinking Regulation for Promoting an Ecologically Based Approach to Sustainability' [2016] C-EENRG Working Papers 1

Morgan J, 'Technological Change and the Development of Liability for Fault in England and Wales' in Martin-Casals M (ed.), *The Development of Liability in Relation to Technological Change*, vol 4 (Cambridge University Press 2010)

Morrow K, 'After the Honeymoon: The Uneasy Marriage of Human Rights and the Environment under the European Convention on Human Rights and in UK Law under the Human Rights Act 1998' (2013) 43 *Revue générale de droit* 317

'The ECHR, Environment-Based Human Rights Claims and the Search for Standards' in Turner SJ and others (eds.), *Environmental Rights: The Development of Standards* (Cambridge University Press 2019)

Murcott M and Van der Westhuizen W, 'The Ebb and Flow of the Application of the Principle of Subsidiarity – Critical Reflections on Motau and My Vote Counts' (2015) 7 *Constitutional Court Review* 43

Murray HM, *In Departmental Committee on Compensation for Industrial Diseases, 1906, Minutes of Evidence* (HMSO (Her Majesty's Stationery Office) 1907)

Murray R, 'Asbestos: A Chronology of Its Origins and Health Effects' (1990) 47 *British Journal of Industrial Medicine* 361

Murswiek D, 'Artikel 20a' in Sachs M (ed.), *Grundgesetz Kommentar* (7th ed., C. H. Beck 2014)

Naess A, 'The Shallow and the Deep, Long-Range Ecology Movement: A Summary.' (1973) 16 *Inquiry* 95

Neimanis A, 'Alongside the Right to Water, a Posthumanist Feminist Imaginary' (2014) 5 *Journal of Human Rights and the Environment* 5

Nelsen A, 'Dutch Government Ordered to Cut Carbon Emissions in Landmark Ruling' *The Guardian*, (The Hague, 24 June 2015) www.theguardian.com/environment/2015/jun/24/dutch-government-ordered-cut-carbon-emissions-landmark-ruling, accessed 1 April 2021

Newman P and others, 'What Would Have Happened to the Ozone Layer if Chlorofluorocarbons (CFCs) Had Not Been Regulated?' (2009) 9 *Atmospheric Chemistry and Physics* 2113
Nickel JW, 'The Human Right to a Safe Environment: Philosophical Perspectives on Its Scope and Justification' (1993) 18 *Yale Journal of International Law* 281
Making Sense of Human Rights (2nd ed., Blackwell 2007)
Norton BG, 'Environmental Ethics and Weak Anthropocentrism' (1984) 6 *Environmental Ethics* 131
Toward Unity among Environmentalists (Oxford University Press 1991)
O'Neill J, 'The Varieties of Intrinsic Value' (1992) 75 *Monist* 119
O'Regan K, 'Text Matters: Some Reflections on the Forging of a New Constitutional Jurisprudence in South Africa' (2012) 75 *Modern Law Review* 1
Ozonoff D, 'Failed Warnings: Asbestos-Related Disease and Industrial Medicine' in Bayer R (ed.), *Case Studies in the Politics of Professional Responsibility* (Oxford University Press 1988)
Page EA, *Climate Change, Justice and Future Generations* (Edward Elgar Publishing 2006)
Pallemaerts M, 'The Human Right to a Healthy Environment as a Substantive Right' in Déjeant-Pons M and Pallemaerts M (eds.), *Human Rights and the Environment* (Council of Europe 2002)
Park C (ed.), A Dictionary of Environment and Conservation (Oxford University Press 2007)
Passmore J, *Man's Responsibility for Nature* (2nd ed., Duckworth 1980)
Pätzold J, 'Artikel 8 EMRK' in Karpenstein U and Meyer FC (eds.), *Europäische Menschenrechtskonvention – Kommentar* (2nd ed., C. H. Beck 2015)
Pedersen OW, 'From Abundance to Indeterminacy: The Precautionary Principle and Its Two Camps of Custom' (2014) 3 *Transnational Environmental Law* 323
Peeters M, 'Urgenda Foundation and 886 Individuals v. The State of the Netherlands: The Dilemma of More Ambitious Greenhouse Gas Reduction Action by EU Member States' (2016) 25 *Review of European, Comparative & International Environmental Law* 123
Pettit P, *Republicanism: A Theory of Freedom and Government* (Oxford University Press 1997)
Posner EA, 'Climate Change and International Human Rights Litigation: A Critical Appraisal' (2007) 155 *University of Pennsylvania Law Review* 1925
Preston BJ, 'Internalizing Ecocentrism in Environmental Law' in Maloney MM and Burdon P (eds.), *Wild Law: In Practice* (Routledge 2014)
Proctor RN, 'The History of the Discovery of the Cigarette–Lung Cancer Link: Evidentiary Traditions, Corporate Denial, Global Toll' (2012) 21 *Tobacco Control* 87
Raaschou-Nielsen O and others, 'Air Pollution and Lung Cancer Incidence in 17 European Cohorts: Prospective Analyses from the European Study of Cohorts for Air Pollution Effects (ESCAPE)' (2013) 14 *The Lancet Oncology* 813
Rawls J, *Political Liberalism* (Columbia University Press 1993)
Raz J, *The Authority of Law: Essays on Law and Morality* (Oxford University Press 1979)
Ethics in the Public Domain: Essays in the Morality of Law and Politics (Clarendon 1994)

The Morality of Freedom (Oxford University Press 1986)
Redgwell C, 'Life, the Universe and Everything: A Critique of Anthropocentric Rights' in Anderson MR and Boyle AE (eds.), *Human Rights Approaches to Environmental Protection* (Clarendon 1996)
Regan T, *The Case for Animal Rights* (Routledge 1983)
Regan T and Singer P, *Animal Rights and Human Obligations* (Prentice-Hall 1976)
Rehbinder E, 'States between Economic Deregulation and Environmental Responsibility' in Bosselmann K and Richardson B (eds.), *Environmental Justice and Market Mechanisms* (Kluwer Law International 1999)
Rest A, 'Europäischer Menschenrechtsschutz als Katalysator für ein verbessertes Umweltrecht' (1997) 19 *Natur und Recht* 209
Rice J, 'The Global Reorganization and Revitalization of the Asbestos Industry, 1970–2007' (2011) 41 *International Journal of Health Services* 239
Rincon P, 'Car Makers 'Failing' on Climate' *BBC News* (20 April 2006) http://news.bbc.co.uk/1/hi/sci/tech/4922614.stm, accessed 1 April 2021
Rivas I and others, 'Source Apportionment of Particle Number Size Distribution in Urban Background and Traffic Stations in Four European Cities' (2020) 135 *Environment International* 105345
Rolston H, *A New Environmental Ethics: The Next Millennium for Life on Earth* (Routledge 2012)
Rosenthal SB and Buchholz RA, 'How Pragmatism Is an Environmental Ethic' in Katz E and Light A (eds.), *Environmental Pragmatism* (Routledge 1996)
Sadurski W, 'Partnering with Strasbourg: Constitutionalisation of the European Court of Human Rights, the Accession of Central and East European States to the Council of Europe, and the Idea of Pilot Judgments' 9 *Human Rights Law Review* 397
Saloojee Y and Dagil E, 'Tobacco Industry Tactics for Resisting Public Policy on Health' (2000) 78 *Bulletin of the World Health Organization* 902
Sand PH, 'The Evolution of International Environmental Law' in Bodansky D, Brunnée J, and Hey E (eds.), *The Oxford Handbook of International Environmental Law* (Oxford University Press 2007)
Sands P and others, *Principles of International Environmental Law* (3rd ed., Cambridge University Press 2012)
Principles of International Environmental Law (4th ed., Cambridge University Press 2018)
Scalia A, *A Matter of Interpretation: Federal Courts and the Law* (Gutmann A ed., Princeton University Press 1997)
Scanlon J, Cassar A, and Nemes N, 'Water as a Human Right?' [2004] IUCN Environmental Policy and Law Paper
Schabas WA, *The European Convention on Human Rights – A Commentary* (Oxford University Press 2015)
Schäfer P, 'Artikel 34 EMRK' in Karpenstein U and Meyer FC (eds.), *Europäische Menschenrechtskonvention – Kommentar* (2nd ed., C. H. Beck 2015)
Schmahl S, 'Piloturteile des EGMR als Mittel der Verfahrensbeschleunigung' [2008] *Europäische Grundrechtezeitschrift* 369
Scholz R, 'Artikel 20a' in Maunz T and others (eds.), *Grundgesetz Loseblatt Kommentar* (76th ed., C. H. Beck 2016)

Schröder M, 'Precautionary Approach/Principle' in Wolfrum R (ed.), *Max Planck Encyclopedia of Public International Law* (Oxford University Press 2014)
Schulze-Fielitz H, 'Artikel 20a' in Dreier H (ed.), *Grundgesetz Kommentar*, vol 2 (3rd ed., Mohr Siebeck 2015)
Sekrafi H and Sghaier A, 'Examining the Relationship between Corruption, Economic Growth, Environmental Degradation, and Energy Consumption: A Panel Analysis in MENA Region' [2016] *Journal of the Knowledge Economy* 1
Sen A, 'Elements of a Theory of Human Rights' (2004) 32 *Philosophy & Public Affairs* 315
Shelton D, 'Environmental Rights' in Alston P (ed.), *Peoples' Rights* (Oxford University Press 2001)
'Human Rights and the Environment: Substantive Rights' in Fitzmaurice M, Ong DM, and Merkouris P (eds.), *Research Handbook on International Environmental Law* (Edward Elgar 2010)
'Human Rights and the Environment: What Specific Environmental Rights Have Been Recognized' (2006) 35 *Denver Journal of International Law and Policy* 129
'Human Rights, Environmental Rights, and the Right to Environment' (1991) 28 *Stanford Journal of International Law* 103
'Resolving Conflicts between Human Rights and Environmental Protection: Is There a Hierarchy?' in De Wet E and Vidmar J (eds.), *Hierarchy in International Law: The Place of Human Rights* (Oxford University Press 2012)
Shue H, 'Changing Images of Climate Change: Human Rights and Future Generations' (2014) 5 *Journal of Human Rights and the Environment* 50
Singer P, *Animal Liberation: A New Ethics for Our Treatment of Animals* (Harper Collins 1976)
Smith L, *Green Belt – Briefing Paper 00934* (House of Commons Library 2016)
Smith RJ and others, 'Governance and the Loss of Biodiversity' (2003) 426 *Nature* 67
Smith RKM, *Textbook on International Human Rights* (6th ed., Oxford University Press 2014)
Sobotta C, 'EuGH: Neue Verfahren im Umweltrecht' (2017) 22 *Zeitschrift für Umweltrecht* 241
Solomon S and others, 'Emergence of Healing in the Antarctic Ozone Layer' (2016) 353 *Science* 269
Solvang O, 'Russia and the European Court of Human Rights: The Price of Non-Cooperation' (2008) 15 *Human Rights Brief* 14
Stapleton J, 'Factual Causation and Asbestos Cancers' (2010) 126 *Law Quarterly Review* 351
'The Two Explosive Proof-of-Causation Doctrines Central to Asbestos Claims' (2008) 74 *Brooklyn Law Review* 1011
Stephens T, *International Courts and Environmental Protection* (Cambridge University Press 2009)
Stone CD, 'Is There a Precautionary Principle?' (2001) 31 *Environmental Law Reporter News and Analysis* 10790
'Should Trees have Standing?' (1972) 45 *Southern California Law Review* 491
Sunstein CR, *Laws of Fear: Beyond the Precautionary Principle* (Cambridge University Press 2005)
Why Nudge? The Politics of Libertarian Paternalism (Yale University Press 2014)

Taylor PE, 'From Environmental to Ecological Human Rights: A New Dynamic in International Law' (1997) 10 *Georgetown International Environmental Law Review* 309

Theil S, 'Excavating landmarks – Empirical Contributions to Doctrinal Analysis' (2020) 32 *Journal of Environmental Law* 221

'Is the "Living Instrument" Approach of the European Court of Human Rights Compatible with the ECHR and International Law?' (2017) 23 *European Public Law* 587

'The Problem with the Normative Content of Section 24 of the Constitution of South Africa' (2019) 37 *Nordic Journal of Human Rights* 105

Thienel T, 'Artikel 53 EMRK' in Karpenstein U and Meyer FC (eds.), *Europäische Menschenrechtskonvention – Kommentar* (2nd ed., C. H. Beck 2015)

'The Burden and Standard of Proof in the European Court of Human Rights' (2007) 50 *German Yearbook of International Law* 543

Thornton D, Kagan RA, and Gunningham N, 'When Social Norms and Pressures Are Not Enough: Environmental Performance in the Trucking Industry' (2009) 43 *Law & Society Review* 405

Tietge U and others, *From Laboratory to Road – A 2016 Update of Official and 'Real-World' Fuel Consumption and CO_2 Values for Passenger Cars in Europe* (International Council on Clean Transportation 2016)

Tomuschat C, *Human Rights – Between Idealism and Realism* (2nd ed., Oxford University Press 2008)

Tossavainen A, 'Asbestos, Asbestosis, and Cancer: The Helsinki Criteria for Diagnosis and Attribution' (2010) 23 *Scandinavian Journal of Work, Environment & Health* 311

Tweedale G, *Magic Mineral to Killer Dust: Turner & Newall and the Asbestos Hazard* (Oxford University Press 2000)

Viljoen F, *International Human Rights Law in Africa* (2nd ed., Oxford University Press 2012)

Viñuales JE, 'The Contribution of the International Court of Justice to the Development of International Environmental Law: A Contemporary Assessment' (2008) 32 *Fordham International Law Journal* 232

'A Human Rights Approach to Extraterritorial Environmental Protection? An Assessment' in Bhuta N (ed.), *The Frontiers of Human Rights: Extraterritoriality and its Challenges* (Oxford University Press 2016)

'Law and the Anthropocence' [2016] C-EENRG Working Papers 1

'Legal Techniques for Dealing with Scientific Uncertainty in Environmental Law' (2010) 43 *Vanderbilt Journal of Transnational Law* 437

Vosgerau U, 'Zur Kollision von Grundrechtsfunktionen: Ein zentrales Problem der Grundrechtsdogmatik' (2008) 133 *Archiv des öffentlichen Rechts* 346

Wainwright O, 'Inside Beijing's Airpocalypse – A City Made "Almost Uninhabitable" by Pollution' *The Guardian* (16 December 2014) www.theguardian.com/cities/2014/dec/16/beijing-airpocalypse-city-almost-uninhabitable-pollution-china, accessed 1 April 2021

Wakefield J, 'Tomorrow's cities: Stockholm turns green' *BBC News* (8 March 2017) www.bbc.co.uk/news/technology-38404884, accessed 1 April 2021

Waldron J, 'The Core of the Case Against Judicial Review' (2006) 115 *Yale Law Journal* 1346

Walton H and others, *Understanding the Health Impacts of Air Pollution in London* (King's College London – Environmental Research Group 2015)

Wansley MT, 'Regulation of Emerging Risks' (2016) 69 *Vanderbilt Law Review* 401

Wegener B, 'Zwangshaft für Markus Söder? Von der Ungemütlichkeit an den Grenzen des Rechtsstaats' (*Verfassungsblog*, 13 December 2018) https://verfassungsblog.de/zwangshaft-fuer-markus-soeder-von-der-ungemuetlichkeit-an-den-grenzen-des-rechtsstaats/, accessed 1 April 2021

Weiss EB, 'Our Rights and Obligations to Future Generations for the Environment' (1990) 84 *American Journal of International Law* 198

Weiss EB and Jacobson HK, *Engaging Countries: Strengthening Compliance with International Environmental Accords* (Massachusetts Institute of Technology Press 1998)

Weitz N and others, *Sustainable Development Goals for Sweden: Insights on Setting a National Agenda* (Stockholm Environment Institute, Working Paper 2015-10, 2015)

Wenzel N, 'Artikel 38 EMRK' in Karpenstein U and Meyer FC (eds.), *Europäische Menschenrechtskonvention – Kommentar* (2nd ed., C. H. Beck 2015)

Weston BH and Bollier D, *Green Governance: Ecological Survival, Human Rights, and the Law of the Commons* (Cambridge University Press 2013)

'Toward a Recalibrated Human Right to a Clean and Healthy Environment: Making the Conceptual Transition' (2013) 4 *Journal of Human Rights and the Environment* 116

Westphal S, 'Artikel 20a GG – Staatsziel "Umweltschutz"' [2000] *Juristische Schulung* 339

Wiener JB, 'Precaution' in Bodansky D, Brunnée J, and Hey E (eds.), *The Oxford Handbook of International Environmental Law* (Oxford University Press 2007)

Wiener JB and Rogers MD, 'Comparing Precaution in the United States and Europe' (2002) 5 *Journal of Risk Research* 317

Wikeley N, 'The Asbestos Regulations 1931: A Licence to Kill?' (1992) 19 *Journal of Law and Society* 365

Wilberg H and Elliott M (eds.), *The Scope and Intensity of Substantive Review: Traversing Taggart's Rainbow* (Hart 2015)

Wolfrum R, 'The Taking and Assessment of Evidence by the European Court of Human Rights' in Breitenmoser S (ed.), *Human Rights, Democracy and the Rule of Law: Liber Amicorum Luzius Wildhaber* (Nomos 2007)

Woolf H and others, *De Smith's Judicial Review* (7th ed., Sweet & Maxwell 2013)

Woolf L, *Access to Justice Final Report* (Department for Constitutional Affairs 1996)

Wynberg R and Fig D, 'Realising Environmental Rights' in Langford M and others (eds.), *Socio-Economic Rights in South Africa: Symbols or Substance?* (Cambridge University Press 2014)

Young N and others, 'Community Water Fluoridation and Health Outcomes in England: A Cross-Sectional Study' (2015) 43 *Community Dentistry and Oral Epidemiology* 550 van Zeben J, 'Establishing a Governmental Duty of Care for Climate Change Mitigation: Will Urgenda Turn the Tide?' (2015) 4 *Transnational Environmental Law* 339

Zhang M, Song Y, and Cai X, 'A Health-Based Assessment of Particulate Air Pollution in Urban Areas of Beijing in 2000–2004' (2007) 376 *Science of The Total Environment* 100

Index

Printed in the United Kingdom by TJ Clays Ltd.